AF615676

Linear Models:
A Mean Model Approach

This is a volume in
PROBABILITY AND MATHEMATICAL STATISTICS

Z. W. Birnbaum, founding editor
David Aldous, Y. L. Tong, series editors

A list of titles in this series appears at the end of this volume.

Linear Models:
A Mean Model Approach

Barry Kurt Moser
Department of Statistics
Oklahoma State University
Stillwater, Oklahoma

Academic Press
San Diego Boston New York
London Sydney Tokyo Toronto

This book is printed on acid-free paper. ♾

Academic Press, Inc.
525 B Street, Suite 1900, San Diego, California 92101-4495, USA
http://www.apnet.com

Academic Press Limited
24-28 Oval Road, London NW1 7DX, UK
http://www.hbuk.co.uk/ap/

Library of Congress Cataloging-in-Publication Data

Moser, Barry Kurt.
Linear models : a mean model approach / by Barry Kurt Moser.
p. cm. -- (Probability and mathematical statistics)
Includes bibliographical references and index.
ISBN 0-12-508465-X (alk. paper)
1. Linear models (Statistics) I. Title. II. Series.
QA279.M685 1996
519.5'35--dc20 96-33930
CIP

PRINTED IN THE UNITED STATES OF AMERICA
96 97 98 99 00 01 BC 9 8 7 6 5 4 3 2 1

To my three precious ones.

Contents

Preface

Linear models is a broad and diversified subject area. Because the subject area is so vast, no attempt was made in this text to cover all possible linear models topics. Rather, the objective of this book is to cover a series of introductory topics that give a student a solid foundation in the study of linear models.

The text is intended for graduate students who are interested in linear statistical modeling. It has been my experience that students in this group enter a linear models course with some exposure to mathematical statistics, linear algebra, normal distribution theory, linear regression, and design of experiments. The attempt here is to build on that experience and to develop these subject areas within the linear models framework.

The early chapters of the text concentrate on the linear algebra and normal distribution theory needed for a linear models study. Examples of experiments with complete, balanced designs are introduced early in the text to give the student a familiar foundation on which to build. Chapter 4 of the text concentrates entirely on complete, balanced models. This early dedication to complete, balanced models is intentional. It has been my experience that students are generally more comfortable learning structured material. Therefore, the structured rules that apply to complete, balanced designs give the student a set of learnable tools on which to build confidence. Later chapters of the text then expand the discussion to more complicated incomplete, unbalanced and mixed models. The same tools learned for the balanced, complete models are simply expanded to apply to these more complicated cases. The hope is that the text progresses in an orderly manner with one topic building on the next.

I thank all the people who contributed to this text. First, I thank Virgil Anderson for introducing me to the wonders of statistics. Special thanks go to Julie Sawyer, Laura Coombs, and David Weeks. Julie and Laura helped edit the text. Julie also contributed heavily to the development of Chapters 4 and 8. David has generally served as a sounding board during the writing process. He has listened to my ideas and contributed many of his own. Finally, and most importantly, I thank my wife, Diane, for her generosity and support.

1 Linear Algebra and Related Introductory Topics

A summary of relevant linear algebra concepts is presented in this chapter. Throughout the text boldfaced letters such as **A**, **U**, **T**, **X**, **Y**, **t**, **g**, **u** are used to represent matrices and vectors, italicized capital letters such as Y, U, T, E, F are used to represent random variables, and lowercase italicized letters such as r, s, t, n, c are used as constants.

1.1 ELEMENTARY MATRIX CONCEPTS

The following list of definitions provides a brief summary of some useful matrix operations.

Definition 1.1.1 *Matrix*: An $r \times s$ matrix $\mathbf{A}$ is a rectangular array of elements with r rows and s columns. An $r \times 1$ vector $\mathbf{Y}$ is a matrix with r rows and 1 column. Matrix elements are restricted to real numbers throughout the text.

Definition 1.1.2 *Transpose*: If $\mathbf{A}$ is an $n \times s$ matrix, then the transpose of $\mathbf{A}$, denoted by $\mathbf{A}'$, is an $s \times n$ matrix formed by interchanging the rows and columns of $\mathbf{A}$.

Definition 1.1.3 *Identity Matrix, Matrix of Ones and Zeros*: $\mathbf{I}_n$ represents an $n \times n$ identity matrix, $\mathbf{J}_n$ is an $n \times n$ matrix of ones, $\mathbf{1}_n$ is an $n \times 1$ vector of ones, and $\mathbf{0}_{m \times n}$ is an $m \times n$ matrix of zeros.

Definition 1.1.4 *Multiplication of Matrices*: Let a_{ij} represent the ij^{th} element of an $r \times s$ matrix $\mathbf{A}$ with $i = 1, \ldots, r$ rows and $j = 1, \ldots, s$ columns. Likewise, let b_{jk} represent the jk^{th} element of an $s \times t$ matrix $\mathbf{B}$ with $j = 1, \ldots, s$ rows and $k = 1, \ldots, t$ columns. The matrix multiplication of $\mathbf{A}$ and $\mathbf{B}$ is represented by $\mathbf{AB} = \mathbf{C}$ where $\mathbf{C}$ is an $r \times t$ matrix whose ik^{th} element $c_{ik} = \sum_{j=1}^{s} a_{ij} b_{jk}$. If the $r \times s$ matrix $\mathbf{A}$ is multiplied by a scalar d, then the resulting $r \times s$ matrix $d\mathbf{A}$ has ij^{th} element da_{ij}.

Example 1.1.1 The following matrix multiplications commonly occur.

$$\mathbf{1}_n'\mathbf{1}_n = n$$

$$\mathbf{1}_n\mathbf{1}_n' = \mathbf{J}_n$$

$$\mathbf{J}_n\mathbf{J}_n = n\mathbf{J}_n$$

$$\mathbf{1}_n'\left(\mathbf{I}_n - \frac{1}{n}\mathbf{J}_n\right) = \mathbf{0}_{1 \times n}$$

$$\mathbf{J}_n'\left(\mathbf{I}_n - \frac{1}{n}\mathbf{J}_n\right) = \mathbf{0}_{n \times n}$$

$$\left(\mathbf{I}_n - \frac{1}{n}\mathbf{J}_n\right)\left(\mathbf{I}_n - \frac{1}{n}\mathbf{J}_n\right) = \left(\mathbf{I}_n - \frac{1}{n}\mathbf{J}_n\right).$$

Definition 1.1.5 *Addition of Matrices*: The sum of two $r \times s$ matrices $\mathbf{A}$ and $\mathbf{B}$ is represented by $\mathbf{A} + \mathbf{B} = \mathbf{C}$ where $\mathbf{C}$ is the $r \times s$ matrix whose ij^{th} element $c_{ij} = a_{ij} + b_{ij}$.

Definition 1.1.6 *Inverse of a Matrix*: An $n \times n$ matrix $\mathbf{A}$ has an inverse if $\mathbf{A}\mathbf{A}^{-1} = \mathbf{A}^{-1}\mathbf{A} = \mathbf{I}_n$ where the $n \times n$ inverse matrix is denoted by $\mathbf{A}^{-1}$.

Definition 1.1.7 *Singularity*: If an $n \times n$ matrix $\mathbf{A}$ has an inverse then $\mathbf{A}$ is a nonsingular matrix. If $\mathbf{A}$ does not have an inverse then $\mathbf{A}$ is a singular matrix.

Definition 1.1.8 *Diagonal Matrix*: Let a_{ii} be the ith diagonal element of an $n \times n$ matrix $\mathbf{A}$. Let a_{ij} be the ij^{th} off-diagonal element of $\mathbf{A}$ for $i \neq j$. Then $\mathbf{A}$ is a diagonal matrix if all the off-diagonal elements a_{ij} equal zero.

Definition 1.1.9 *Trace of a Square Matrix*: The trace of an $n \times n$ matrix $\mathbf{A}$, denoted by $\text{tr}(\mathbf{A})$, is the sum of the diagonal elements of $\mathbf{A}$. That is, $\text{tr}(\mathbf{A}) = \sum_{i=1}^{n} a_{ii}$.

It is assumed that the reader is familiar with the definition of the determinant of a square matrix. Therefore, a rigorous definition is omitted. The next definition actually provides the notation used for a determinant.

Definition 1.1.10 *Determinant of a Square Matrix*: Let $\det(\mathbf{A}) = |\mathbf{A}|$ denote the determinant of an $n \times n$ matrix $\mathbf{A}$. Note $\det(\mathbf{A}) = 0$ if $\mathbf{A}$ is singular.

Definition 1.1.11 *Symmetric Matrix*: An $n \times n$ matrix $\mathbf{A}$ is symmetric if $\mathbf{A} = \mathbf{A}'$.

Definition 1.1.12 *Linear Dependence and the Rank of a Matrix*: Let $\mathbf{A}$ be an $n \times s$ matrix ($s \leq n$) where $\mathbf{a}_1, \ldots, \mathbf{a}_s$ represent the s $n \times 1$ column vectors of $\mathbf{A}$. The s vectors $\mathbf{a}_1, \ldots, \mathbf{a}_s$ are linearly dependent provided there exists s elements $k_1, \ldots, k_s$, not all zero, such that $k_1\mathbf{a}_1 + \cdots + k_s\mathbf{a}_s = 0$. Otherwise, the s vectors are linearly independent. Furthermore, if there are exactly $r \leq s$ vectors of the set $\mathbf{a}_1, \ldots, \mathbf{a}_s$ which are linearly independent, while the remaining $s - r$ can be expressed as a linear combination of these r vectors, then the rank of $\mathbf{A}$, denoted by rank ($\mathbf{A}$), is r.

The following list shows the results of the preceding definitions and are stated without proof:

Result 1.1: Let $\mathbf{A}$ and $\mathbf{B}$ each be $n \times n$ nonsingular matrices. Then $(\mathbf{AB})^{-1} = \mathbf{B}^{-1}\mathbf{A}^{-1}$.

Result 1.2: Let $\mathbf{A}$ and $\mathbf{B}$ be any two matrices such that $\mathbf{AB}$ is defined. Then $(\mathbf{AB})' = \mathbf{B}'\mathbf{A}'$.

Result 1.3: Let $\mathbf{A}$ be any matrix. The $\mathbf{A}'\mathbf{A}$ and $\mathbf{AA}'$ are symmetric.

Result 1.4: Let $\mathbf{A}$ and $\mathbf{B}$ each be $n \times n$ matrices. Then $\det(\mathbf{AB}) = [\det(\mathbf{A})][\det(\mathbf{B})]$.

Result 1.5: Let $\mathbf{A}$ and $\mathbf{B}$ be $m \times n$ and $n \times m$ matrices, respectively. Then $\text{tr}(\mathbf{AB}) = \text{tr}(\mathbf{BA})$.

Quadratic forms play a key role in linear model theory. The following definitions introduce quadratic forms.

Definition 1.1.13 *Quadratic Forms*: A function $f(x_1, \ldots, x_n)$ is a quadratic form if $f(x_1, \ldots, x_n) = \sum_{i=1}^{n} \sum_{j=1}^{n} a_{ij}x_ix_j = \mathbf{X}'\mathbf{AX}$ where $\mathbf{X} = (x_1, \ldots, x_n)'$ is an $n \times 1$ vector and $\mathbf{A}$ is an $n \times n$ symmetric matrix whose ij^{th} element is a_{ij}.

Example 1.1.2 Let $f(x_1, x_2, x_3) = x_1^2 + 3x_2^2 + 4x_3^2 + x_1x_2 + 2x_2x_3$. Then $f(x_1, x_2, x_3) = \mathbf{X}'\mathbf{A}\mathbf{X}$ is a quadratic form with $\mathbf{X} = (x_1, x_2, x_3)'$ and

$$\mathbf{A} = \begin{bmatrix} 1 & 0.5 & 0 \\ 0.5 & 3 & 1 \\ 0 & 1 & 4 \end{bmatrix}.$$

The symmetric matrix $\mathbf{A}$ is constructed by setting a_{ij} and a_{ji} equal to one-half the coefficient on the x_ix_j term for $i \neq j$.

Example 1.1.3 Quadratic forms are very useful for defining sums of squares. For example, let

$$f(x_1, \ldots, x_n) = \sum_{i=1}^{n} x_i^2 = \mathbf{X}'\mathbf{X} = \mathbf{X}'\mathbf{I}_n\mathbf{X}$$

where the $n \times 1$ vector $\mathbf{X} = (x_1, \ldots, x_n)'$. The sum of squares around the sample mean is another common example. Let

$$\begin{aligned} f(x_1, \ldots, x_n) &= \sum_{i=1}^{n} (x_i - \overline{x})^2 \\ &= \sum_{i=1}^{n} x_i^2 - n\overline{x}^2 \\ &= \mathbf{X}'\mathbf{X} - \frac{1}{n}\left(\sum_{i=1}^{n} x_i\right)\left(\sum_{i=1}^{n} x_i\right) \\ &= \mathbf{X}'\mathbf{X} - \frac{1}{n}(\mathbf{X}'\mathbf{1}_n)(\mathbf{1}_n'\mathbf{X}) \\ &= \mathbf{X}'\left[\mathbf{I}_n - \frac{1}{n}\mathbf{1}_n\mathbf{1}_n'\right]\mathbf{X} \\ &= \mathbf{X}'\left[\mathbf{I}_n - \frac{1}{n}\mathbf{J}_n\right]\mathbf{X} \end{aligned}$$

Definition 1.1.14 *Orthogonal Matrix*: An $n \times n$ matrix $\mathbf{P}$ is orthogonal if and only if $\mathbf{P}^{-1} = \mathbf{P}'$. Therefore, $\mathbf{P}\mathbf{P}' = \mathbf{P}'\mathbf{P} = \mathbf{I}_n$. If $\mathbf{P}$ is written as $(\mathbf{p}_1, \mathbf{p}_2, \ldots, \mathbf{p}_n)$ where $\mathbf{p}_i$ is an $n \times 1$ column vector of $\mathbf{P}$ for $i = 1, \ldots, n$, then necessary and sufficient conditions for $\mathbf{P}$ to be orthogonal are

(i) $\mathbf{p}_i'\mathbf{p}_i = 1$ for each $i = 1, \ldots, n$ and

(ii) $\mathbf{p}_i'\mathbf{p}_j = 0$ for any $i \neq j$.

Example 1.1.4 Let the $n \times n$ matrix

$$\mathbf{P} = \begin{bmatrix} 1/\sqrt{n} & 1/\sqrt{2} & 1/\sqrt{6} & \cdots & 1/\sqrt{n(n-1)} \\ 1/\sqrt{n} & -1/\sqrt{2} & 1/\sqrt{6} & \cdots & 1/\sqrt{n(n-1)} \\ 1/\sqrt{n} & 0 & -2/\sqrt{6} & \cdots & 1/\sqrt{n(n-1)} \\ 1/\sqrt{n} & 0 & 0 & \cdots & 1/\sqrt{n(n-1)} \\ \vdots & \vdots & \vdots & \ddots & \vdots \\ 1/\sqrt{n} & 0 & 0 & \cdots & -\sqrt{(n-1)/n} \end{bmatrix},$$

where $\mathbf{PP}' = \mathbf{P}'\mathbf{P} = \mathbf{I}_n$. The columns of $\mathbf{P}$ are created as follows:

$$\mathbf{p}_1 = \left(1/\sqrt{1^2 + 1^2 + \cdots + 1^2}\right)(1, 1, 1, 1, \ldots, 1)' = \left(1/\sqrt{n}\right)\mathbf{1}_n$$

$$\mathbf{p}_2 = \left(1/\sqrt{1^2 + (-1)^2}\right)(1, -1, 0, 0, \ldots, 0)'$$

$$\mathbf{p}_3 = \left(1/\sqrt{1^2 + 1^2 + (-2)^2}\right)(1, 1, -2, 0, \ldots, 0)'$$

$$\vdots$$

$$\mathbf{p}_n = \left(1/\sqrt{1^2 + 1^2 + \cdots + 1^2 + (-(n-1))^2}\right)(1, 1, \ldots, 1, -(n-1))'.$$

The matrix $\mathbf{P}'$ in Example 1.1.4 is generally referred to as an n-dimensional Helmert matrix. The Helmert matrix has some interesting properties. Write $\mathbf{P}$ as $\mathbf{P} = (\mathbf{p}_1|\mathbf{P}_n)$ where the $n \times 1$ vector $\mathbf{p}_1 = (1/\sqrt{n})\mathbf{1}_n$ and the $n \times (n-1)$ matrix $\mathbf{P}_n = (\mathbf{p}_2, \mathbf{p}_3, \ldots, \mathbf{p}_n)$ then

$$\begin{aligned} \mathbf{p}_1\mathbf{p}_1' &= \frac{1}{n}\mathbf{J}_n \\ \mathbf{p}_1'\mathbf{p}_1 &= 1 \\ \mathbf{p}_1'\mathbf{P}_n &= \mathbf{0}_{1\times(n-1)} \\ \mathbf{P}_n\mathbf{P}_n' &= \left(\mathbf{I}_n - \frac{1}{n}\mathbf{J}_n\right) \\ \mathbf{P}_n'\mathbf{P}_n &= \mathbf{I}_{n-1}. \end{aligned}$$

The $(n-1)\times n$ matrix $\mathbf{P}_n'$ will be referred to as the lower portion of an n-dimensional Helmert matrix.

If $\mathbf{X}$ is an $n \times 1$ vector and $\mathbf{A}$ is an $n \times n$ matrix, then $\mathbf{AX}$ defines n linear combinations of the elements of $\mathbf{X}$. Such transformations from $\mathbf{X}$ to $\mathbf{AX}$ are very useful in linear models. Of particular interest are transformations of the vector $\mathbf{X}$ that produce multiples of $\mathbf{X}$. That is, we are interested in transformations that satisfy the relationship

$$\mathbf{AX} = \lambda\mathbf{X}$$

where λ is a scalar multiple. The above relationship holds if and only if

$$|\lambda\mathbf{I}_n - \mathbf{A}| = 0.$$

But the determinant of $\lambda\mathbf{I}_n - \mathbf{A}$ is an nth degree polynomial in λ. Thus, there are exactly n values of λ that satisfy $|\lambda\mathbf{I}_n - \mathbf{A}| = 0$. These n values of λ are called the n eigenvalues of the matrix $\mathbf{A}$. They are denoted by $\lambda_1, \lambda_2, \ldots, \lambda_n$. Corresponding to each eigenvalue λ_i there is an $n \times 1$ vector $\mathbf{X}_i$ that satisfies

$$\mathbf{AX}_i = \lambda_i\mathbf{X}_i$$

where $\mathbf{X}_i$ is called the ith eigenvector of the matrix $\mathbf{A}$ corresponding to the eigenvalue λ_i.

Example 1.1.5 Find the eigenvalues and vectors of the 3×3 matrix $\mathbf{A} = 0.6\mathbf{I}_3 + 0.4\mathbf{J}_3$. First, set $|\lambda\mathbf{I}_3 - \mathbf{A}| = 0$. This relationship produces the cubic equation

$$\lambda^3 - 3\lambda^2 + 2.52\lambda - 0.648 = 0$$

or

$$(\lambda - 1.8)(\lambda - 0.6)(\lambda - 0.6) = 0.$$

Therefore, the eigenvalues of $\mathbf{A}$ are $\lambda_1 = 1.8, \lambda_2 = \lambda_3 = 0.6$. Next, find vectors $\mathbf{X}_i$ that satisfy $(\mathbf{A} - \lambda_i\mathbf{I}_3)\mathbf{X}_i = \mathbf{0}_{3\times 1}$ for each $i = 1, 2, 3$. For $\lambda_1 = 1.8$, $(\mathbf{A} - 1.8\mathbf{I}_3)\mathbf{X}_1 = \mathbf{0}_{3\times 1}$ or $(-1.2\mathbf{I}_3 + 0.4\mathbf{J}_3)\mathbf{X}_1 = \mathbf{0}_{3\times 1}$. The vector $\mathbf{X}_1 = (1/\sqrt{3})\mathbf{I}_3$ satisfies this relationship. For $\lambda_2 = \lambda_3 = 0.6$, $(\mathbf{A} - 0.6\mathbf{I}_3)\mathbf{X}_i = \mathbf{0}_{3\times 1}$ or $\mathbf{1}_3'\mathbf{X}_i = \mathbf{0}_{3\times 1}$ for $i = 2, 3$. The vectors $\mathbf{X}_2 = (1/\sqrt{2}, -1/\sqrt{2}, 0)'$ and $\mathbf{X}_3 = (1/\sqrt{6}, 1/\sqrt{6}, -2/\sqrt{6})'$ satisfy this condition. Note that vectors $\mathbf{X}_1, \mathbf{X}_2, \mathbf{X}_3$ are normalized and orthogonal since $\mathbf{X}_1'\mathbf{X}_1 = \mathbf{X}_2'\mathbf{X}_2 = \mathbf{X}_3'\mathbf{X}_3 = 1$ and $\mathbf{X}_1'\mathbf{X}_2 = \mathbf{X}_1'\mathbf{X}_3 = \mathbf{X}_2'\mathbf{X}_3 = 0$.

The following theorems address the uniqueness or nonuniqueness of the eigenvector associated with each eigenvalue.

Theorem 1.1.1 *There exists at least one eigenvector corresponding to each eigenvalue.*

Theorem 1.1.2 *If an $n \times n$ matrix $\mathbf{A}$ has n distinct eigenvalues, then there exist exactly n linearly independent eigenvectors, one associated with each eigenvalue.*

In the next theorem and corollary a symmetric matrix is defined in terms of its eigenvalues and eigenvectors.

Theorem 1.1.3 *Let* $\mathbf{A}$ *be an* $n \times n$ *symmetric matrix. There exists an* $n \times n$ *orthogonal matrix* $\mathbf{P}$ *such that* $\mathbf{P}'\mathbf{A}\mathbf{P} = \mathbf{D}$ *where* $\mathbf{D}$ *is a diagonal matrix whose diagonal elements are the eigenvalues of* $\mathbf{A}$ *and where the columns of* $\mathbf{P}$ *are the orthogonal, normalized eigenvectors of* $\mathbf{A}$. *The* i^{th} *column of* $\mathbf{P}$ *(i.e., the* i^{th} *eigenvectors of* $\mathbf{A}$*) corresponds to the* i^{th} *diagonal element of* $\mathbf{D}$ *for* $i = 1, \ldots, n$.

Example 1.1.6 Let $\mathbf{A}$ be the 3×3 matrix from Example 1.1.5. Then $\mathbf{P}'\mathbf{A}\mathbf{P} = \mathbf{D}$ or

$$\begin{bmatrix} 1/\sqrt{3} & 1/\sqrt{3} & 1/\sqrt{3} \\ 1/\sqrt{2} & -1/\sqrt{2} & 0 \\ 1/\sqrt{6} & 1/\sqrt{6} & -2/\sqrt{6} \end{bmatrix} \begin{bmatrix} 1 & 0.4 & 0.4 \\ 0.4 & 1 & 0.4 \\ 0.4 & 0.4 & 1 \end{bmatrix} \begin{bmatrix} 1/\sqrt{3} & 1/\sqrt{2} & 1/\sqrt{6} \\ 1/\sqrt{3} & -1/\sqrt{2} & 1/\sqrt{6} \\ 1/\sqrt{3} & 0 & -2/\sqrt{6} \end{bmatrix}$$

$$= \begin{bmatrix} 1.8 & 0 & 0 \\ 0 & 0.6 & 0 \\ 0 & 0 & 0.6 \end{bmatrix}$$

Theorem 1.1.3 can be used to relate the trace and determinant of a symmetric matrix to its eigenvalues.

Theorem 1.1.4 *If* $\mathbf{A}$ *is an* $n \times n$ *symmetric matrix then* $tr(\mathbf{A}) = \sum_{i=1}^{n} \lambda_i$ *and* $det(\mathbf{A}) = \prod_{i=1}^{n} \lambda_i$.

Proof: Let $\mathbf{P}'\mathbf{A}\mathbf{P} = \mathbf{D}$ then

$$\text{tr}(\mathbf{A}) = \text{tr}(\mathbf{A}\mathbf{P}\mathbf{P}') = \text{tr}(\mathbf{P}'\mathbf{A}\mathbf{P}) = \text{tr}(\mathbf{D}) = \sum_{i=1}^{n} \lambda_i.$$

$$\det(\mathbf{A}) = \det(\mathbf{A}\mathbf{P}\mathbf{P}') = \det(\mathbf{P}'\mathbf{A}\mathbf{P}) = \det(\mathbf{D}) \prod_{i=1}^{n} \lambda_i. \qquad \blacksquare$$

The number of times an eigenvalue occurs is the multiplicity of the value. This idea is formalized in the next definition.

Definition 1.1.15 *Multiplicity*: The $n \times n$ matrix $\mathbf{A}$ has eigenvalue λ^* with multiplicity $m \leq n$ if m of the eigenvalues of $\mathbf{A}$ equal λ^*.

Example 1.1.7 All the n eigenvalues of the identity matrix $\mathbf{I}_n$ equal 1. Therefore, $\mathbf{I}_n$ has eigenvalue 1 with multiplicity n.

Example 1.1.8 Find the eigenvalues and eigenvectors of the $n \times n$ matrix $\mathbf{G} = (a - b)\mathbf{I}_n + b\mathbf{J}_n$. First, note that

$$\mathbf{G}[(1/\sqrt{n})\mathbf{1}_n] = [a + (n-1)b](1/\sqrt{n})\mathbf{1}_n.$$

Therefore, $a+(n-1)b$ is an eigenvalue of matrix $\mathbf{G}$ with corresponding normalized eigenvector $(1\sqrt{n})\mathbf{1}_n$. Next, take any $n \times 1$ vector $\mathbf{X}$ such that $\mathbf{1}_n'\mathbf{X} = 0$. (One set of $n-1$ vectors that satisfies $\mathbf{1}_n'\mathbf{X} = 0$ are the column vectors $\mathbf{p}_2, \mathbf{p}_3, \ldots, \mathbf{p}_n$ from Example 1.1.4.) Rewrite $\mathbf{G} = (a-b)\mathbf{I}_n + b\mathbf{1}_n\mathbf{1}_n'$. Therefore,

$$\mathbf{GX} = (a-b)\mathbf{X} + b\mathbf{1}_n\mathbf{1}_n'\mathbf{X} = (a-b)\mathbf{X}$$

and matrix $\mathbf{G}$ has eigenvalue $a-b$. Furthermore,

$$\begin{aligned} |\lambda\mathbf{I}_n - \mathbf{G}| &= |(\lambda - a + b)\mathbf{I}_n - b\mathbf{J}_n| \\ &= [a + (n-1)b](a-b)^{n-1}. \end{aligned}$$

Therefore, eigenvalue $a + (n-1)b$ has multiplicity 1 and eigenvalue $a-b$ has multiplicity $n-1$. Note that the 3×3 matrix $\mathbf{A}$ in Example 1.1.5 is a special case of matrix $\mathbf{G}$ with $a = 1$, $b = 0.4$, and $n = 3$.

It will be convenient at times to separate a matrix into its submatrix components. Such a separation is called partitioning.

Definition 1.1.16 *Partitioning a Matrix*: If $\mathbf{A}$ is an $m \times n$ matrix then $\mathbf{A}$ can be separated or partitioned as

$$\mathbf{A} = \begin{bmatrix} \mathbf{A}_{11} & \mathbf{A}_{12} \\ \mathbf{A}_{21} & \mathbf{A}_{22} \end{bmatrix}$$

where $\mathbf{A}_{ij}$ is an $m_i \times n_j$ matrix for $i, j = 1, 2$, $m = m_1 + m_2$ and $n = n_1 + n_2$.

Most of the square matrices used in this text are either positive definite or positive semidefinite. These two general matrix types are described in the following definitions.

Definition 1.1.17 *Positive Semidefinite Matrix*: An $n \times n$ matrix $\mathbf{A}$ is positive semidefinite if

(i) $\mathbf{A} = \mathbf{A}'$,

(ii) $\mathbf{Y}'\mathbf{A}\mathbf{Y} \geq 0$ for all $n \times 1$ real vectors $\mathbf{Y}$, and

(iii) $\mathbf{Y}'\mathbf{A}\mathbf{Y} = 0$ for at least one $n \times 1$ nonzero real vector $\mathbf{Y}$.

Example 1.1.9 The matrix $\mathbf{J}_n$ is positive semidefinite because $\mathbf{J}_n = \mathbf{J}_n'$, $\mathbf{Y}'\mathbf{J}_n\mathbf{Y} = (\mathbf{1}_n'\mathbf{Y})'(\mathbf{1}_n'\mathbf{Y}) = (\sum_{i=1}^n y_i)^2 \geq 0$ for $\mathbf{Y} = (y_1, \ldots, y_n)'$ and $\mathbf{Y}'\mathbf{J}_n\mathbf{Y} = 0$ for $\mathbf{Y} = (1, -1, 0, \ldots, 0)'$.

Definition 1.1.18 *Positive Definite Matrix*: An $n \times n$ matrix $\mathbf{A}$ is positive definite if

(i) $\mathbf{A} = \mathbf{A}'$ and

(ii) $\mathbf{Y}'\mathbf{A}\mathbf{Y} > 0$ for all nonzero $n \times 1$ real vectors $\mathbf{Y}$.

Example 1.1.10 The $n \times n$ identity matrix $\mathbf{I}_n$ is positive definite because $\mathbf{I}_n$ is symmetric and $\mathbf{Y}'\mathbf{I}_n\mathbf{Y} > 0$ for all nonzero $n \times 1$ real vectors $\mathbf{Y}$.

Theorem 1.1.5 *Let $\mathbf{A}$ be an $n \times n$ positive definite matrix. Then*

(i) there exists an $n \times n$ matrix $\mathbf{B}$ of rank n such that $\mathbf{A} = \mathbf{B}\mathbf{B}'$ and

(ii) the eigenvalues of $\mathbf{A}$ are all positive.

The following example demonstrates how the matrix $\mathbf{B}$ in Theorem 1.1.5 can be constructed.

Example 1.1.11 Let $\mathbf{A}$ be an $n \times n$ positive definite matrix. Thus, $\mathbf{A} = \mathbf{A}'$ and by Theorem 1.1.3 there exists $n \times n$ matrices $\mathbf{P}$ and $\mathbf{D}$ such that $\mathbf{P}'\mathbf{A}\mathbf{P} = \mathbf{D}$ where $\mathbf{P}$ is the orthogonal matrix whose columns are the eigenvectors of $\mathbf{A}$, and $\mathbf{D}$ is the corresponding diagonal matrix of eigenvalues. Therefore, $\mathbf{A} = \mathbf{P}\mathbf{D}\mathbf{P}' = \mathbf{P}\mathbf{D}^{1/2}\mathbf{D}^{1/2}\mathbf{P}' = \mathbf{B}\mathbf{B}'$ where $\mathbf{D}^{1/2}$ is an $n \times n$ diagonal matrix whose i^{th} diagonal element is $\lambda_i^{1/2}$ and $\mathbf{B} = \mathbf{P}\mathbf{D}^{1/2}$.

Certain square matrices have the characteristic that $\mathbf{A}^2 = \mathbf{A}$. For example, let $\mathbf{A} = \mathbf{I}_n - \frac{1}{n}\mathbf{J}_n$. Then

$$\begin{aligned}\mathbf{A}^2 = \left(\mathbf{I}_n - \frac{1}{n}\mathbf{J}_n\right)^2 &= \left(\mathbf{I}_n - \frac{1}{n}\mathbf{J}_n\right)\left(\mathbf{I}_n - \frac{1}{n}\mathbf{J}_n\right)\\ &= \mathbf{I}_n - \frac{1}{n}\mathbf{J}_n - \frac{1}{n}\mathbf{J}_n + \frac{1}{n}\mathbf{J}_n\\ &= \mathbf{A}.\end{aligned}$$

Matrices of this type are introduced in the next definition.

Definition 1.1.19 *Idempotent Matrices*. Let $\mathbf{A}$ be an $n \times n$ matrix. Then

(i) $\mathbf{A}$ is idempotent if $\mathbf{A}^2 = \mathbf{A}$ and

(ii) $\mathbf{A}$ is symmetric, idempotent if $\mathbf{A} = \mathbf{A}^2$ and $\mathbf{A} = \mathbf{A}'$.

Note that if $\mathbf{A}$ is idempotent of rank n then $\mathbf{A} = \mathbf{I}_n$.

In linear model applications, idempotent matrices generally occur in the context of quadratic forms. Since the matrix in a quadratic form is symmetric, we generally restrict our attention to symmetric, idempotent matrices.

Theorem 1.1.6 *Let* $\mathbf{B}$ *be an* $n \times n$ *symmetric, idempotent matrix of rank* $r < n$. *Then* $\mathbf{B}$ *is positive semidefinite.*

The next theorem will prove useful when examining sums of squares in ANOVA problems.

Theorem 1.1.7 *Let* $\mathbf{A}_1, \ldots, \mathbf{A}_m$ *be* $n \times n$ *symmetric matrices where rank* $(\mathbf{A}_s) = n_s$ *for* $s = 1, \ldots, m$ *and* $\sum_{s=1}^{m} \mathbf{A}_s = \mathbf{I}_n$. *If* $\sum_{s=1}^{m} n_s = n$, *then*

(i) $\mathbf{A}_r \mathbf{A}_s = \mathbf{0}_{n \times n}$ *for* $r \neq s$, $r, s = 1, \ldots, m$ *and*

(ii) $\mathbf{A}_s = \mathbf{A}_s^2$ *for* $s = 1, \ldots, m$.

The eigenvalues of the matrix $\mathbf{I}_n - \frac{1}{n}\mathbf{J}_n$ are derived in the next example.

Example 1.1.12 The symmetric, idempotent matrix $\mathbf{I}_n - \frac{1}{n}\mathbf{J}_n$ takes the form $(a - b)\mathbf{I}_n + b\mathbf{J}_n$ with $a = 1 - \frac{1}{n}$ and $b = -\frac{1}{n}$. Therefore, by Example 1.1.8, the eigenvalues of $\mathbf{I}_n - \frac{1}{n}\mathbf{J}_n$ are $a + (n - 1)b = (1 - \frac{1}{n}) + (n - 1)(-\frac{1}{n}) = 0$ with multiplicity 1 and $a - b = (1 - \frac{1}{n}) - (-\frac{1}{n}) = 1$ with multiplicity $n - 1$.

The result that the eigenvalues of an idempotent matrix are all zeros and ones is generalized in the next theorem.

Theorem 1.1.8 *The eigenvalues of an* $n \times n$ *symmetric matrix* $\mathbf{A}$ *of rank* $r \leq n$ *are 1, multiplicity* r *and 0, multiplicity* $n - r$ *if and only if* $\mathbf{A}$ *is idempotent.*

Proof: The proof is given by Graybill (1976, p. 39). ■

The following theorem relates the trace and the rank of a symmetric, idempotent matrix.

Theorem 1.1.9 *If* $\mathbf{A}$ *is an* $n \times n$ *symmetric, idempotent matrix then tr(*$\mathbf{A}$*)* $=$ *rank(*$\mathbf{A}$*).*

Proof: The proof is left to the reader. ■

In the next example the matrix $\mathbf{I}_n - \frac{1}{n}\mathbf{J}_n$ is written as a function of $n - 1$ of its eigenvalues.

Example 1.1.13 The $n \times n$ matrix $\mathbf{I}_n - \frac{1}{n}\mathbf{J}_n$ takes the form $(a - b)\mathbf{I}_n + b\mathbf{J}_n$ with $a = 1 - \frac{1}{n}$ and $b = -\frac{1}{n}$. By Examples 1.1.8 and 1.1.12, the $n - 1$ eigenvalues equal to 1 have corresponding eigenvectors equal to the $n - 1$ columns of $\mathbf{P}_n$ where $\mathbf{P}'_n$ is the $(n - 1) \times n$ lower portion of an n-dimensional Helmert matrix. Further, $\mathbf{I}_n - \frac{1}{n}\mathbf{J}_n = \mathbf{P}_n\mathbf{P}'_n$.

This representation of a symmetric idempotent matrix is generalized in the next theorem.

Theorem 1.1.10 *If* $\mathbf{A}$ *is an* $n \times n$ *symmetric, idempotent matrix of rank* r *then* $\mathbf{A} = \mathbf{PP}'$ *where* $\mathbf{P}$ *is an* $n \times r$ *matrix whose columns are the eigenvectors of* $\mathbf{A}$ *associated with the* r *eigenvalues equal to 1.*

Proof: Let $\mathbf{A}$ be an $n \times n$ symmetric, idempotent matrix of rank r. By Theorem 1.1.3,

$$\mathbf{R}'\mathbf{AR} = \begin{bmatrix} \mathbf{I}_r & \mathbf{0} \\ \mathbf{0} & \mathbf{0} \end{bmatrix}$$

where $\mathbf{R} = [\mathbf{P}|\mathbf{Q}]$ is the $n \times n$ matrix of eigenvectors of $\mathbf{A}$, $\mathbf{P}$ is the $n \times r$ matrix whose r columns are the eigenvectors associated with the r eigenvalues 1, and $\mathbf{Q}$ is the $n \times (n-r)$ matrix of eigenvectors associated with the $(n-r)$ eigenvalues 0. Therefore,

$$\mathbf{A} = [\mathbf{P}\ \mathbf{Q}] \begin{bmatrix} \mathbf{I}_r & \mathbf{0} \\ \mathbf{0} & \mathbf{0} \end{bmatrix} \begin{bmatrix} \mathbf{P}' \\ \mathbf{Q}' \end{bmatrix} = \mathbf{PP}'.$$

Furthermore, $\mathbf{P}'\mathbf{P} = \mathbf{I}_r$ because $\mathbf{P}$ is an $n \times r$ matrix of orthogonal eigenvectors. ■

If $\mathbf{X}'\mathbf{AX}$ is a quadratic form with $n \times 1$ vector $\mathbf{X}$ and $n \times n$ symmetric matrix $\mathbf{A}$, then $\mathbf{X}'\mathbf{AX}$ is a quadratic form constructed from an n-dimensional vector. The following example uses Theorem 1.1.10 to show that if $\mathbf{A}$ is an $n \times n$ symmetric, idempotent matrix of rank $r \leq n$ then the quadratic form $\mathbf{X}'\mathbf{AX}$ can be rewritten as a quadratic form constructed from an r-dimensional vector.

Example 1.1.14 Let $\mathbf{X}'\mathbf{AX}$ be a quadratic form with $n \times 1$ vector $\mathbf{X}$ and $n \times n$ symmetric, idempotent matrix $\mathbf{A}$ of rank $r \leq n$. By Theorem 1.1.10, $\mathbf{X}'\mathbf{AX} = \mathbf{X}'\mathbf{PP}'\mathbf{X} = \mathbf{Z}'\mathbf{Z}$ where $\mathbf{P}$ is an $n \times r$ matrix of eigenvectors of $\mathbf{A}$ associated with the eigenvalues 1 and $\mathbf{Z} = \mathbf{P}'\mathbf{X}$ is an $r \times 1$ vector. For a more specific example, note that $\mathbf{X}'(\mathbf{I}_n - \frac{1}{n}\mathbf{J}_n)\mathbf{X} = \mathbf{X}'\mathbf{P}_n\mathbf{P}'_n\mathbf{X} = \mathbf{Z}'\mathbf{Z}$ where $\mathbf{P}'_n$ is the $(n-1) \times n$ lower portion of an n dimensional Helmert matrix and $\mathbf{Z} = \mathbf{P}'_n\mathbf{X}$ is an $(n-1) \times 1$ vector.

Later sections of the text cover covariance matrices and quadratic forms within the context of complete, balanced data structures. A data set is complete if all combinations of the levels of the factors contain data. A data set is balanced if the number of observations in each level of any factor is constant. Kronecker product notation will prove very useful when discussing covariance matrices and quadratic forms for balanced data structures. The next section of the text therefore provides some useful Kronecker product results.

1.2 KRONECKER PRODUCTS

Kronecker products will be used extensively in this text. In this section the Kronecker product operation is defined and a number of related theorems are listed without proof.

Definition 1.2.1 *Kronecker Product*: If $\mathbf{A}$ is an $r \times s$ matrix with ij^{th} element a_{ij} for $i = 1, \ldots, r$ and $j = 1, \ldots, s$, and $\mathbf{B}$ is any $t \times v$ matrix, then the Kronecker product of $\mathbf{A}$ and $\mathbf{B}$, denoted by $\mathbf{A} \otimes \mathbf{B}$, is the $rt \times sv$ matrix formed by multiplying each a_{ij} element by the entire matrix $\mathbf{B}$. That is,

$$\mathbf{A} \otimes \mathbf{B} = \begin{bmatrix} a_{11}\mathbf{B} & a_{12}\mathbf{B} & \cdots & a_{1s}\mathbf{B} \\ a_{21}\mathbf{B} & a_{22}\mathbf{B} & \cdots & a_{2s}\mathbf{B} \\ \vdots & \vdots & \ddots & \vdots \\ a_{r1}\mathbf{B} & a_{r2}\mathbf{B} & \cdots & a_{rs}\mathbf{B} \end{bmatrix}.$$

Theorem 1.2.1 *Let $\mathbf{A}$ and $\mathbf{B}$ be any matrices. Then $(\mathbf{A} \otimes \mathbf{B})' = \mathbf{A}' \otimes \mathbf{B}'$.*

Example 1.2.1 $[\mathbf{1}_a \otimes (2, 1, 4)]' = \mathbf{1}'_a \otimes (2, 1, 4)'$.

Theorem 1.2.2 *Let $\mathbf{A}, \mathbf{B}$, and $\mathbf{C}$ be any matrices and let a be a scalar. Then $a\mathbf{A} \otimes \mathbf{B} \otimes \mathbf{C} = a(\mathbf{A} \otimes \mathbf{B}) \otimes \mathbf{C} = \mathbf{A} \otimes (a\mathbf{B} \otimes \mathbf{C})$.*

Example 1.2.2 $\frac{1}{a}\mathbf{J}_a \otimes \mathbf{J}_b \otimes \mathbf{J}_c = \frac{1}{a}[(\mathbf{J}_a \otimes \mathbf{J}_b) \otimes \mathbf{J}_c] = \mathbf{J}_a \otimes (\frac{1}{a}\mathbf{J}_b \otimes \mathbf{J}_c)$.

Theorem 1.2.3 *Let $\mathbf{A}$ and $\mathbf{B}$ be any square matrices. Then $tr(\mathbf{A} \otimes \mathbf{B}) = [tr(\mathbf{A})][tr(\mathbf{B})]$.*

Example 1.2.3 $\text{tr}[(\mathbf{I}_a - \frac{1}{a}\mathbf{J}_a) \otimes (\mathbf{I}_n - \frac{1}{n}\mathbf{J}_n)] = \text{tr}[(\mathbf{I}_a - \frac{1}{a}\mathbf{J}_a)]\text{tr}[(\mathbf{I}_n - \frac{1}{n}\mathbf{J}_n)] = (a-1)(n-1)$.

Theorem 1.2.4 *Let $\mathbf{A}$ be an $r \times s$ matrix, $\mathbf{B}$ be a $t \times u$ matrix, $\mathbf{C}$ be an $s \times v$ matrix, and $\mathbf{D}$ be a $u \times w$ matrix. Then $(\mathbf{A} \otimes \mathbf{B})(\mathbf{C} \otimes \mathbf{D}) = \mathbf{AC} \otimes \mathbf{BD}$.*

Example 1.2.4 $[\mathbf{I}_a \otimes \mathbf{J}_n][(\mathbf{I}_a - \frac{1}{a}\mathbf{J}_a) \otimes \frac{1}{n}\mathbf{J}_n] = \mathbf{I}_a(\mathbf{I}_a - \frac{1}{a}\mathbf{J}_a) \otimes \frac{1}{n}\mathbf{J}_n\mathbf{J}_n = (\mathbf{I}_a - \frac{1}{a}\mathbf{J}_a) \otimes \mathbf{J}_n$.

Theorem 1.2.5 *Let $\mathbf{A}$ and $\mathbf{B}$ be $m \times m$ and $n \times n$ nonsingular matrices, respectively. Then the inverse of $\mathbf{A} \otimes \mathbf{B}$ is $(\mathbf{A} \otimes \mathbf{B})^{-1} = \mathbf{A}^{-1} \otimes \mathbf{B}^{-1}$.*

Example 1.2.5 $[(\mathbf{I}_a + \alpha\mathbf{J}_a) \otimes (\mathbf{I}_n + \beta\mathbf{J}_n)]^{-1} = [\mathbf{I}_a - (\alpha/(1 + a\alpha))\mathbf{J}_a] \otimes [\mathbf{I}_n - (\beta/(1 + n\beta))\mathbf{J}_n]$ for $0 < \alpha, \beta$.

Theorem 1.2.6 *Let* $\mathbf{A}$ *and* $\mathbf{B}$ *be* $m \times n$ *matrices and let* $\mathbf{C}$ *be a* $p \times q$ *matrix. Then* $(\mathbf{A} + \mathbf{B}) \otimes \mathbf{C} = (\mathbf{A} \otimes \mathbf{C}) + (\mathbf{B} \otimes \mathbf{C})$.

Example 1.2.6 $(\mathbf{I}_a - \frac{1}{a}\mathbf{J}_a) \otimes (\mathbf{I}_n - \frac{1}{n}\mathbf{J}_n) = [\mathbf{I}_a \otimes (\mathbf{I}_n - \frac{1}{n}\mathbf{J}_n)] - [\frac{1}{a}\mathbf{J}_a \otimes (\mathbf{I}_n - \frac{1}{n}\mathbf{J}_n)] = \mathbf{I}_a \otimes \mathbf{I}_n - \mathbf{I}_a \otimes \frac{1}{n}\mathbf{J}_n - \frac{1}{a}\mathbf{J}_a \otimes \mathbf{I}_n + \frac{1}{a}\mathbf{J}_a \frac{1}{n}\mathbf{J}_n$.

Theorem 1.2.7 *Let* $\mathbf{A}$ *be an* $m \times m$ *matrix with eigenvalues* $\alpha_1, \ldots, \alpha_m$ *and let* $\mathbf{B}$ *be an* $n \times n$ *matrix with eigenvalues* $\beta_1, \ldots, \beta_n$. *Then the eigenvalues of* $\mathbf{A} \otimes \mathbf{B}$ *(or* $\mathbf{B} \otimes \mathbf{A}$*) are the* mn *values* $\alpha_i \beta_j$ *for* $i = 1, \ldots, m$ *and* $j = 1, \ldots, n$.

Example 1.2.7 From Example 1.1.8, the eigenvalues of $\mathbf{I}_n - \frac{1}{n}\mathbf{J}_n$ are 1 with multiplicity $n - 1$ and 0 with multiplicity 1. Likewise, $\mathbf{I}_a - \frac{1}{a}\mathbf{J}_a$ has eigenvalues 1 with multiplicity $a - 1$ and 0 with multiplicity 1. Therefore, the eigenvalues of $(\mathbf{I}_a - \frac{1}{a}\mathbf{J}_a) \otimes (\mathbf{I}_n - \frac{1}{n}\mathbf{J}_n)$ are 1 with multiplicity $(a - 1)(n - 1)$ and 0 with multiplicity $a + n - 1$.

Theorem 1.2.8 *Let* $\mathbf{A}$ *be an* $m \times n$ *matrix of rank* r *and let* $\mathbf{B}$ *be a* $p \times q$ *matrix of rank* s. *Then* $\mathbf{A} \otimes \mathbf{B}$ *has rank* rs.

Example 1.2.8 $\text{rank}[(\mathbf{I}_a - \frac{1}{a}\mathbf{J}_a) \otimes \mathbf{I}_n] = [\text{rank}\,(\mathbf{I}_a - \frac{1}{a}\mathbf{J}_a)]\,[\text{rank}\,(\mathbf{I}_n)] = (a-1)n$.

Theorem 1.2.9 *Let* $\mathbf{A}$ *be an* $m \times m$ *symmetric, idempotent matrix of rank* r *and let* $\mathbf{B}$ *be an* $n \times n$ *symmetric, idempotent matrix of rank* s. *Then* $\mathbf{A} \otimes \mathbf{B}$ *is an* $mn \times mn$ *symmetric, idempotent matrix where* $tr(\mathbf{A} \otimes \mathbf{B}) = rank\,(\mathbf{A} \otimes \mathbf{B}) = rs$.

Example 1.2.9 $\text{tr}[(\mathbf{I}_a - \frac{1}{a}\mathbf{J}_a) \otimes \mathbf{I}_n] = \text{rank}[(\mathbf{I}_a - \frac{1}{a}\mathbf{J}_a) \otimes \mathbf{I}_n] = (a - 1)n$.

The following example demonstrates that Kronecker products are useful for describing sums of squares in complete, balanced ANOVA problems.

Example 1.2.10 Consider a one-way classification where there are r replicate observations nested in each of the t levels of a fixed factor. Let y_{ij} represent the jth replicate observation in the ith level of the fixed factor for $i = 1, \ldots, t$ and $j = 1, \ldots, r$. Define the $tr \times 1$ vector of observations $\mathbf{Y} = (y_{11}, \ldots, y_{1r}, \ldots, y_{t1}, \ldots, y_{tr})'$. The layout for this experiment is given in Figure 1.2.1. The ANOVA table is presented in Table 1.2.1. Note that the sums of squares are written in summation notation and as quadratic forms, $\mathbf{Y}'\mathbf{A}_m\mathbf{Y}$, for $m = 1, \ldots, 4$. The objective is to demonstrate that the $tr \times tr$ matrices $\mathbf{A}_m$ can be expressed as Kronecker products. Each matrix $\mathbf{A}_m$ is derived later. Note

$$\bar{y}_{..} = [1/(rt)]\mathbf{1}'_{tr}\mathbf{Y} = [1/(rt)](\mathbf{1}_t \otimes \mathbf{1}_r)'\mathbf{Y}$$

	Fixed Factor (i)			
Replicates (j)	1	2	$\cdots$	t
1	y_{11}	y_{21}	$\cdots$	y_{t1}
2	y_{12}	y_{22}	$\cdots$	y_{t2}
$\vdots$	$\vdots$	$\vdots$	$\ddots$	$\vdots$
r	y_{1r}	y_{2r}	$\cdots$	y_{tr}

Figure 1.2.1 One-Way Layout.

Table 1.2.1
One-Way ANOVA Table

Source	df	SS
Mean	1	$\sum_{i=1}^t \sum_{j=1}^r \bar{y}^2.. \quad = \mathbf{Y}'\mathbf{A}_1\mathbf{Y}$
Fixed factor	$t-1$	$\sum_{i=1}^t \sum_{j=1}^r (\bar{y}_{i.} - \bar{y}..)^2 = \mathbf{Y}'\mathbf{A}_2\mathbf{Y}$
Nested replicates	$t(r-1)$	$\sum_{i=1}^t \sum_{j=1}^r (y_{ij} - \bar{y}_{i.})^2 = \mathbf{Y}'\mathbf{A}_3\mathbf{Y}$
Total	tr	$\sum_{i=1}^t \sum_{j=1}^r y_{ij}^2 \quad = \mathbf{Y}'\mathbf{A}_4\mathbf{Y}$

and

$$(\overline{y}_{1.}, \ldots, \overline{y}_{t.})' = \left(\mathbf{I}_t \otimes \frac{1}{r}\mathbf{1}'_r\right)\mathbf{Y}.$$

Therefore, the sum of squares due to the mean is given by

$$\begin{aligned}\sum_{i=1}^{t}\sum_{j=1}^{r}\overline{y}^2.. &= rt\overline{y}^2..\\ &= rt\{[1/(rt)](\mathbf{1}_t \otimes \mathbf{1}_r)'\mathbf{Y}\}'\{[1/(rt)](\mathbf{1}_t \otimes \mathbf{1}_r)'\mathbf{Y}\}\\ &= \mathbf{Y}'\left[\frac{1}{t}\mathbf{J}_t \otimes \frac{1}{r}\mathbf{J}_r\right]\mathbf{Y}\\ &= \mathbf{Y}'\mathbf{A}_1\mathbf{Y}\end{aligned}$$

where the $tr \times tr$ matrix $\mathbf{A}_1 = \frac{1}{t}\mathbf{J}_t \otimes \frac{1}{r}\mathbf{J}_r$. The sum of squares due to the fixed factor is

$$\begin{aligned}\sum_{i=1}^{t}\sum_{j=1}^{r}(\bar{y}_{i.}-\bar{y}_{..})^2 &= r\sum_{i=1}^{t}\bar{y}_{i.}^2 - tr\bar{y}^2_{..}\\ &= r\left[\left(\mathbf{I}_t\otimes\frac{1}{r}\mathbf{1}'_r\right)\mathbf{Y}\right]'\left[\left(\mathbf{I}_t\otimes\frac{1}{r}\mathbf{1}'_r\right)\mathbf{Y}\right]-\mathbf{Y}'\left[\frac{1}{t}\mathbf{J}_t\otimes\frac{1}{r}\mathbf{J}_r\right]\mathbf{Y}\\ &= \mathbf{Y}'\left[\mathbf{I}_t\otimes\frac{1}{r}\mathbf{J}_r\right]\mathbf{Y}-\mathbf{Y}'\left[\frac{1}{t}\mathbf{J}_t\otimes\frac{1}{r}\mathbf{J}_r\right]\mathbf{Y}\\ &= \mathbf{Y}'\left[\left(\mathbf{I}_t-\frac{1}{t}\mathbf{J}_t\right)\otimes\frac{1}{r}\mathbf{J}_r\right]\mathbf{Y}\\ &= \mathbf{Y}'\mathbf{A}_2\mathbf{Y}\end{aligned}$$

where the $tr \times tr$ matrix $\mathbf{A}_2 = (\mathbf{I}_t - \frac{1}{t}\mathbf{J}_t) \otimes \frac{1}{r}\mathbf{J}_r$. The sum of squares due to the nested replicates is

$$\begin{aligned}\sum_{i=1}^{t}\sum_{j=1}^{r}(y_{ij}-\bar{y}_{i.})^2 &= \sum_{i=1}^{t}\sum_{j=1}^{r}y_{ij}^2 - r\sum_{i=1}^{t}\bar{y}_{i.}^2\\ &= \mathbf{Y}'[\mathbf{I}_t\otimes\mathbf{I}_r]\mathbf{Y}-\mathbf{Y}'\left[\mathbf{I}_t\otimes\frac{1}{r}\mathbf{J}_r\right]\mathbf{Y}\\ &= \mathbf{Y}'\left[\mathbf{I}_t\otimes\left(\mathbf{I}_r-\frac{1}{r}\mathbf{J}_r\right)\right]\mathbf{Y}\\ &= \mathbf{Y}'\mathbf{A}_3\mathbf{Y}\end{aligned}$$

where the $tr \times tr$ matrix $\mathbf{A}_3 = \mathbf{I}_t \otimes (\mathbf{I}_r - \frac{1}{r}\mathbf{J}_r)$. Finally, the sum of squares total is

$$\sum_{i=1}^{t}\sum_{j=1}^{r}y_{ij}^2 = \mathbf{Y}'[\mathbf{I}_t\otimes\mathbf{I}_r]\mathbf{Y} = \mathbf{Y}'\mathbf{A}_4\mathbf{Y}$$

where the $tr \times tr$ matrix $\mathbf{A}_4 = \mathbf{I}_t \otimes \mathbf{I}_r$.

The derivations of the sums of squares matrices $\mathbf{A}_m$ can be tedious. In Chapter 4 an algorithm is provided for determining the sums of squares matrices for complete, balanced designs with any number of main effects, interactions, or nested factors. This algorithm makes the calculation of sums of squares matrices $\mathbf{A}_m$ very simple.

This section concludes with a matrix operator that will prove useful in Chapter 8.

Definition 1.2.2 *BIB Product*: If $\mathbf{B}$ is a $c \times d$ matrix and $\mathbf{A}$ is an $a \times b$ matrix where each column of $\mathbf{A}$ has $c \leq a$ nonzero elements and $a - c$ zero elements, then the BIB product of matrices $\mathbf{A}$ and $\mathbf{B}$, denoted by $\mathbf{A} \square \mathbf{B}$, is the $a \times bd$ matrix formed by multiplying each zero element in the ith column of $\mathbf{A}$ by a $1 \times d$ row vector of zeros and multiplying the jth nonzero element in the ith column of $\mathbf{A}$ by the jth row of $\mathbf{B}$ for $i = 1, \ldots, b$ and $j = 1, \ldots, c$.

Example 1.2.11 Let the 3×3 matrix $\mathbf{A} = \mathbf{J}_3 - \mathbf{I}_3$ and the 2×2 matrix $\mathbf{B} = \mathbf{I}_2 - \frac{1}{2}\mathbf{J}_2$. Then the 3×6 BIB product matrix

$$\mathbf{A} \square \mathbf{B} = \begin{bmatrix} 0 & 1 & 1 \\ 1 & 0 & 1 \\ 1 & 1 & 0 \end{bmatrix} \square \begin{bmatrix} 1/2 & -1/2 \\ -1/2 & 1/2 \end{bmatrix}$$

$$= \begin{bmatrix} 0 & 0 & 1/2 & -1/2 & 1/2 & -1/2 \\ 1/2 & -1/2 & 0 & 0 & -1/2 & 1/2 \\ -1/2 & 1/2 & -1/2 & 1/2 & 0 & 0 \end{bmatrix}.$$

Theorem 1.2.10 *If* $\mathbf{A}$ *is an* $a \times b$ *matrix with* $c \le a$ *nonzero elements per column and* $a - c$ *zeros per column;* $\mathbf{B}_1$, $\mathbf{B}_2$, *and* $\mathbf{B}$ *are each* $c \times d$ *matrices;* $\mathbf{D}$ *is a* $b \times b$ *diagonal matrix of rank* b*;* $\mathbf{Z}$ *is a* $d \times 1$ *vector; and* $\mathbf{Y}$ *is a* $c \times 1$ *vector, then*

(i) $[\mathbf{A} \square \mathbf{B}][\mathbf{D} \otimes \mathbf{Z}] = \mathbf{AD} \square \mathbf{BZ}$

(ii) $[\mathbf{A} \square \mathbf{Y}][\mathbf{D} \otimes \mathbf{Z}'] = \mathbf{AD} \square \mathbf{YZ}'$

(iii) $\mathbf{D} \square \mathbf{Y}' = \mathbf{D} \otimes \mathbf{Y}'$

(iv) $[\mathbf{A} \square (\mathbf{B}_1 + \mathbf{B}_2)] = [\mathbf{A} \square \mathbf{B}_1] + [\mathbf{A} \square \mathbf{B}_2]$.

Example 1.2.12 Let $\mathbf{A}$ be any 3×3 matrix with two nonzero elements per column and let $\mathbf{B} = \mathbf{I}_2 - \frac{1}{2}\mathbf{J}_2$. Then

$$\begin{aligned}
[\mathbf{A} \square \mathbf{B}][\mathbf{I}_3 \otimes \mathbf{J}_2] &= \left[\mathbf{A} \square \left(\mathbf{I}_2 - \frac{1}{2}\mathbf{J}_2\right)\right] [\mathbf{I}_3 \otimes \mathbf{1}_2][\mathbf{I}_3 \otimes \mathbf{1}_2'] \\
&= \left[\mathbf{A} \square \left(\left(\mathbf{I}_2 - \frac{1}{2}\mathbf{J}_2\right) \mathbf{1}_2\right)\right] [\mathbf{I}_3 \otimes \mathbf{1}_2'] \\
&= [\mathbf{A} \square \mathbf{0}_{2\times 1}][\mathbf{I}_3 \otimes \mathbf{1}_2'] \\
&= \mathbf{0}_{3\times 6}.
\end{aligned}$$

1.3 RANDOM VECTORS

Let the $n \times 1$ random vector $\mathbf{Y} = (Y_1, Y_2, \ldots, Y_n)'$ where Y_i is a random variable for $i = 1, \ldots, n$. The vector $\mathbf{Y}$ is a random entity. Therefore, $\mathbf{Y}$ has an expectation; each element of $\mathbf{Y}$ has a variance; and any two elements of $\mathbf{Y}$ have a covariance (assuming the expectations, variances, and covariances exist). The following definitions and theorems describe the structure of random vectors.

Definition 1.3.1 *Joint Probability Distribution*: The probability distribution of the $n \times 1$ random vector $\mathbf{Y} = (Y_1, \ldots, Y_n)'$ equals the joint probability distribution of $Y_1, \ldots, Y_n$. Denote the distribution of $\mathbf{Y}$ by $f_{\mathbf{Y}}(y) - f_{\mathbf{Y}}(y_1, \ldots, y_n)$.

Definition 1.3.2 *Expectation of a Random Vector*: The expected value of the $n \times 1$ random vector $\mathbf{Y} = (Y_1, \ldots, Y_n)'$ is given by $\mathrm{E}(\mathbf{Y}) = [\mathrm{E}(Y_1), \ldots, \mathrm{E}(Y_n)]'$.

Definition 1.3.3 *Covariance Matrix of a Random Vector* $\mathbf{Y}$: The $n \times 1$ random vector $\mathbf{Y} = (Y_1, \ldots, Y_n)'$ has $n \times n$ covariance matrix given by

$$\mathrm{cov}(\mathbf{Y}) = \mathrm{E}\{[\mathbf{Y} - \mathrm{E}(\mathbf{Y})][\mathbf{Y} - \mathrm{E}(\mathbf{Y})]'\}.$$

The ij^{th} element of $\mathrm{cov}(\mathbf{Y})$ equals $\mathrm{E}\{[Y_i - \mathrm{E}(Y_i)][Y_j - \mathrm{E}(Y_j)]\}$ for $i, j = 1, \ldots, n$.

Definition 1.3.4 *Linear Transformations of a Random Vector* $\mathbf{Y}$: If $\mathbf{B}$ is an $m \times n$ matrix of constants and $\mathbf{Y}$ is an $n \times 1$ random vector, then the $m \times 1$ random vector $\mathbf{BY}$ represents m linear transformations of $\mathbf{Y}$.

The following theorem provides the covariance matrix of linear transformations of a random vector.

Theorem 1.3.1 *If* $\mathbf{B}$ *is an* $m \times n$ *matrix of constants,* $\mathbf{Y}$ *is an* $n \times 1$ *random vector, and* $\mathrm{cov}(\mathbf{Y})$ *is the* $n \times n$ *covariance matrix of* $\mathbf{Y}$*, then the* $m \times 1$ *random vector* $\mathbf{BY}$ *has an* $m \times m$ *covariance matrix given by* $\mathbf{B}[\mathrm{cov}(\mathbf{Y})]\mathbf{B}'$.

Proof:

$$\begin{aligned}
\mathrm{cov}(\mathbf{BY}) &= \mathrm{E}\{[\mathbf{BY} - \mathrm{E}(\mathbf{BY})][\mathbf{BY} - \mathrm{E}(\mathbf{BY})]'\} \\
&= \mathrm{E}\{\mathbf{B}[\mathbf{Y} - \mathrm{E}(\mathbf{Y})][\mathbf{Y} - \mathrm{E}(\mathbf{Y})]'\mathbf{B}'\} \\
&= \mathbf{B}\{\mathrm{E}\{[\mathbf{Y} - \mathrm{E}(\mathbf{Y})][\mathbf{Y} - \mathrm{E}(\mathbf{Y})]'\}\}\mathbf{B}' \\
&= \mathbf{B}[\mathrm{cov}(\mathbf{Y})]\mathbf{B}'. \quad \blacksquare
\end{aligned}$$

The next theorem provides the expected value of a quadratic form.

Theorem 1.3.2 *Let* $\mathbf{Y}$ *be an* $n \times 1$ *random vector with mean vector* $\mu = \mathrm{E}(\mathbf{Y})$ *and* $n \times n$ *covariance matrix* $\mathbf{\Sigma} = \mathrm{cov}(\mathbf{Y})$ *then* $\mathrm{E}(\mathbf{Y}'\mathbf{AY}) = \mathrm{tr}(\mathbf{A\Sigma}) + \mu'\mathbf{A}\mu$ *where* $\mathbf{A}$ *is any* $n \times n$ *symmetric matrix of constants.*

Proof: Since $(\mathbf{Y}-\mu)'\mathbf{A}(\mathbf{Y}-\mu)$ is a scalar and using Result 1.5,

$$(\mathbf{Y}-\mu)'\mathbf{A}(\mathbf{Y}-\mu) = \text{tr}[(\mathbf{Y}-\mu)'\mathbf{A}(\mathbf{Y}-\mu)] = \text{tr}[\mathbf{A}(\mathbf{Y}-\mu)(\mathbf{Y}-\mu)'].$$

Therefore,

$$\begin{aligned}
\text{E}[\mathbf{Y}'\mathbf{A}\mathbf{Y}] &= \text{E}[(\mathbf{Y}-\mu)'\mathbf{A}(\mathbf{Y}-\mu) + 2\mathbf{Y}'\mathbf{A}\mu - \mu'\mathbf{A}\mu] \\
&= \text{E}\{\text{tr}[\mathbf{A}(\mathbf{Y}-\mu)(\mathbf{Y}-\mu)']\} + 2\text{E}(\mathbf{Y}'\mathbf{A}\mu) - \mu'\mathbf{A}\mu \\
&= \text{tr}[\mathbf{A}\text{E}\{(\mathbf{Y}-\mu)(\mathbf{Y}-\mu)'\}] + \mu'\mathbf{A}\mu \\
&= \text{tr}[\mathbf{A}\mathbf{\Sigma}] + \mu'\mathbf{A}\mu. \qquad \blacksquare
\end{aligned}$$

The moment generating function of a random vector is used extensively in the next chapter. The following definitions and theorems provide some general moment generating function results.

Definition 1.3.5 *Moment Generating Function (MGF) of a Random Vector* $\mathbf{Y}$: The MGF of an $n \times 1$ random vector $\mathbf{Y}$ is given by

$$m_{\mathbf{Y}}(\mathbf{t}) = \text{E}(e^{\mathbf{t}'\mathbf{Y}})$$

where the $n \times 1$ vector of constants $\mathbf{t} = (t_1, \ldots, t_n)'$ if the expectation exists for $-h < t_i < h$ where $h > 0$ and $i = 1, \ldots, n$.

There is a one-to-one correspondence between the probability distribution of $\mathbf{Y}$ and the MGF of $\mathbf{Y}$, if the MGF exists. Therefore, the probability distribution of $\mathbf{Y}$ can be identified if the MGF of $\mathbf{Y}$ can be found. The following two theorems and corollary are used to derive the MGF of a random vector $\mathbf{Y}$.

Theorem 1.3.3 *Let the $n \times 1$ random vector $\mathbf{Y} = (\mathbf{Y}_1', \mathbf{Y}_2', \ldots, \mathbf{Y}_m')'$ where $\mathbf{Y}_i$ is an $n_i \times 1$ random vector for $i = 1, \ldots, m$ and $n = \sum_{i=1}^m n_i$. Let $m_{\mathbf{Y}}(.), m_{\mathbf{Y}_1}(.), \ldots, m_{\mathbf{Y}_m}(.)$ represent the MGFs of $\mathbf{Y}, \mathbf{Y}_1, \ldots, \mathbf{Y}_m$ respectively. The vectors $\mathbf{Y}_1, \ldots, \mathbf{Y}_m$ are mutually independent if and only if*

$$m_{\mathbf{Y}}(\mathbf{t}) = m_{\mathbf{Y}_1}(\mathbf{t}_1)m_{\mathbf{Y}_2}(\mathbf{t}_2)\ldots m_{\mathbf{Y}_m}(\mathbf{t}_m)$$

for all $\mathbf{t} = (\mathbf{t}_1', \ldots .\mathbf{t}_m')'$ on the open rectangle around 0.

Theorem 1.3.4 *If $\mathbf{Y}$ is an $n \times 1$ random vector, $\mathbf{g}$ is an $n \times 1$ vector of constants, and c is a scalar constant, then*

$$m_{\mathbf{g}'\mathbf{Y}}(c) = m_{\mathbf{Y}}(c\mathbf{g}).$$

Proof: $m_{\mathbf{g}'\mathbf{Y}}(c) = \text{E}[e^{c\mathbf{g}'\mathbf{Y}}] = \text{E}[e^{(c\mathbf{g})'\mathbf{Y}}] = m_{\mathbf{Y}}(c\mathbf{g}).$ $\blacksquare$

Corollary 1.3.4 *Let $m_{\mathbf{Y}_1}(.), \ldots, m_{\mathbf{Y}_m}(.)$ represent the MGFs of the independent random variables $Y_1, \ldots, Y_m$, respectively. If $\mathbf{Z} = \sum_{i=1}^m Y_i$ then the MGF of $\mathbf{Z}$ is given by*

$$m_{\mathbf{Z}}(s) = \prod_{i=1}^m m_{Y_i}(s).$$

Proof:

$$\begin{aligned} m_{\mathbf{Z}}(s) = m_{\mathbf{1}'_m\mathbf{Y}}(s) &= m_{\mathbf{Y}}(s\mathbf{1}_m) \quad \text{by Theorem 1.3.4} \\ &= \prod_{i=1}^m m_{Y_i}(s) \quad \text{by Theorem 1.3.3.} \quad \blacksquare \end{aligned}$$

Moment generating functions are used in the next example to derive the distribution of the sum of independent chi-square random variables.

Example 1.3.1 Let $Y_1, \ldots, Y_m$ be m independent central chi-square random variables where Y_i and n_i degrees of freedom for $i = 1, \ldots, m$. For any i

$$\begin{aligned} m_{Y_i}(t) &= \mathrm{E}(e^{tY_i}) \\ &= \int_0^\infty e^{ty_i} \left[\Gamma(n_i/2)2^{n_i/2}\right]^{-1} y_i^{n_i/2-1} e^{-y_i/2} dy_i \\ &= \left[\Gamma(n_i/2)2^{n_i/2}\right]^{-1} \int_0^\infty y_i^{n_i/2-1} e^{-y_i/(1/2-t)^{-1}} dy_i \\ &= \left[\Gamma(n_i/2)2^{n_i/2}\right]^{-1} \left[\Gamma(n_i/2)(1/2-t)^{-n_i/2}\right] \\ &= (1-2t)^{-n_i/2}. \end{aligned}$$

Let $\mathbf{Z} = \sum_{i=1}^m Y_i$. By Corollary 1.3.4,

$$\begin{aligned} m_{\mathbf{Z}}(t) &= \prod_{i=1}^m m_{Y_i}(t) \\ &= (1-2t)^{-\sum_{i=1}^m n_i/2}. \end{aligned}$$

Therefore, $\sum_{i=1}^m Y_i$ is distributed as a central chi-square random variable with $\sum_{i=1}^m n_i$ degrees of freedom.

The next theorem is useful when dealing with functions of independent random vectors.

Theorem 1.3.5 *Let $g_1(\mathbf{Y}_1), \ldots . g_m(\mathbf{Y}_m)$ be m functions of the random vectors $\mathbf{Y}_1, \ldots, \mathbf{Y}_m$, respectively. If $\mathbf{Y}_1, \ldots, \mathbf{Y}_m$ are mutually independent, then $g_1, \ldots, g_m$ are mutually independent.*

The next example demonstrates that the sum of squares of n independent $\mathrm{N}_1(0, 1)$ random variables has a central chi-square distribution with n degrees of freedom.

Example 1.3.2 Let $Z_1, \ldots, Z_n$ be a random sample of normally distributed random variables with mean 0 and variance 1. Let $Y_i = Z_i^2$ for $i = 1, \ldots, n$. The moment generating function of Y_i is

$$\begin{aligned} m_{Y_i}(t) = m_{Z_i^2}(t) &= \mathrm{E}(e^{tZ_i^2}) \\ &= \int_{-\infty}^{\infty} (2\pi)^{-1/2} e^{tz_i^2 - z_i^2/2} dz_i \\ &= \int_{-\infty}^{\infty} (2\pi)^{-1/2} e^{-(1-2t)z_i^2/2} dz_i \\ &= (1-2t)^{-1/2}. \end{aligned}$$

That is, each Y_i has a central chi-square distribution with one degree of freedom. Furthermore, by Theorem 1.3.5, the Y_i's are independent random variables. Therefore, by Example 1.3.1, $\sum_{i=1}^n Y_i = \sum_{i=1}^n Z_i^2$ is a central chi-square random variable with n degrees of freedom.

EXERCISES

1. Find an $mn \times (n-1)$ matrix $\mathbf{P}$ such that $\mathbf{PP}' = \frac{1}{m}\mathbf{J}_m \otimes (\mathbf{I}_n - \frac{1}{n}\mathbf{J}_n)$.

2. Let $S_1 = \sum_{i=1}^n U_i(V_i - \overline{V})$, $S_2 = \sum_{i=1}^n (U_i - \overline{U})^2$, and $S_3 = \sum_{i=1}^n (V_i - \overline{V})^2$. If $\mathbf{A} = (\mathbf{I}_n - \frac{1}{n}\mathbf{J}_n)(\mathbf{I}_n - \mathbf{VV}')(\mathbf{I}_n - \frac{1}{n}\mathbf{J}_n)$, $\mathbf{U} = (U_1, \ldots, U_n)'$, and $\mathbf{V} = (V_1, \ldots, V_n)'$, is the statement $S_2 - (S_1^2/S_3) = \mathbf{U}'\mathbf{AU}$ true or false? If the statement is true, verify that $\mathbf{A}$ is correct. If the statement $\mathbf{A}$ is false, find the correct form of $\mathbf{A}$.

3. Let $\mathbf{V} = \mathbf{I}_m \otimes [\sigma_1^2\mathbf{I}_n + \sigma_2^2\mathbf{J}_n]$, $\mathbf{A}_1 = (\mathbf{I}_m - \frac{1}{m}\mathbf{J}_m) \otimes \frac{1}{n}\mathbf{J}_n$, and $\mathbf{A}_2 = \mathbf{I}_m \otimes (\mathbf{I}_n - \frac{1}{n}\mathbf{J}_n)$.

   (a) Show $\mathbf{A}_1\mathbf{A}_2 = \mathbf{0}_{mn \times mn}$.

   (b) Define an $mn \times mn$ matrix $\mathbf{C}$ such that $\mathbf{I}_{mn} = \mathbf{A}_1 + \mathbf{A}_2 + \mathbf{C}$.

(c) Let the $mn \times 1$ vector $\mathbf{Y} = (Y_{11}, \ldots, Y_{mn})'$ and let $\mathbf{C}$ be defined as in part b. Is the following statement true or false?: $\mathbf{Y}'\mathbf{C}\mathbf{Y} = [\sum_{i=1}^{m} \sum_{j=1}^{n} Y_{ij}]^2/(mn)$. If the statement is true, verify it. If the statement is false, redefine $\mathbf{Y}'\mathbf{C}\mathbf{Y}$ in terms of the Y_{ij}'s.

(d) Define constants k_1, k_2, and $mn \times mn$ idempotent matrices $\mathbf{C}_1$ and $\mathbf{C}_2$ such that $\mathbf{A}_1\mathbf{V} = k_1\mathbf{C}_1$ and $\mathbf{A}_2\mathbf{V} = k_2\mathbf{C}_2$. Verify that $\mathbf{C}_1$ and $\mathbf{C}_2$ are idempotent.

4. Find the inverse of the matrix $0.4\mathbf{I}_4 + 0.6\mathbf{J}_4$.

5. Show that the inverse of the matrix $(\mathbf{I}_n + \mathbf{V}\mathbf{V}')$ is $\left(\frac{\mathbf{I}_n - \mathbf{V}\mathbf{V}'}{1+\mathbf{V}'\mathbf{V}}\right)$ where $\mathbf{V}$ is an $n \times 1$ vector.

6. Use the result in Exercise 5 to find the inverse of matrix $(a-b)\mathbf{I}_n + b\mathbf{J}_n$ where a and b are positive constants.

7. Let $\mathbf{V} = \mathbf{I}_n \otimes [(1-\rho)\mathbf{I}_2 + \rho\mathbf{J}_2]$ where $-1 < \rho < 1$.

   (a) Find the $2n$ eigenvalues of $\mathbf{V}$.

   (b) Find a nonsingular $2n \times 2n$ matrix $\mathbf{Q}$ such that $\mathbf{V} = \mathbf{Q}\mathbf{Q}'$.

8. Let $\mathbf{A}_1 = \frac{1}{m}\mathbf{J}_m \otimes \frac{1}{n}\mathbf{J}_n$, $\mathbf{A}_2 = (\mathbf{I}_m - \frac{1}{m}\mathbf{J}_m) \otimes \frac{1}{n}\mathbf{J}_n$, and $\mathbf{A}_3 = \mathbf{I}_m \otimes (\mathbf{I}_n - \frac{1}{n}\mathbf{J}_n)$. Find $\sum_{i=1}^{3} \mathbf{A}_i$, $\mathbf{A}_i\mathbf{A}_j$ for all $i, j = 1, 2, 3$ and $\sum_{i=1}^{3} c_i\mathbf{A}_i$ where $c_1 = c_2 = 1 + (m-1)b$ and $c_3 = 1 - b$ for $-\frac{1}{(m-1)} < b < 1$.

9. Define $n \times n$ matrices $\mathbf{P}$ and $\mathbf{D}$ such that $(a-b)\mathbf{I}_n + b\mathbf{J}_n = \mathbf{P}\mathbf{D}\mathbf{P}'$ where $\mathbf{D}$ is a diagonal matrix and a and b are constants.

10. Let

$$\mathbf{B} = \begin{bmatrix} \mathbf{I}_{n_1} & -\boldsymbol{\Sigma}_{12}\boldsymbol{\Sigma}_{22}^{-1} \\ \mathbf{0} & \mathbf{I}_{n_2} \end{bmatrix} \quad \text{and} \quad \boldsymbol{\Sigma} = \begin{bmatrix} \boldsymbol{\Sigma}_{11} & \boldsymbol{\Sigma}_{12} \\ \boldsymbol{\Sigma}_{21} & \boldsymbol{\Sigma}_{22} \end{bmatrix}$$

where $\boldsymbol{\Sigma}$ is a symmetric matrix and the $\boldsymbol{\Sigma}_{ij}$ are $n_i \times n_j$ matrices. Find $\mathbf{B}\boldsymbol{\Sigma}\mathbf{B}'$.

11. Let $\mathbf{A}_1 = \frac{1}{m}\mathbf{J}_m \otimes \frac{1}{n}\mathbf{J}_n$, $\mathbf{A}_2 = \mathbf{X}(\mathbf{X}'\mathbf{X})^{-1}\mathbf{X}'$, $\mathbf{A}_3 = \mathbf{I}_m \otimes (\mathbf{I}_n - \frac{1}{n}\mathbf{J}_n)$ and $\mathbf{X} = \mathbf{X}^+ \otimes \mathbf{1}_n$ where $\mathbf{X}^+$ is an $m \times p$ matrix such that $\mathbf{1}_m'\mathbf{X}^+ = \mathbf{0}_{1\times p}$. Find the $mn \times mn$ matrix $\mathbf{A}_4$ such that $\mathbf{I}_m \otimes \mathbf{I}_n = \sum_{i=1}^{4} \mathbf{A}_i$. Express $\mathbf{A}_4$ in its simplest form.

12. Is the matrix $\mathbf{A}_4$ in Exercise 11 idempotent?

13. Let $\mathbf{Y}$ be an $n \times 1$ random vector with $n \times n$ covariance matrix $\text{cov}(\mathbf{Y}) = \sigma_1^2\mathbf{I}_n + \sigma_2^2\mathbf{J}_n$. Define $\mathbf{Z} = \mathbf{P}'\mathbf{Y}$ where the $n \times n$ matrix $\mathbf{P} = (\mathbf{1}_n|\mathbf{P}_n)$ and the $n \times (n-1)$ matrix $\mathbf{P}_n$ are defined in Example 1.1.4. Find $\text{cov}(\mathbf{P}'\mathbf{Y})$.

14. Let $\mathbf{Y}$ be a $bt \times 1$ random vector with $\mathrm{E}(\mathbf{Y}) = \mathbf{1}_b \otimes (\mu_1, \ldots, \mu_t)'$ and $\mathrm{cov}(\mathbf{Y}) = \sigma_B^2[\mathbf{I}_b \otimes \mathbf{J}_t] + \sigma_{BT}^2[\mathbf{I}_b \otimes (\mathbf{I}_t - \frac{1}{t}\mathbf{J}_t)]$. Define $\mathbf{A}_1 = \frac{1}{b}\mathbf{J}_b \otimes \frac{1}{t}\mathbf{J}_t$, $\mathbf{A}_2 = (\mathbf{I}_b - \frac{1}{b}\mathbf{J}_b) \otimes \frac{1}{t}\mathbf{J}_t$, $\mathbf{A}_3 = (\mathbf{I}_b - \frac{1}{b}\mathbf{J}_b) \otimes (\mathbf{I}_t - \frac{1}{t}\mathbf{J}_t)$. Find $\mathrm{E}(\mathbf{Y}'\mathbf{A}_i\mathbf{Y})$ for $i = 1, 2, 3$.

15. Let the $btr \times 1$ vector $\mathbf{Y} = (Y_{111}, \ldots, Y_{11r}, Y_{121}, \ldots, Y_{12r}, \ldots, Y_{bt1}, \ldots, Y_{btr})'$ and let

$$S_1 = \sum_{i=1}^{b}\sum_{j=1}^{t}\sum_{k=1}^{r}\overline{Y}^2_{\ldots},$$

$$S_2 = \sum_{i=1}^{b}\sum_{j=1}^{t}\sum_{k=1}^{r}(\overline{Y}_{i..} - \overline{Y}_{\ldots})^2,$$

$$S_3 = \sum_{i=1}^{b}\sum_{j=1}^{t}\sum_{k=1}^{r}(\overline{Y}_{.j.} - \overline{Y}_{\ldots})^2,$$

$$S_4 = \sum_{i=1}^{b}\sum_{j=1}^{t}\sum_{k=1}^{r}(\overline{Y}_{ij.} - \overline{Y}_{i..} - \overline{Y}_{.j.} + \overline{Y}_{\ldots})^2,$$

$$S_5 = \sum_{i=1}^{b}\sum_{j=1}^{t}\sum_{k=1}^{r}(Y_{ijk} - \overline{Y}_{ij.})^2.$$

Derive $btr \times btr$ matrices $\mathbf{A}_m$ for $m = 1, \ldots, 5$ where $S_m = \mathbf{Y}'\mathbf{A}_m\mathbf{Y}$.

2 Multivariate Normal Distribution

In later chapters we will investigate linear models with normally distributed error structures. Therefore, this chapter concentrates on some important concepts related to the multivariate normal distribution.

2.1 MULTIVARIATE NORMAL DISTRIBUTION FUNCTION

Let $Z_1, \ldots, Z_n$ be independent, identically distributed normal random variables with mean 0 and variance 1. The marginal distribution of Z_i is

$$f_{Z_i}(z_i) = (2\pi)^{-1/2} e^{-z_i^2/2} \qquad -\infty < z_i < \infty$$

for $i = 1, \ldots, n$. Since the Z_i's are independent random variables, the joint probability distribution of the $n \times 1$ random vector $\mathbf{Z} = (Z_1, \ldots, Z_n)'$ is

$$\begin{aligned} f_{\mathbf{Z}}(\mathbf{z}) &= (2\pi)^{-n/2} e^{-\sum_{i=1}^{n} z_i^2/2} \\ &= (2\pi)^{-n/2} e^{-\mathbf{z}'\mathbf{z}/2} \qquad -\infty < z_i < \infty \end{aligned}$$

for $i = 1, \ldots, n$. Let the $n \times 1$ vector $\mathbf{Y} = \mathbf{GZ} + \boldsymbol{\mu}$ where $\mathbf{G}$ is an $n \times n$ nonsingular matrix and $\boldsymbol{\mu}$ is an $n \times 1$ vector. The joint distribution of the $n \times 1$ random vector $\mathbf{Y}$ is

$$f_{\mathbf{Y}}(\mathbf{y}) = |\boldsymbol{\Sigma}|^{-1/2}(2\pi)^{-n/2}e^{-\{(\mathbf{y}-\boldsymbol{\mu})'\boldsymbol{\Sigma}^{-1}(\mathbf{y}-\boldsymbol{\mu})\}/2}.$$

where $\boldsymbol{\Sigma} = \mathbf{GG}'$ is an $n \times n$ positive definite matrix and the Jacobian for the transformation $\mathbf{Z} = \mathbf{G}^{-1}(\mathbf{Y} - \mu)$ is $|\mathbf{GG}'|^{-1/2} = |\boldsymbol{\Sigma}|^{-1/2}$.

The function $f_{\mathbf{Y}}(\mathbf{y})$ is the multivariate normal distribution of an $n \times 1$ random vector $\mathbf{Y}$ with $n \times 1$ mean vector $\boldsymbol{\mu}$ and $n \times n$ positive definite covariance matrix $\boldsymbol{\Sigma}$. The following notation will be used to represent this distribution: the $n \times 1$ random vector $\mathbf{Y} \sim \mathrm{N}_n(\boldsymbol{\mu}, \boldsymbol{\Sigma})$.

The moment generating function of an n-dimensional multivariate normal random vector is provided in the next theorem.

Theorem 2.1.1 *Let the $n \times 1$ random vector $\mathbf{Y} \sim \mathrm{N}_n(\boldsymbol{\mu}, \boldsymbol{\Sigma})$. The MGF of $\mathbf{Y}$ is*

$$m_{\mathbf{Y}}(\mathbf{t}) = e^{\mathbf{t}'\boldsymbol{\mu}+\mathbf{t}'\boldsymbol{\Sigma}\mathbf{t}/2}$$

where the $n \times 1$ vector $\mathbf{t} = (t_1, \ldots, t_n)'$ for $-h < t_i < h, h > 0$, and $i = 1, \ldots, n$.

Proof: Let the $n \times 1$ random vector $\mathbf{Z} = (Z_1, \ldots, Z_n)'$ where Z_i are independent, identically distributed $\mathrm{N}_1(0, 1)$ random variables. The $n \times 1$ random vector $\mathbf{Y} = \mathbf{GZ} + \boldsymbol{\mu} \sim \mathrm{N}_n(\boldsymbol{\mu}, \boldsymbol{\Sigma})$ where $\boldsymbol{\Sigma} = \mathbf{GG}'$. Therefore,

$$\begin{aligned}
m_{\mathbf{Y}}(\mathbf{t}) &= \mathrm{E}_{\mathbf{Y}}[e^{\mathbf{t}'\mathbf{Y}}] \\
&= \mathrm{E}_{\mathbf{Z}} = [e^{\mathbf{t}'(\mathbf{GZ}+\boldsymbol{\mu})}] \\
&= \int \cdots \int (2\pi)^{-n/2} e^{\mathbf{t}'(\mathbf{Gz}+\boldsymbol{\mu})} e^{-\mathbf{z}'\mathbf{z}/2} d\mathbf{z} \\
&= e^{\mathbf{t}'\boldsymbol{\mu}+\mathbf{t}'\boldsymbol{\Sigma}\mathbf{t}/2} \int \cdots \int (2\pi)^{-n/2} e^{\{-(\mathbf{z}-\mathbf{G}'\mathbf{t})'(\mathbf{z}-\mathbf{G}'\mathbf{t})\}/2} d\mathbf{z} \\
&= e^{\mathbf{t}'\boldsymbol{\mu}+\mathbf{t}'\boldsymbol{\Sigma}\mathbf{t}/2}. \qquad \blacksquare
\end{aligned}$$

We will now consider distributions of linear transformations of the random vector $\mathbf{Y}$ when $\mathbf{Y} \sim \mathrm{N}_n(\boldsymbol{\mu}, \boldsymbol{\Sigma})$. The following theorem provides the joint probability distribution of the $m \times 1$ random vector $\mathbf{BY} + \mathbf{b}$ where $\mathbf{B}$ is an $m \times n$ matrix of constants and $\mathbf{b}$ is an $m \times 1$ vector of constants.

Theorem 2.1.2 *If $\mathbf{Y}$ is an $n \times 1$ random vector distributed $\mathrm{N}_n(\boldsymbol{\mu}, \boldsymbol{\Sigma})$, $\mathbf{B}$ is an $m \times n$ matrix of constants with $m \leq n$, and $\mathbf{b}$ is an $m \times 1$ vector of constants, then the $m \times 1$ random vector $\mathbf{BY} + \mathbf{b} \sim \mathrm{N}_m(\mathbf{B}\boldsymbol{\mu} + \mathbf{b}, \mathbf{B}\boldsymbol{\Sigma}\mathbf{B}')$.*

Proof:

$$\begin{aligned} m_{\mathbf{BY}+\mathbf{b}}(\mathbf{t}) = \mathrm{E}_{\mathbf{Y}}[e^{\mathbf{t}'(\mathbf{BY}+\mathbf{b})}] &= e^{\mathbf{t}'\mathbf{b}}\mathrm{E}_{\mathbf{Y}}[e^{(\mathbf{B}'\mathbf{t})'\mathbf{Y}}] \\ &= e^{\mathbf{t}'\mathbf{b}}e^{(\mathbf{B}'\mathbf{t})'\boldsymbol{\mu}+(\mathbf{B}'\mathbf{t})'\boldsymbol{\Sigma}(\mathbf{B}'\mathbf{t})/2} \\ &= e^{\mathbf{t}'(\mathbf{B}\boldsymbol{\mu}+\mathbf{b})+\mathbf{t}'(\mathbf{B}\boldsymbol{\Sigma}\mathbf{B}')\mathbf{t}/2}. \end{aligned}$$

The MGF of $\mathbf{BY}+\mathbf{b}$ takes the form of a multivariate normal random vector with dimension m, mean vector $\mathbf{B}\boldsymbol{\mu}+\mathbf{b}$ and covariance matrix $\mathbf{B}\boldsymbol{\Sigma}\mathbf{B}'$ and the proof is complete. ∎

Example 2.1.1 Find the distribution of $\bar{Y} = (1/n)\mathbf{1}_n'\mathbf{Y}$ when the $n \times 1$ random vector $\mathbf{Y} = (Y_1, \ldots, Y_n)' \sim \mathrm{N}_n(\alpha\mathbf{1}_n, \sigma^2\mathbf{I}_n)$. By Theorem 2.1.2 with $1 \times n$ matrix $\mathbf{B} = (1/n)\mathbf{1}_n'$ and scalar $\mathrm{b} = 0$, $\bar{Y} \sim \mathrm{N}_1(\alpha, \sigma^2/n)$ since

$$\begin{aligned} \mathbf{B}(\alpha\mathbf{1}_n) + \mathbf{b} &= (1/n)\mathbf{1}_n'(\alpha\mathbf{1}_n) + 0 = \alpha \quad \text{and} \\ \mathbf{B}(\sigma^2\mathbf{I}_n)\mathbf{B}' &= (1/n)\mathbf{1}_n'(\sigma^2\mathbf{I}_n)(\mathbf{1}_n'/n)' = \sigma^2/n \end{aligned}$$

Example 2.1.2 Let the $n \times 1$ random vector $\mathbf{Y}$ be defined as in Example 2.1.1. Find the distribution of the $(n-1) \times 1$ random vector $\mathbf{U} = (U_1, \ldots, U_{n-1})' = (1/\sigma)\mathbf{P}_n'\mathbf{Y}$ where $\mathbf{P}_n'$ is the $(n-1) \times n$ lower portion of an n-dimensional Helmert matrix. By Theorem 2.1.2 with $(n-1) \times n$ matrix $\mathbf{B} = (1/\sigma)\mathbf{P}_n'$ and $(n-1) \times 1$ vector $\mathbf{b} = \mathbf{0}_{(n-1)\times 1}$, $\mathbf{U} \sim \mathrm{N}_{n-1}(\mathbf{0}, \mathbf{I}_{n-1})$ since

$$\begin{aligned} \mathbf{B}(\alpha\mathbf{1}_n) + \mathbf{b} &= (1/\sigma)\mathbf{P}_n'(\alpha\mathbf{1}_n) = \mathbf{0}_{n-1\times 1} \quad \text{and} \\ \mathbf{B}(\sigma^2\mathbf{I}_n)\mathbf{B}' &= (1/\sigma)\mathbf{P}_n'(\sigma^2\mathbf{I}_n)[(1/\sigma)\mathbf{P}_n']' = \mathbf{I}_{n-1}. \end{aligned}$$

The $n \times 1$ random vector $\mathbf{Y}$ can be partitioned as $\mathbf{Y} = (\mathbf{Y}_1', \mathbf{Y}_2')'$ where $\mathbf{Y}_i$ is an $n_i \times 1$ vector for $i = 1, 2$ and $n = n_1 + n_2$. Theorem 2.1.2 is used to derive the marginal distributions of the $n_i \times 1$ random vectors $\mathbf{Y}_i$.

Theorem 2.1.3 *Let the $n \times 1$ random vector $\mathbf{Y} = (\mathbf{Y}_1', \mathbf{Y}_2')' \sim \mathrm{N}_n(\boldsymbol{\mu}, \boldsymbol{\Sigma})$ where $\boldsymbol{\mu} = (\boldsymbol{\mu}_1', \boldsymbol{\mu}_2')'$ is the $n \times 1$ mean vector,*

$$\boldsymbol{\Sigma} = \begin{bmatrix} \boldsymbol{\Sigma}_{11} & \boldsymbol{\Sigma}_{12} \\ \boldsymbol{\Sigma}_{21} & \boldsymbol{\Sigma}_{22} \end{bmatrix}$$

is the $n \times n$ covariance matrix, $\mathbf{Y}_i$ and $\boldsymbol{\mu}_i$ are $n_i \times 1$ vectors, $\boldsymbol{\Sigma}_{ij}$ is an $n_i \times n_j$ matrix for $i, j = 1, 2$ and $n = n_1 + n_2$. The marginal distribution of the $n_i \times 1$ random vector $\mathbf{Y}_i$ is $\mathrm{N}_{n_i}(\boldsymbol{\mu}_i, \boldsymbol{\Sigma}_{ii})$.

Proof: By Theorem 2.1.2 with $n_1 \times n$ matrix $\mathbf{B} = [\mathbf{I}_{n_1}|\mathbf{0}_{n_1\times n_2}]$ and $n_1 \times 1$ vector $\mathbf{b} = \mathbf{0}_{n_1\times 1}$, $\mathbf{Y}_1 = \mathbf{BY} + \mathbf{b} \sim \mathrm{N}_{n_1}(\boldsymbol{\mu}_1, \boldsymbol{\Sigma}_{11})$. The marginal distribution of $\mathbf{Y}_2$ is derived in the same way with $n_2 \times 1$ matrix $\mathbf{B} = [\mathbf{0}_{n_2 \times n_1}|\mathbf{I}_{n_2}]$ and $n_2 \times 1$ vector $\mathbf{b} = \mathbf{0}_{n_2} \times 1$. ■

The results of Theorem 2.1.3 are generalized as follows. If the $n \times 1$ random vector $\mathbf{Y} \sim \mathrm{N}_n(\boldsymbol{\mu}, \boldsymbol{\Sigma})$, then any subset of elements of $\mathbf{Y}$ has a multivariate normal distribution where the mean vector of the subset is obtained by choosing the corresponding elements of $\boldsymbol{\mu}$, and the covariance matrix of the subset is obtained by choosing the corresponding rows and columns of $\boldsymbol{\Sigma}$.

Two normally distributed random vectors have the unique characteristic that the two vectors are independent if and only if they are uncorrelated. This result is given in the next theorem.

Theorem 2.1.4 *Let the $n \times 1$ random vector $\mathbf{Y} = (\mathbf{Y}_1', \ldots, \mathbf{Y}_m')' \sim \mathrm{N}_n(\boldsymbol{\mu}, \boldsymbol{\Sigma})$ where $\boldsymbol{\mu} = (\boldsymbol{\mu}_1', \ldots, \boldsymbol{\mu}_m')'$ is the $n \times 1$ mean vector,*

$$\boldsymbol{\Sigma} = \begin{bmatrix} \boldsymbol{\Sigma}_{11} & \boldsymbol{\Sigma}_{12} & \cdots & \boldsymbol{\Sigma}_{1m} \\ \boldsymbol{\Sigma}_{21} & \boldsymbol{\Sigma}_{22} & \cdots & \boldsymbol{\Sigma}_{2m} \\ \vdots & \vdots & \ddots & \vdots \\ \boldsymbol{\Sigma}_{m1} & \boldsymbol{\Sigma}_{m2} & \cdots & \boldsymbol{\Sigma}_{mm} \end{bmatrix}$$

is the $n \times n$ covariance matrix, $\mathbf{Y}_i$ and $\boldsymbol{\mu}_i$ are $n_i \times 1$ vectors, $\boldsymbol{\Sigma}_{ij}$ is an $n_i \times n_j$ matrix for $i, j = 1, \ldots, m$ and $n = \sum_{i=1}^m n_i$. The random vectors $\mathbf{Y}_1, \ldots, \mathbf{Y}_m$ are independent if and only if $\Sigma_{ij} = \mathbf{0}_{n_i \times n_j}$ for all $i \neq j$.

Proof: First assume $\boldsymbol{\Sigma}_{ij} = \mathbf{0}_{n_i \times n_j}$ for all $i \neq j$. The moment generating function of $\mathbf{Y}$ is

$$\begin{aligned} m_{\mathbf{Y}}(\mathbf{t}) = \mathrm{E}[e^{\mathbf{t}'\mathbf{Y}}] &= e^{\mathbf{t}'\boldsymbol{\mu} + \mathbf{t}'\boldsymbol{\Sigma}\mathbf{t}/2} \\ &= e^{\mathbf{t}'\boldsymbol{\mu} + \sum_{i=1}^m \sum_{j=1}^m \mathbf{t}_i'\boldsymbol{\Sigma}_{ij}\mathbf{t}_j/2} \\ &= e^{\mathbf{t}'\boldsymbol{\mu} + \sum_{i=1}^m \mathbf{t}_i'\boldsymbol{\Sigma}_{ii}\mathbf{t}_i/2} \\ &= \prod_{i=1}^m e^{\mathbf{t}_i'\boldsymbol{\mu}_i + \mathbf{t}_i'\boldsymbol{\Sigma}_{ii}\mathbf{t}_i/2} \\ &= \prod_{i=1}^m m_{\mathbf{Y}_i}(\mathbf{t}_i) \end{aligned}$$

where $\mathbf{t} = (\mathbf{t}_1', \ldots, \mathbf{t}_m')'$ with $n_i \times 1$ vector $\mathbf{t}_i$ for $i = 1, \ldots, m$. Therefore, by Theorem 1.3.3, the vectors $\mathbf{Y}_i$ are mutually independent. Now assume the vectors $\mathbf{Y}_i$ are mutually independent. For any $i \neq j$,

$$\boldsymbol{\Sigma}_{ij} = \mathrm{E}\{[\mathbf{Y}_i - \boldsymbol{\mu}_i][\mathbf{Y}_j - \boldsymbol{\mu}_j]'\} = \mathrm{E}[\mathbf{Y}_i - \boldsymbol{\mu}_i]\mathrm{E}[\mathbf{Y}_j - \boldsymbol{\mu}_j]' = \mathbf{0}_{n_i \times n_j}. \quad ■$$

In the following examples mean vectors and covariance matrices are derived for a few common problems.

Example 2.1.3 Let $Y_1, \ldots, Y_n$ be independent, identically distributed $\mathrm{N}_1(\alpha, \sigma^2)$ random variables. By Theorem 2.1.4, $\mathrm{cov}(Y_i, Y_j) = 0$ for $i \neq j$. Furthermore, $\mathrm{E}(Y_i) = \alpha$ and the $\mathrm{var}(Y_i) = \sigma^2$ for all $i = 1, \ldots, n$. Therefore, the $n \times 1$ random vector $\mathbf{Y} = (Y_1, \ldots, Y_n)' \sim \mathrm{N}_n(\alpha \mathbf{1}_n, \sigma^2 \mathbf{I}_n)$.

Example 2.1.4 Consider the one-way classification described in Example 1.2.10. Let Y_{ij} be a random variable representing the jth replicate observation in the ith level of the fixed factor for $i = 1, \ldots, t$ and $j = 1, \ldots, r$. Let the $tr \times 1$ random vector $\mathbf{Y} = Y_{11}, \ldots, Y_{1r}, \ldots, Y_{t1}, \ldots, Y_{tr})'$ where the Y_{ij}'s are assumed to be independent, normally distributed random variables with $\mathrm{E}(Y_{ij}) = \mu_i$ and $\mathrm{var}(Y_{ij}) = \sigma^2$. This experiment can be characterized with the model

$$Y_{ij} = \mu_i + R(T)_{(i)j}$$

where the $R(T)_{(i)j}$ are independent, identically distributed normal random variables with mean 0 and variance σ^2. The letter R signifies replicates and the letter T signifies the fixed factor or fixed treatments. Therefore, $R(T)$ represents the effect of the random replicates nested in the fixed treatment levels. The parentheses around T identify the nesting. By Theorems 2.1.2 and 2.1.4, $\mathbf{Y} \sim \mathrm{N}_{tr}(\boldsymbol{\mu}, \boldsymbol{\Sigma})$ where the $tr \times 1$ mean vector $\boldsymbol{\mu}$ is given by

$$\begin{aligned}
\boldsymbol{\mu} &= [\mathrm{E}(Y_{11}), \ldots, \mathrm{E}(Y_{1r}), \ldots, \mathrm{E}(Y_{t1}), \ldots, \mathrm{E}(Y_{tr})]' \\
&= [\mu_1, \ldots, \mu_1, \ldots, \mu_t, \ldots, \mu_t]' \\
&= [\mu_1 \mathbf{1}'_r, \ldots, \mu_t \mathbf{1}'_r]' \\
&= (\mu_1, \ldots, \mu_t)' \otimes \mathbf{1}_r
\end{aligned}$$

and using Definition 1.3.3 the elements of the $tr \times tr$ covariance matrix $\boldsymbol{\Sigma}$ are

$$\begin{aligned}
\mathrm{cov}(Y_{ij}, Y_{i'j'}) &= \mathrm{E}[(Y_{ij} - \mathrm{E}(Y_{ij}))(Y_{i'j'} - \mathrm{E}(Y_{i'j'}))] \\
&= \mathrm{E}[(\mu_i + R(T)_{(i)j} - \mu_i)(\mu_{i'} + R(T)_{(i')j'} - \mu_{i'})] \\
&= \mathrm{E}[R(T)_{(i)j} R(T)_{(i')j'}] \\
&= \begin{cases} \sigma^2 & \text{if } i = i' \text{ and } j = j' \\ 0 & \text{otherwise.} \end{cases}
\end{aligned}$$

That is, $\boldsymbol{\Sigma} = \sigma^2 \mathbf{I}_t \otimes \mathbf{I}_r$.

The preceding model is composed of two parts: a fixed portion represented by the t fixed constants μ_i and a random portion represented by the tr random variables $R(T)_{(i)j}$. In models of this type, each constant in the fixed portion equals the expected value of observations in particular combinations of the levels of the fixed factor(s). Models whose fixed portions are represented in this way are called *mean models*. Numerous examples of mean models are presented in this text and a specific discussion on mean models is provided in Chapters 9 and 10.

Example 2.1.5 Consider a two-way cross classification where both factors are random. Let Y_{ij} be a random variable representing the observation in the ith level of the first random factor S and the jth level of the second random factor T for $i = 1, \ldots, s$ and $j = 1, \ldots, t$. Let the $st \times 1$ random vector $\mathbf{Y} = Y_{11}, \ldots, Y_{1t}, \ldots, Y_{s1}, \ldots, Y_{st})'$. This experiment can be characterized with the model

$$Y_{ij} = \alpha + S_i + T_j + ST_{ij}$$

where α is a constant representing the overall mean; the random variable S_i represents the effect of the ith level of the first random factor; the random variable T_j represents the effect of the jth level of the second random factor; and the random variable ST_{ij} represents the interaction effect of the ith level of factor S and the jth level of factor T. Furthermore, $S_1, \ldots, S_s, T_1, \ldots, T_t$ and $ST_{11}, \ldots, ST_{st}$ are assumed to be independent normal random variables with zero expectations and variances given by $\text{var}(S_i) = \sigma_S^2$, $\text{var}(T_j) = \sigma_T^2$, and $\text{var}[ST_{ij}] = \sigma_{ST}^2$ for $i = 1, \ldots, s$ and $j = 1, \ldots, t$. Therefore, by Theorems 2.1.2 and 2.1.4, $\mathbf{Y} \sim \text{N}_{st}(\boldsymbol{\mu}, \boldsymbol{\Sigma})$ where the $st \times 1$ mean vector $\boldsymbol{\mu}$ is given by

$$\begin{aligned}
\boldsymbol{\mu} &= [\text{E}(Y_{11}), \ldots, \text{E}(Y_{1t}), \ldots, \text{E}(Y_{s1}), \ldots, \text{E}(Y_{st})]' \\
&= [\alpha, \ldots, \alpha, \ldots, \alpha, \ldots, \alpha]' \\
&= \alpha \mathbf{1}_s \otimes \mathbf{1}_t
\end{aligned}$$

and the elements of the $st \times st$ covariance matrix $\boldsymbol{\Sigma}$ are

$$\begin{aligned}
\text{cov}(Y_{ij}, Y_{i'j'}) &= \text{E}[(Y_{ij} - \text{E}(Y_{ij}))(Y_{i'j'} - \text{E}(Y_{i'j'}))] \\
&= \text{E}[(\alpha + S_i + T_j + ST_{ij} - \alpha)(\alpha + S_{i'} + T_{j'} + ST_{i'j'} - \alpha)] \\
&= \text{E}[S_i S_{i'}] + \text{E}[T_j T_{j'}] + \text{E}[ST_{ij} ST_{i'j'}] \\
&= \begin{cases} \sigma_S^2 + \sigma_T^2 + \sigma_{ST}^2 & \text{if } i = i' \text{ and } j = j' \\ \sigma_S^2 & \text{if } i = i' \text{ but } j \neq j' \\ \sigma_T^2 & \text{if } j = j' \text{ but } i \neq i' \\ 0 & \text{otherwise.} \end{cases}
\end{aligned}$$

That is, $\boldsymbol{\Sigma} = \sigma_S^2(\mathbf{I}_s \otimes \mathbf{J}_t) + \sigma_T^2(\mathbf{J}_s \otimes \mathbf{I}_t) + \sigma_{ST}^2(\mathbf{I}_s \otimes \mathbf{I}_t)$.

Direct derivation of the covariance matrix $\boldsymbol{\Sigma}$ can be difficult even for balanced design structures. In Chapter 4 a simple algorithm is provided for determining the covariance matrix for complete, balanced designs with any number of fixed or random main effects, interactions, and nested factors.

2.2 CONDITIONAL DISTRIBUTIONS OF MULTIVARIATE NORMAL RANDOM VECTORS

In this section conditional multivariate normal distributions are discussed.

Theorem 2.2.1 *Let the $n \times 1$ random vector $\mathbf{Y} = (\mathbf{Y}_1', \mathbf{Y}_2')' \sim \mathrm{N}_n(\boldsymbol{\mu}, \boldsymbol{\Sigma})$ where $\boldsymbol{\mu} = (\boldsymbol{\mu}_1', \boldsymbol{\mu}_2')'$ is the $n \times 1$ mean vector,*

$$\boldsymbol{\Sigma} = \begin{bmatrix} \boldsymbol{\Sigma}_{11} & \boldsymbol{\Sigma}_{12} \\ \boldsymbol{\Sigma}_{21} & \boldsymbol{\Sigma}_{22} \end{bmatrix}$$

is the $n \times n$ positive definite covariance matrix, $\mathbf{Y}_i$ and $\boldsymbol{\mu}_i$ are $n_i \times 1$ vectors, $\boldsymbol{\Sigma}_{ij}$ is an $n_i \times n_j$ matrix for $i, j = 1, 2$ and $n = n_1 + n_2$. The conditional distribution of the $n_1 \times 1$ random vector $\mathbf{Y_1}$ given the $n_2 \times 1$ vector of constants $\mathbf{Y}_2 = \mathbf{c_2}$ is $\mathrm{N}_{n_1}[\boldsymbol{\mu}_1 + \boldsymbol{\Sigma}_{12}\boldsymbol{\Sigma}_{22}^{-1}(\mathbf{c}_2 - \boldsymbol{\mu}_2), \boldsymbol{\Sigma}_{11} - \boldsymbol{\Sigma}_{12}\boldsymbol{\Sigma}_{22}^{-1}\boldsymbol{\Sigma}_{21}]$.

Proof: Define the $n_1 \times 1$ vector $\mathbf{V}_1$ and the $n_2 \times 1$ vector $\mathbf{V}_2$ as

$$\begin{bmatrix} \mathbf{V}_1 \\ \mathbf{V}_2 \end{bmatrix} = \begin{bmatrix} \mathbf{Y}_1 - \boldsymbol{\Sigma}_{12}\boldsymbol{\Sigma}_{22}^{-1}\mathbf{Y}_2 \\ \mathbf{Y}_2 \end{bmatrix} = \begin{bmatrix} \mathbf{I}_{n_1} & -\boldsymbol{\Sigma}_{12}\boldsymbol{\Sigma}_{22}^{-1} \\ \mathbf{0} & \mathbf{I}_{n_2} \end{bmatrix} \mathbf{Y}.$$

By Theorem 2.1.2 with the $n \times 1$ vector $\mathbf{b} = \mathbf{0}_{n\times 1}$ and the $n \times n$ matrix

$$\mathbf{B} = \begin{bmatrix} \mathbf{I}_{n_1} & -\boldsymbol{\Sigma}_{12}\boldsymbol{\Sigma}_{22}^{-1} \\ \mathbf{0} & \mathbf{I}_{n_2} \end{bmatrix}.$$

the $n \times 1$ random vector $(\mathbf{V}_1', \mathbf{V}_2')' \sim \mathrm{N}_n(\boldsymbol{\mu}^*, \boldsymbol{\Sigma}^*)$ where the $n \times 1$ mean vector $\boldsymbol{\mu}^*$ is

$$\boldsymbol{\mu}^* = \mathbf{B}\begin{bmatrix} \boldsymbol{\mu}_1 \\ \boldsymbol{\mu}_2 \end{bmatrix} = \begin{bmatrix} \boldsymbol{\mu}_1 - \boldsymbol{\Sigma}_{12}\boldsymbol{\Sigma}_{22}^{-1}\boldsymbol{\mu}_2 \\ \boldsymbol{\mu}_2 \end{bmatrix}.$$

and the $n \times n$ covariance matrix $\boldsymbol{\Sigma}^*$ is

$$\begin{aligned} \boldsymbol{\Sigma}^* &= \mathbf{B}\boldsymbol{\Sigma}\mathbf{B}' \\ &= \begin{bmatrix} \mathbf{I}_{n_1} & -\boldsymbol{\Sigma}_{12}\boldsymbol{\Sigma}_{22}^{-1} \\ \mathbf{0} & \mathbf{I}_{n_2} \end{bmatrix} \begin{bmatrix} \boldsymbol{\Sigma}_{11} & \boldsymbol{\Sigma}_{12} \\ \boldsymbol{\Sigma}_{21} & \boldsymbol{\Sigma}_{22} \end{bmatrix} \begin{bmatrix} \mathbf{I}_{n_1} & \mathbf{0} \\ \boldsymbol{\Sigma}_{22}^{-1}\boldsymbol{\Sigma}_{21} & \mathbf{I}_{n_2} \end{bmatrix} \\ &= \begin{bmatrix} \boldsymbol{\Sigma}_{11} - \boldsymbol{\Sigma}_{12}\boldsymbol{\Sigma}_{22}^{-1}\boldsymbol{\Sigma}_{21} & \mathbf{0} \\ \mathbf{0} & \boldsymbol{\Sigma}_{22} \end{bmatrix}. \end{aligned}$$

Thus, $\mathbf{V}_1$ and $\mathbf{V}_2$ are independent multivariate normal random vectors. That is, the joint distribution of $\mathbf{V}_1$ and $\mathbf{V}_2$ can be written as the product of the two marginal distributions

$$f_{\mathbf{V}_1,\mathbf{V}_2}(\mathbf{v}_1, \mathbf{v}_2) = f_{\mathbf{V}_1}(\mathbf{v}_1) f_{\mathbf{V}_2}(\mathbf{v}_2)$$

where $f_{\mathbf{V}_1}(\mathbf{v}_1)$ is a $\mathrm{N}_{n_1}(\boldsymbol{\mu}_1 - \boldsymbol{\Sigma}_{12}\boldsymbol{\Sigma}_{22}^{-1}\boldsymbol{\mu}_2, \boldsymbol{\Sigma}_{11} - \boldsymbol{\Sigma}_{12}\boldsymbol{\Sigma}_{22}^{-1}\boldsymbol{\Sigma}_{21})$ distribution and $f_{\mathbf{V}_2}(\mathbf{v}_2)$ is a $\mathrm{N}_{n_2}(\boldsymbol{\mu}_2, \boldsymbol{\Sigma}_{22})$ distribution. The conditional distribution of $\mathbf{Y}_1|\mathbf{Y}_2 = \mathbf{c}_2$ is derived by utilizing the transformation from $\mathbf{Y}_1, \mathbf{Y}_2$ to $\mathbf{V}_1, \mathbf{V}_2$ and noting that the Jacobian of this transformation is 1.

$$\begin{aligned} f_{\mathbf{Y}_1|\mathbf{Y}_2=\mathbf{c}_2}(\mathbf{y}_1|\mathbf{y}_2 = \mathbf{c}_2) &= f_{\mathbf{Y}_1,\mathbf{Y}_2}(\mathbf{y}_1, \mathbf{c}_2)/f_{\mathbf{Y}_2}(\mathbf{c}_2) \\ &= f_{\mathbf{V}_1,\mathbf{V}_2}(\mathbf{v}_1 + \boldsymbol{\Sigma}_{12}\boldsymbol{\Sigma}_{22}^{-1}\mathbf{c}_2, \mathbf{c}_2)/f_{\mathbf{V}_2}(\mathbf{c}_2) \\ &= f_{\mathbf{V}_1}(\mathbf{v}_1 + \boldsymbol{\Sigma}_{12}\boldsymbol{\Sigma}_{22}^{-1}\mathbf{c}_2) f_{\mathbf{V}_2}(\mathbf{c}_2)/f_{\mathbf{V}_2}(\mathbf{c}_2) \\ &= f_{\mathbf{V}_1}(\mathbf{v}_1 + \boldsymbol{\Sigma}_{12}\boldsymbol{\Sigma}_{22}^{-1}\mathbf{c}_2). \end{aligned}$$

But the distribution of $f_{\mathbf{V}_1}(\mathbf{v}_1 + \boldsymbol{\Sigma}_{12}\boldsymbol{\Sigma}_{22}^{-1}\mathbf{c}_2)$ is the distribution of $\mathbf{V}_1$ plus the constant vector $\boldsymbol{\Sigma}_{12}\boldsymbol{\Sigma}_{22}^{-1}\mathbf{c}_2)$. By Theorem 2.1.2 with $\mathbf{B} = \mathbf{I}_{n_1}$ and $\mathbf{b} = \boldsymbol{\Sigma}_{12}\boldsymbol{\Sigma}_{22}^{-1}\mathbf{c}_2$, the proof is complete. ■

Theorem 2.2.1 is applied in the next few examples.

Example 2.2.1 Let $\mathbf{Y} = (Y_1, Y_2, Y_3)' \sim \mathrm{N}_3(\boldsymbol{\mu}, \boldsymbol{\Sigma})$ where $\boldsymbol{\mu} = (1, 4, -2)'$ and

$$\boldsymbol{\Sigma} = \begin{bmatrix} 1 & 0.5 & 0.4 \\ 0.5 & 1 & 0.2 \\ 0.4 & 0.2 & 1 \end{bmatrix}.$$

Then by Theorem 2.2.1 the conditional distribution of $Y_1, Y_2|Y_3 = 1$ is $\mathrm{N}_2(\boldsymbol{\mu}_c, \boldsymbol{\Sigma}_c)$ where the 2×1 conditional mean vector $\boldsymbol{\mu}_c$ is given by

$$\boldsymbol{\mu}_c = \begin{bmatrix} 1 \\ 4 \end{bmatrix} + \begin{bmatrix} 0.4 \\ 0.2 \end{bmatrix} (1)^{-1}[1 - (-2)] = \begin{bmatrix} 2.2 \\ 4.6 \end{bmatrix}$$

and the 2×2 conditional covariance matrix $\boldsymbol{\Sigma}_c$ is given by

$$\boldsymbol{\Sigma}_c = \begin{bmatrix} 1 & 0.5 \\ 0.5 & 1 \end{bmatrix} - \begin{bmatrix} 0.4 \\ 0.2 \end{bmatrix} (1)^{-1}[0.4, 0.2] = \begin{bmatrix} 0.84 & 0.42 \\ 0.42 & 0.96 \end{bmatrix}.$$

Example 2.2.2 Use the distribution of $\mathbf{Y} = (Y_1, Y_2, Y_3)'$ from Example 2.2.1 to find the conditional distribution of $2Y_1 + Y_2|Y_1 + 2Y_2 + 3Y_3 = 2$. First, let the 2×1 vector $\mathbf{b} = \mathbf{0}_{2\times 1}$ and let the 2×3 matrix

$$\mathbf{B} = \begin{bmatrix} 2 & 1 & 0 \\ 1 & 2 & 3 \end{bmatrix}.$$

By Theorem 2.1.2, the joint distribution of

$$\mathbf{BY} = \begin{bmatrix} 2Y_1 + Y_2 \\ Y_1 + 2Y_2 + 3Y_3 \end{bmatrix}$$

is $N_2(\boldsymbol{\mu}^*, \Sigma^*)$ where the 2×1 mean vector $\boldsymbol{\mu}^*$ is

$$\boldsymbol{\mu}^* = \mathbf{B}\boldsymbol{\mu} = \begin{bmatrix} 2(1) + 1(4) + 0(-2) \\ 1(1) + 2(4) + 3(-2) \end{bmatrix} = \begin{bmatrix} 6 \\ 3 \end{bmatrix}$$

and the 2×2 covariance matrix $\boldsymbol{\Sigma}^*$ is

$$\begin{aligned} \boldsymbol{\Sigma}^* &= \mathbf{B\Sigma B}' \\ &= \begin{bmatrix} 2 & 1 & 0 \\ 1 & 2 & 3 \end{bmatrix} \begin{bmatrix} 1 & 0.5 & 0.4 \\ 0.5 & 1 & 0.2 \\ 0.4 & 0.2 & 1 \end{bmatrix} \begin{bmatrix} 2 & 1 \\ 1 & 2 \\ 0 & 3 \end{bmatrix} = \begin{bmatrix} 7 & 9.5 \\ 9.5 & 20.8 \end{bmatrix} \end{aligned}$$

where $\boldsymbol{\mu}$ and $\boldsymbol{\Sigma}$ are given in Example 2.2.1. Applying Theorem 2.2.1 to the distribution of $\mathbf{BY}$, the conditional distribution of $2Y_1 + Y_2 | Y_1 + 2Y_2 + 3Y_3 = 2$ is $\mathrm{N}_1(\mu_c, \sigma_c^2)$ where the conditional mean μ_c is

$$\mu_c = 6 + 9.5\ (20.8)^{-1}\ (2 - 3) = 5.54$$

and the conditional variance σ_c^2 is given by

$$\sigma_c^2 = 7 - 9.5\ (20.8)^{-1}\ 9.5 = 2.66.$$

Example 2.2.3 Let the $n \times 1$ random vector $\mathbf{Y} = (Y_1, \ldots, Y_n)' \sim \mathrm{N}_n(\alpha \mathbf{1}_n, \sigma^2 \mathbf{I}_n)$. Find the conditional distribution of $Y_1, \ldots, Y_{n-1} | \bar{Y} = \bar{y}$. By Theorem 2.1.2 with $n \times 1$ vector $\mathbf{b} = \mathbf{0}_{n \times 1}$ and $n \times n$ matrix

$$\mathbf{B} = \begin{bmatrix} \mathbf{I}_{n-1} & \mathbf{0} \\ (1/n)\mathbf{1}'_{n-1} & 1/n \end{bmatrix},$$

the $n \times 1$ random vector $\mathbf{BY} = (Y_1, \ldots, Y_{n-1}, \bar{Y})' \sim \mathrm{N}_n(\boldsymbol{\mu}^*, \boldsymbol{\Sigma}^*)$ where the $n \times 1$ mean vector $\boldsymbol{\mu}^*$ is

$$\boldsymbol{\mu}^* = \mathbf{B}(\alpha \mathbf{1}_n) = \begin{bmatrix} \mathbf{I}_{n-1} & \mathbf{0} \\ (1/n)\mathbf{1}'_{n-1} & 1/n \end{bmatrix} (\alpha \mathbf{1}_n) - \alpha \mathbf{1}_n$$

and the $n \times n$ covariance matrix $\boldsymbol{\Sigma}^*$ is

$$\begin{aligned} \boldsymbol{\Sigma}^* &= \mathbf{B}(\sigma^2 \mathbf{I}_n)\mathbf{B}' \\ &= \sigma^2 \begin{bmatrix} \mathbf{I}_{n-1} & \mathbf{0} \\ (1/n)\mathbf{1}'_{n-1} & 1/n \end{bmatrix} \begin{bmatrix} \mathbf{I}_{n-1} & (1/n)\mathbf{1}_{n-1} \\ \mathbf{0} & 1/n \end{bmatrix} \\ &= \sigma^2 \begin{bmatrix} \mathbf{I}_{n-1} & (1/n)\mathbf{1}_{n-1} \\ (1/n)\mathbf{1}'_{n-1} & 1/n \end{bmatrix}. \end{aligned}$$

Applying Theorem 2.2.1 to the distribution of $\mathbf{BY}$, the conditional distribution of $Y_1, \ldots, Y_{n-1}|\bar{Y} = \bar{y}$ is $\mathrm{N}_{n-1}(\boldsymbol{\mu}_c, \boldsymbol{\Sigma}_c)$ where the $(n-1) \times 1$ conditional mean vector $\boldsymbol{\mu}_c$ is

$$\boldsymbol{\mu}_c = \alpha \mathbf{1}_{n-1} + (1/n)\mathbf{1}_{n-1}(1/n)^{-1}(\bar{y} - \alpha) = \bar{y}\mathbf{1}_{n-1}$$

and the $(n-1) \times (n-1)$ conditional covariance matrix $\boldsymbol{\Sigma}_c$ is

$$\begin{aligned}\boldsymbol{\Sigma}_c &= \sigma^2[\mathbf{I}_{n-1} - (1/n)\mathbf{1}_{n-1}(1/n)^{-1}(1/n)\mathbf{1}'_{n-1}] \\ &= \sigma^2 \left[\mathbf{I}_{n-1} - \frac{1}{n}\mathbf{J}_{n-1}\right].\end{aligned}$$

Applying Theorem 2.1.2 [with scalar $b = 0$ and $1 \times (n-1)$ matrix $\mathbf{B} = (1, 0, \ldots, 0)$] to the conditional distribution of $Y_1, \ldots, Y_{n-1}|\bar{Y} = \bar{y}$, the conditional distribution of $Y_1|\bar{Y} = \bar{y}$ is $\mathrm{N}_1(\alpha_c, \sigma_c^2)$ where the conditional mean α_c is

$$\alpha_c = \mathbf{B}\boldsymbol{\mu}_c = (1, 0, \ldots, 0)(\bar{y}\mathbf{1}_{n-1}) = \bar{y}$$

and the conditional variance σ_c^2 is

$$\begin{aligned}\sigma_c^2 &= \mathbf{B}\boldsymbol{\Sigma}_c\mathbf{B}' = (1, 0, \ldots, 0)\sigma^2 \left[\mathbf{I}_{n-1} - \frac{1}{n}\mathbf{J}_{n-1}\right](1, 0, \ldots, 0)' \\ &= \sigma^2(n-1)/n.\end{aligned}$$

The distribution theory of quadratic forms is presented in Chapter 3. However, a number of interesting quadratic form problems are solved in the next section before the general distribution theory is developed.

2.3 DISTRIBUTIONS OF CERTAIN QUADRATIC FORMS

The distributions of several quadratic forms are derived in the following examples.

Example 2.3.1 Let the $n \times 1$ random vector $\mathbf{Y} = (Y_1, \ldots, Y_n)' \sim \mathrm{N}_n(\alpha\mathbf{1}_n, \sigma^2\mathbf{I}_n)$. Define $U = \sum_{i=1}^n (Y_i - \bar{Y})^2/\sigma^2$ and $V = n(\bar{Y} - \alpha)^2/\sigma^2$ where $\bar{Y} = (1/n)\mathbf{1}'_n\mathbf{Y}$. Find the distributions of U and V and show these two random variables are independent. First, note $\sqrt{n}(\bar{Y} - \alpha)/\sigma = [1/(\sigma\sqrt{n})]\mathbf{1}'_n\mathbf{Y} - (\alpha\sqrt{n})/\sigma$. By Theorem 2.1.2 with $1 \times n$ matrix $\mathbf{B} = [1/(\sigma\sqrt{n})]\mathbf{1}'_n$ and scalar $b = -(\alpha\sqrt{n})/\sigma$, $\sqrt{n}(\bar{Y} - \alpha)/\sigma \sim \mathrm{N}_1(0, 1)$ since

$$\mathbf{B}(\alpha\mathbf{1}_n) + b = [\alpha/(\sigma\sqrt{n})]\mathbf{1}'_n\mathbf{1}_n - \alpha\sqrt{n}/\sigma = 0$$

and

$$\mathbf{B}(\sigma^2\mathbf{I}_n)\mathbf{B}' = \sigma^2\{[1/(\sigma\sqrt{n})]\mathbf{1}'_n\}\{[1/(\sigma\sqrt{n})]\mathbf{1}'_n\}' = 1.$$

Therefore, by Example 1.3.2, $n(\bar{Y}-\alpha)^2/\sigma^2$ is distributed as a central chi-square random variable with 1 degree of freedom. Next, rewrite U as

$$\begin{aligned} U &= \sum_{i=1}^{n}(Y_i-\bar{Y})^2/\sigma^2 = (1/\sigma^2)\mathbf{Y}'\left(\mathbf{I}_n - \frac{1}{n}\mathbf{J}_n\right)\mathbf{Y} \\ &= \mathbf{Y}'(1/\sigma)\mathbf{P}_n[(1/\sigma)\mathbf{P}_n']\mathbf{Y} \\ &= \mathbf{X}'\mathbf{X} \\ &= \sum_{i=1}^{n-1} X_i^2 \end{aligned}$$

where the $(n-1)\times 1$ vector $\mathbf{X} = [(1/\sigma)\mathbf{P}_n']\mathbf{Y} = (X_1, \ldots, X_{n-1})'$ and the $(n-1)\times n$ matrix $\mathbf{P}_n'$ is the lower portion of an n-dimensional Helmert matrix with $\mathbf{P}_n\mathbf{P}_n' = (\mathbf{I}_n - \frac{1}{n}\mathbf{J}_n)$, $\mathbf{P}_n'\mathbf{P}_n = \mathbf{I}_{n-1}$ and $\mathbf{1}_n'\mathbf{P}_n = \mathbf{0}_{1\times(n-1)}$. By Theorem 2.1.2 with $(n-1)\times n$ matrix $\mathbf{B} = (1/\sigma)\mathbf{P}_n'$ and $(n-1)\times 1$ vector $\mathbf{b} = \mathbf{0}_{(n-1)\times 1}$, $\mathbf{X} \sim \mathrm{N}_{n-1}(\mathbf{0}, \mathbf{I}_{n-1})$ since

$$\mathbf{B}(\alpha\mathbf{1}_n) = (1/\sigma)\mathbf{P}_n'(\alpha\mathbf{1}_n) = \mathbf{0}_{(n-1)\times 1}$$

and

$$\mathbf{B}(\sigma^2\mathbf{I}_n)\mathbf{B}' = (1/\sigma)\mathbf{P}_n'(\sigma^2\mathbf{I}_n)[(1/\sigma)\mathbf{P}_n']' = \mathbf{I}_{n-1}.$$

Therefore, U equals the sum of squares of $n-1$ independent $\mathrm{N}_1(0,1)$ random variables. By Example 1.3.2, U has a central chi-square distribution with $n-1$ degrees of freedom. Finally, by Theorem 2.1.2 with $n\times n$ matrix $\mathbf{B}$ and $n\times 1$ vector $\mathbf{b}$ given by

$$\mathbf{B} = \begin{bmatrix} 1/(\sigma\sqrt{n})\mathbf{1}_n' \\ (1/\sigma)\mathbf{P}_n' \end{bmatrix} \quad \text{and} \quad \mathbf{b} = \begin{bmatrix} -\sqrt{n}\alpha/\sigma \\ \mathbf{0}_{(n-1)\times 1} \end{bmatrix},$$

the $n\times 1$ vector $(\sqrt{n}(\bar{Y}-\alpha)/\sigma, \mathbf{X}')' = \mathbf{BY} + \mathbf{b} \sim \mathrm{N}_n(\mathbf{0}, \mathbf{I}_n)$ since

$$\mathbf{B}(\alpha\mathbf{1}_n) + \mathbf{b} = \begin{bmatrix} [\alpha/(\sigma\sqrt{n})]\mathbf{1}_n'\mathbf{1}_n \\ (\alpha/\sigma)\mathbf{P}_n'\mathbf{1}_n \end{bmatrix} + \begin{bmatrix} -\sqrt{n}\alpha/\sigma \\ \mathbf{0}_{(n-1)\times 1} \end{bmatrix} = \mathbf{0}_{n\times 1}$$

and

$$\mathbf{B}(\sigma^2\mathbf{I}_n)\mathbf{B}' = \begin{bmatrix} [1/(\sigma\sqrt{n})]\mathbf{1}_n' \\ (1/\sigma)\mathbf{P}_n' \end{bmatrix} (\sigma^2\mathbf{I}_n)\{[1/(\sigma\sqrt{n})]\mathbf{1}_n, (1/\sigma)\mathbf{P}_n\} = \mathbf{I}_n.$$

By Theorem 2.1.4, $\sqrt{n}(\bar{Y}-\alpha)/\sigma$ and $\mathbf{X}$ are independent. Therefore, by Theorem 1.3.5, U and V are independent.

		Factor S (i)			
		1	2	$\cdots$	s
Factor T (j)	1	Y_{11}	Y_{21}	$\cdots$	Y_{s1}
	2	Y_{12}	Y_{22}	$\cdots$	Y_{s2}
	$\vdots$	$\vdots$	$\vdots$	$\ddots$	$\vdots$
	t	Y_{1t}	Y_{2t}	$\cdots$	Y_{st}

Figure 2.3.1 Two-Way Layout.

Example 2.3.2 Consider the two-way cross classification described in Example 2.1.5 where the $st \times 1$ random vector

$$\begin{aligned}\mathbf{Y} &= (Y_{11}, \ldots, Y_{1t}, \ldots, Y_{s1}, \ldots, Y_{st})' \\ &\sim \mathrm{N}_{st}(\alpha \mathbf{1}_s \otimes \mathbf{1}_t, \sigma_S^2 \mathbf{I}_s \otimes \mathbf{J}_t + \sigma_T^2 \mathbf{J}_s \otimes \mathbf{I}_t + \sigma_{ST}^2 \mathbf{I}_s \otimes \mathbf{I}_t).\end{aligned}$$

The layout for this experiment is given in Figure 2.3.1 and the ANOVA table is provided in Table 2.3.1, where the sums of squares are written in summation notation and as quadratic forms, $\mathbf{Y}'\mathbf{A}_m\mathbf{Y}$, for $m = 1, \ldots, 5$. The matrices $\mathbf{A}_1$, $\mathbf{A}_2$, and $\mathbf{A}_5$ for the sums of squares due to the mean, random factor S, and the total, respectively, were already derived in Example 1.2.10. Note that $(\bar{Y}_{.1}, \ldots, \bar{Y}_{.t})' = [(1/s)\mathbf{1}_s' \otimes \mathbf{I}_t]\mathbf{Y}$. Therefore, the sum of squares due to the random factor T is

$$\begin{aligned}\sum_{i=1}^{s}\sum_{j=1}^{t}(\bar{Y}_{.j} - \bar{Y}_{..})^2 &= \sum_{i=1}^{s}\sum_{j=1}^{t}\bar{Y}_{.j}^2 - st\bar{Y}^2_{..} \\ &= s\{[(1/s)\mathbf{1}_s' \otimes \mathbf{I}_t]\mathbf{Y}\}'\{[(1/s)\mathbf{1}_s' \otimes \mathbf{I}_t]\mathbf{Y}\} - \mathbf{Y}'\left[\frac{1}{s}\mathbf{J}_s \otimes \frac{1}{t}\mathbf{J}_t\right]\mathbf{Y}\end{aligned}$$

Table 2.3.1
Two-Way ANOVA Table

Source	df	SS	
Mean	1	$\sum_{i=1}^{s}\sum_{j=1}^{t}\bar{Y}^2_{..}$	$= \mathbf{Y}'\mathbf{A}_1\mathbf{Y}$
Factor S	$s-1$	$\sum_{i=1}^{s}\sum_{j=1}^{t}(\bar{Y}_{i.} - \bar{Y}_{..})^2$	$= \mathbf{Y}'\mathbf{A}_2\mathbf{Y}$
Factor T	$t-1$	$\sum_{i=1}^{s}\sum_{j=1}^{t}(\bar{Y}_{.j} - \bar{Y}_{..})^2$	$= \mathbf{Y}'\mathbf{A}_3\mathbf{Y}$
Interaction ST	$(s-1)(t-1)$	$\sum_{i=1}^{s}\sum_{j=1}^{t}(Y_{ij} - \bar{Y}_{i.} - \bar{Y}_{.j} + \bar{Y}_{..})^2$	$= \mathbf{Y}'\mathbf{A}_4\mathbf{Y}$
Total	st	$\sum_{i=1}^{s}\sum_{j=1}^{t}Y_{ij}^2$	$= \mathbf{Y}'\mathbf{A}_5\mathbf{Y}$

$$= \mathbf{Y}'\left[\frac{1}{s}\mathbf{J}_s \otimes \left(\mathbf{I}_t - \frac{1}{t}\mathbf{J}_t\right)\right]\mathbf{Y}$$

$$= \mathbf{Y}'\mathbf{A}_3\mathbf{Y}$$

where the $st \times st$ matrix $\mathbf{A}_3 = [\frac{1}{s}\mathbf{J}_s \otimes (\mathbf{I}_t - \frac{1}{t}\mathbf{J}_t)]$. The matrix $\mathbf{A}_4$ for the sum of squares due to the random interaction can be solved by subtraction, $\mathbf{A}_4 = \mathbf{A}_5 - \mathbf{A}_1 - \mathbf{A}_2 - \mathbf{A}_3$, that is,

$$\mathbf{A}_4 = [\mathbf{I}_s \otimes \mathbf{I}_t] - \left[\frac{1}{s}\mathbf{J}_s \otimes \frac{1}{t}\mathbf{J}_t\right] - \left[\left(\mathbf{I}_s - \frac{1}{s}\mathbf{J}_s\right) \otimes \frac{1}{t}\mathbf{J}_t\right]$$

$$- \left[\frac{1}{s}\mathbf{J}_s \otimes \left(\mathbf{I}_t - \frac{1}{t}\mathbf{J}_t\right)\right]$$

$$= [\mathbf{I}_s \otimes \mathbf{I}_t] - \left[\mathbf{I}_s \otimes \frac{1}{t}\mathbf{J}_t\right] - \left[\frac{1}{s}\mathbf{J}_s \otimes \left(\mathbf{I}_t - \frac{1}{t}\mathbf{J}_t\right)\right]$$

$$= \left[\mathbf{I}_s \otimes \left(\mathbf{I}_t - \frac{1}{t}\mathbf{J}_t\right)\right] - \left[\frac{1}{s}\mathbf{J}_s \otimes \left(\mathbf{I}_t - \frac{1}{t}\mathbf{J}_t\right)\right]$$

$$= \left[\left(\mathbf{I}_s - \frac{1}{s}\mathbf{J}_s\right) \otimes \left(\mathbf{I}_t - \frac{1}{t}\mathbf{J}_t\right)\right].$$

To find the distribution of $\mathbf{Y}'\mathbf{A}_2\mathbf{Y}$, note that $\mathbf{A}_2$ is an idempotent matrix of rank $s - 1$. Thus, $\mathbf{Y}'\mathbf{A}_2\mathbf{Y} = \mathbf{Y}'\mathbf{P}\mathbf{P}'\mathbf{Y} = \mathbf{X}'\mathbf{X}$ where the $(s-1) \times 1$ vector $\mathbf{X} = (X_1, \ldots, X_{s-1})' = \mathbf{P}'\mathbf{Y}$, the $st \times (s-1)$ matrix $\mathbf{P} = \mathbf{P}_s \otimes (1/\sqrt{t})\mathbf{1}_t$, and the $(s-1) \times s$ matrix $\mathbf{P}'_S$ is the lower portion of an s-dimensional Helmert matrix with $\mathbf{P}_s\mathbf{P}'_s = (\mathbf{I}_s - \frac{1}{s}\mathbf{J}_s)$, $\mathbf{1}'_s\mathbf{P}_s = \mathbf{0}_{1\times(s-1)}$, and $\mathbf{P}'_s\mathbf{P}_s = \mathbf{I}_{s-1}$. By Theorem 2.1.2 with $(s-1) \times st$ matrix $\mathbf{B} = \mathbf{P}'$ and $(s-1) \times 1$ vector $\mathbf{b} = \mathbf{0}_{(s-1)\times 1}$, $\mathbf{X} \sim \mathrm{N}_{s-1}(\mathbf{0}, (\sigma^2_{ST} + t\sigma^2_S)\mathbf{I}_{s-1})$ since

$$\mathbf{B}(\alpha\mathbf{1}_s \otimes \mathbf{1}_t) + \mathbf{b} = [\mathbf{P}_s \otimes (1/\sqrt{t})\mathbf{1}_t]'(\alpha\mathbf{1}_s \otimes \mathbf{1}_t) = \mathbf{0}_{(s-1)\times 1}$$

and

$$\mathbf{B}[\sigma^2_S\mathbf{I}_s \otimes \mathbf{J}_t \mid \sigma^2_T\mathbf{J}_s \otimes \mathbf{I}_t \mid \sigma^2_{ST}\mathbf{I}_s \otimes \mathbf{I}_t]\mathbf{B}'$$

$$= \sigma^2_S[\mathbf{P}'_s\mathbf{P}_s \otimes (1/\sqrt{t})\mathbf{1}'_t\mathbf{J}_t((1/\sqrt{t})\mathbf{1}_t)]$$

$$+\sigma^2_T[\mathbf{P}'_s\mathbf{J}_s\mathbf{P}_s \otimes (1/\sqrt{t})\mathbf{1}'_t((1/\sqrt{t})\mathbf{1}_t)]$$

$$+\sigma^2_{ST}[\mathbf{P}'_s\mathbf{P}_s \otimes (1/\sqrt{t})\mathbf{1}'_t((1/\sqrt{t})\mathbf{1}_t)]$$

$$= (\sigma^2_{ST} + t\sigma^2_S)\mathbf{I}_{s-1}.$$

Therefore, $[1/\sqrt{(\sigma^s_{ST} + t\sigma^2_S)}]\mathbf{X} \sim \mathrm{N}_{s-1}(\mathbf{0}, \mathbf{I}_{s-1})$. From Example 1.3.2, $\mathbf{Y}'\mathbf{A}_2\mathbf{Y}/(\sigma^2_{ST} + t\sigma^2_S) = \mathbf{X}'\mathbf{X}/(\sigma^2_{ST} + t\sigma^2_S)$ has a central chi-square distribution with $s - 1$

degrees of freedom. The distributions of $\mathbf{Y}'\mathbf{A}_3\mathbf{Y}$ and $\mathbf{Y}'\mathbf{A}_4\mathbf{Y}$ and the independence of $\mathbf{Y}'\mathbf{A}_2\mathbf{Y}$, $\mathbf{Y}'\mathbf{A}_3\mathbf{Y}$, and $\mathbf{Y}'\mathbf{A}_4\mathbf{Y}$ are left to the reader.

The preceding examples suggest the following general technique for finding the distribution of the quadratic form $\mathbf{Y}'\mathbf{A}\mathbf{Y}$ when $\mathbf{Y} \sim \mathrm{N}_n(\boldsymbol{\mu}, \boldsymbol{\Sigma})$ and $\mathbf{A}$ is an $n \times n$ idempotent matrix of rank r.

1. Set $\mathbf{A} = \mathbf{P}\mathbf{P}'$ where $\mathbf{P}$ is an $n \times r$ matrix of eigenvectors corresponding to the r eigenvalues of $\mathbf{A}$ equal to 1.
2. Let $\mathbf{Y}'\mathbf{A}\mathbf{Y} = \mathbf{Y}'\mathbf{P}\mathbf{P}'\mathbf{Y} = \mathbf{X}'\mathbf{X} = \sum_{i=1}^{r} X_i^2$ where the $r \times 1$ vector $\mathbf{X} = (X_1, \ldots, X_r)' = \mathbf{P}'\mathbf{Y}$. Then find the distribution of $\mathbf{X}$.
3. Find the distribution of $\mathbf{Y}'\mathbf{A}\mathbf{Y}$ using the distribution of $\mathbf{X}$.

This technique is used in the next chapter to prove some general theorems about the distributions of quadratic forms.

EXERCISES

1. Let the 2×1 random vector $\mathbf{Y} \sim \mathrm{N}_2(\boldsymbol{\mu}, \boldsymbol{\Sigma})$ where $\boldsymbol{\mu} = (0, 0)'$ and

$$\boldsymbol{\Sigma} = \begin{bmatrix} 1 & 0.3 \\ 0.3 & 1 \end{bmatrix}.$$

Find a 2×2 triangular matrix $\mathbf{T}$ such that $\mathbf{T}\mathbf{Y} \sim \mathrm{N}_2(\mathbf{0}, \mathbf{I}_2)$.

2. Let the 6×1 random vector $\mathbf{Y} = (Y_1, Y_2, Y_3, Y_4, Y_5, Y_6)' \sim \mathrm{N}_6(\boldsymbol{\mu}, \boldsymbol{\Sigma})$ where $\boldsymbol{\mu} = (\mu_1\mathbf{1}_3', \mu_2\mathbf{1}_2', \mu_3)'$ and

$$\boldsymbol{\Sigma} = \begin{bmatrix} 0.5\mathbf{I}_3 + 0.5\mathbf{J}_3 & \mathbf{0} & \mathbf{0} \\ \mathbf{0} & 0.3\mathbf{I}_2 + 0.7\mathbf{J}_2 & \mathbf{0} \\ \mathbf{0} & \mathbf{0} & 1 \end{bmatrix}$$

Find the distribution of the 3×1 random vector $\bar{\mathbf{Y}} = (\bar{Y}_1, \bar{Y}_2, \bar{Y}_3)'$ where $\bar{Y}_1 = \frac{1}{3}(Y_1 + Y_2 + Y_3)$, $\bar{Y}_2 = \frac{1}{2}(Y_4 + Y_5)$, and $\bar{Y}_3 = Y_6$.

3. Let the $n \times 1$ random vector $\mathbf{Y} = (Y_1, \ldots, Y_n)' \sim \mathrm{N}_n(\mathbf{0}, \boldsymbol{\Sigma})$ where the $n \times n$ covariance matrix

$$\boldsymbol{\Sigma} = \begin{bmatrix} w_1^{-1} & & & \\ & w_2^{-1} & & \mathbf{0} \\ & \mathbf{0} & \ddots & \\ & & & w_n^{-1} \end{bmatrix}$$

and let $Y^* = \sum_{i=1}^{n} w_i Y_i / [\sum_{i=1}^{n} w_i]^{1/2}$.

(a) Find the distribution of Y^*. Rewrite Y^* in terms of the $n \times 1$ vector $\mathbf{W} = (w_1, \ldots, w_n)'$ and the $n \times 1$ vector $\mathbf{Y}$.

(b) Find the distribution of $\sum_{i=1}^{n} w_i Y_i^2$.

4. Let the 9×1 random vector $\mathbf{Y} \sim \mathrm{N}_9(\boldsymbol{\mu}, \boldsymbol{\Sigma})$ where the mean vector $\boldsymbol{\mu} = (\mu_1 \mathbf{1}_2', \mu_2 \mathbf{1}_4', \mu_3 \mathbf{1}_3')'$ and covariance matrix

$$\boldsymbol{\Sigma} = \begin{bmatrix} \boldsymbol{\Sigma}_1 & & \mathbf{0} \\ & \boldsymbol{\Sigma}_2 & \\ \mathbf{0} & & \boldsymbol{\Sigma}_3 \end{bmatrix}$$

with $\boldsymbol{\Sigma}_i = (a - b)\mathbf{I}_{n_i} + b\mathbf{J}_{n_i}$ for $i = 1, 2, 3$ and $n_1 = 2, n_2 = 4$, and $n_3 = 3$. Also let

$$\mathbf{A} = \begin{bmatrix} \mathbf{A}_1 & & \mathbf{0} \\ & \mathbf{A}_2 & \\ \mathbf{0} & & \mathbf{A}_3 \end{bmatrix}$$

where $\mathbf{A}_i = \mathbf{I}_{n_i} - \frac{1}{n_i}\mathbf{J}_{n_i}$ for $i = 1, 2, 3$.

(a) Define a 9×6 matrix $\mathbf{P}$ such that $\mathbf{A} = \mathbf{P}\mathbf{P}'$.

(b) Find the distribution of $\mathbf{P}'\mathbf{Y}$.

(c) Find the distribution of $\mathbf{Y}'\mathbf{A}\mathbf{Y}$.

5. Let $Y_{ijk} = \mu_i + S_{ij} + T_{ijk}$ for $i = 1, \ldots, a$; $j = 1, \ldots, s$; and $k = 1, \ldots, t$ where $S_{ij} \sim iid\ \mathrm{N}_1(0, \sigma_S^2)$ and independent of $T_{ijk} \sim iid\ \mathrm{N}_1(0, \sigma_T^2)$.

(a) Find the distribution of the $a \times 1$ random vector $(\bar{Y}_{1..}, \ldots, \bar{Y}_{a..})'$ where $\bar{Y}_{i..} = \sum_{j=1}^{s} \sum_{k=1}^{t} Y_{ijk}/(st)$.

(b) Find the distribution of the $as \times 1$ random vector $\bar{Y}_{11.} - \bar{Y}_{1..}, \ldots, \bar{Y}_{1s.} - \bar{Y}_{1..}, \ldots, \bar{Y}_{a1.} - \bar{Y}_{a..}, \ldots, \bar{Y}_{as.} - \bar{Y}_{a..})'$ where $\bar{Y}_{ij.} = \sum_{k=1}^{t} Y_{ijk}/t$.

6. Let the $n \times 1$ random vector $\mathbf{Y} = (Y_1, \ldots, Y_n)' \sim \mathrm{N}_n(\boldsymbol{\mu}, \boldsymbol{\Sigma})$ where $\boldsymbol{\mu} = \alpha \mathbf{1}_n$ and $\Sigma = (a - b)\mathbf{I}_n + b\mathbf{J}_n$. Let $\bar{Y}. = \sum_{i=1}^{n} Y_i/n$ and $U = \sum_{i=1}^{n}(Y_i - \bar{Y}.)^2$.

(a) Find the distribution of $\bar{Y}$.

(b) Find the distribution of U.

(c) Are $\bar{Y}.$ and U independent? Prove your answer.

7. Let the 3×1 random vector $\mathbf{Y} = (Y_1, Y_2, Y_3)' \sim \mathrm{N}_3(\boldsymbol{\mu}, \boldsymbol{\Sigma})$ where $\boldsymbol{\mu} = (0, 1, 2)'$ and

$$\boldsymbol{\Sigma} = \begin{bmatrix} 1 & 1 & 2 \\ 1 & 4 & 3 \\ 2 & 3 & 9 \end{bmatrix}.$$

Let $Z = (2, -1, -1)\mathbf{Y}$. Find a random variable $U = \mathbf{b}'\mathbf{Y}$ where $\mathbf{b} = (b_1, b_2, b_3)'$ such that $E(U) = 0$ and U is independent of Z.

8. Let the 9×1 random vector $\mathbf{Y} = (Y_{11}, Y_{12}, Y_{21}, Y_{22}, Y_{23}, Y_{31}, Y_{32}, Y_{33}, Y_{34})' \sim \mathrm{N}_9(\boldsymbol{\mu}, \sigma^2 \mathbf{I}_9)$ where $\boldsymbol{\mu} = (\mu_1 \mathbf{1}_2', \mu_2 \mathbf{1}_3', \mu_3 \mathbf{1}_4')'$. Let $\bar{Y}_i = \sum_{j=1}^{r_i} Y_{ij}/r_i$ for $r_1 = 2, r_2 = 3$, and $r_3 = 4$.

   (a) Find a linear combination of $\bar{Y}_{1.}$, $\bar{Y}_{2.}$, and $\bar{Y}_{3.}$ that is independent of $\bar{Y}_{1.} - \bar{Y}_{2.}$. Prove your result.

   (b) What is the distribution of the linear combination you found in part a?

9. Let the $(n_1+n_2) \times 1$ random vector $\mathbf{Y} = (Y_{11}, \ldots, Y_{1n_1}, \ldots, Y_{21}, \ldots, Y_{2n_2})' \sim \mathrm{N}_{n_1+n_2}(\alpha \mathbf{1}_{n_1+n_2}, \sigma^2 \mathbf{I}_{n_1} + n_2)$.

   (a) Find the distribution of $\bar{Y}_{1.} - \bar{Y}_{2.}$ where $\bar{Y}_{i.} = \sum_{i=1}^{n_i} Y_{ij}/n_i$ for $i = 1, 2$.

   (b) Find the distribution of $(\bar{Y}_{1.} - \bar{Y}_{2.})^2$.

10. Let $\mathbf{Y} = \mathbf{X}\boldsymbol{\beta} + \mathbf{E}$ where $\mathbf{Y}$ is an $n \times 1$ random vector, $\mathbf{X}$ is an $n \times p$ matrix of known constants, $\boldsymbol{\beta}$ is a $p \times 1$ vector of unknown constants, and $\mathbf{E}$ is an $n \times 1$ random vector. Assume $\mathbf{E} \sim \mathrm{N}_n(\mathbf{0}, \sigma^2 \mathbf{I}_n)$.

    (a) Find the distribution of $\hat{\boldsymbol{\beta}} = (\mathbf{X}'\mathbf{X})^{-1}\mathbf{X}'\mathbf{Y}$.

    (b) Find the rank $[\mathrm{cov}(\hat{\mathbf{Y}})]$ where $\hat{\mathbf{Y}} = \mathbf{X}\hat{\boldsymbol{\beta}}$.

11. Let the 3×1 random vector $\mathbf{Y} = (Y_1, Y_2, Y_3)' \sim \mathrm{N}_3(\mathbf{0}, \sigma^2 \mathbf{I}_3)$.

    (a) Find the distribution of $Y_1 + 2Y_2 - Y_3$.

    (b) Show that $Y_1 + 2Y_2 - Y_3$ is independent of $2Y_1^2 + Y_2^2 + 2Y_3^2 - 2Y_1Y_2 + 2Y_2Y_3$.

    (c) Find a constant c such that $c[5Y_1^2 + 2Y_2^2 + 5Y_3^2 - 4Y_1Y_2 + 2Y_1Y_3 + 4Y_2Y_3]$ has a central chi-square distribution.

12. Let the 4×1 random vector $\mathbf{Y} = (Y_1, Y_2, Y_3, Y_4)' \sim \mathrm{N}_4(\boldsymbol{\mu}, \boldsymbol{\Sigma})$ where $\boldsymbol{\mu} = (2, 3, 0, -1)'$ and

$$\boldsymbol{\Sigma} = \begin{bmatrix} 11 & 5 & 1 & -1 \\ 5 & 11 & -1 & 1 \\ 1 & -1 & 11 & 5 \\ -1 & 1 & 5 & 11 \end{bmatrix}.$$

(a) Find the conditional distribution of $Y_1 + Y_2 | Y_3 + Y_4 = 1$.

(b) Show that the conditional mean of $Y_3 | Y_1 = y_1, Y_2 = y_2$ has the form $\beta_0 + \beta_1 y_1 + \beta_2 y_2$ and find the values of the β's.

13. Let the 3×1 random vector $\mathbf{Y} = (Y_1, Y_2, Y_3)' \sim \mathrm{N}_3(\boldsymbol{\mu}, \boldsymbol{\Sigma})$ where $\boldsymbol{\mu} = (1, 2, -2)'$ and

$$\boldsymbol{\Sigma} = \begin{bmatrix} 2 & 0 & -1 \\ 0 & 3 & 1 \\ -1 & 1 & 4 \end{bmatrix}.$$

Find the conditional distribution of $Y_1 + Y_2 + Y_3 | Y_2 + Y_3 = 1$.

14. Let the 4×1 random vector $\mathbf{Y} = (Y_1, Y_2, Y_3, Y_4)' \sim \mathrm{N}_4(\boldsymbol{\mu}, \boldsymbol{\Sigma})$ where $\boldsymbol{\mu} = (1, 0, 2, -1)'$ and

$$\boldsymbol{\Sigma} = \begin{bmatrix} 4 & 1 & 0 & 1 \\ 1 & 2 & 1 & 0 \\ 0 & 1 & 3 & 1 \\ 1 & 0 & 1 & 4 \end{bmatrix}.$$

(a) Find the marginal distribution of $(Y_1, Y_3)'$.

(b) Find the partial (i.e., conditional) correlation of Y_1 and Y_2 given $Y_3 = 2$.

(c) Find the distribution of $4Y_1 - Y_2 - 2$.

(d) Find a normally distributed random variable Z which is a (nontrivial) function of Y_1 and Y_2 and is independent of $4Y_1 - Y_2 - 2$.

15. Let the $n \times 1$ random vector $\mathbf{Y} = (Y_1, \ldots, Y_n)' \sim \mathrm{N}_n(\alpha \mathbf{1}_n, \sigma^2 \mathbf{I}_n)$ for $n \geq 3$.

(a) Find the conditional expectation of $\frac{1}{2}(Y_2 + Y_3)|\bar{Y}$ where $\bar{Y} = \sum_{i=1}^{n} Y_i/n$; that is, find $\mathrm{E}[\frac{1}{2}(Y_2 + Y_3)|\bar{Y}]$.

(b) Show that the variance of $\mathrm{E}[\frac{1}{2}(Y_2 + Y_3)|\bar{Y}]$ is smaller than the variance of $\frac{1}{2}(Y_2 + Y_3)$.

16. Let F and G be independent normal random variables with 0 means and variances σ^2 and $1 - \sigma^2$, respectively, $0 < \sigma^2 < 1$. Find the conditional distribution of F given $F + G = c$.

17. Let the 3×1 random vector $\mathbf{Y} = (Y_1, Y_2, Y_3)' \sim \mathrm{N}_3(\boldsymbol{\mu}, \boldsymbol{\Sigma})$. Show that the variance of Y_1 in the conditional distribution of Y_1 given Y_2 is greater than or equal to the variance of Y_1 in the conditional distribution of Y_1 given Y_2 and Y_3.

3 Distributions of Quadratic Forms

The distribution of the quadratic form $\mathbf{Y}'\mathbf{AY}$ is now derived when $\mathbf{Y} \sim \mathrm{N}_n(\mathbf{0}, \mathbf{I}_n)$. Later, the distribution of $\mathbf{Y}'\mathbf{AY}$ is developed when $\mathbf{Y} \sim \mathrm{N}_n(\mathbf{0}, \mathbf{\Sigma})$ for any positive definite $n \times n$ matrix $\mathbf{\Sigma}$.

3.1 QUADRATIC FORMS OF NORMAL RANDOM VECTORS

A chi-square random variable with n degrees of freedom and the noncentrality parameter λ will be designated by $\chi_n^2(\lambda)$. Therefore, a central chi-square random variable with n degrees of freedom is denoted by $\chi_n^2(\lambda = 0)$ or $\chi_n^2(0)$.

Theorem 3.1.1 *Let the $n \times 1$ random vector $\mathbf{Y} \sim \mathrm{N}_n(\mathbf{0}, \mathbf{I}_n)$ then $\mathbf{Y}'\mathbf{AY} \sim \chi_p^2(\lambda = 0)$ if and only if $\mathbf{A}$ is an $n \times n$ idempotent matrix of rank p.*

Proof: First assume $\mathbf{A}$ is an $n \times n$ idempotent matrix of rank p. By Theorem 1.1.10, $\mathbf{A} = \mathbf{PP}'$ where $\mathbf{P}$ is an $n \times p$ matrix of eigenvectors with $\mathbf{P}'\mathbf{P} = \mathbf{I}_p$. Let the $p \times 1$ random vector $\mathbf{X} = \mathbf{P}'\mathbf{Y}$. By Theorem 2.1.2 with $p \times n$ matrix $\mathbf{B} = \mathbf{P}'$

and $p \times 1$ vector $\mathbf{b} = \mathbf{0}_{p\times 1}$, $\mathbf{X} = (X_1, \ldots, X_p)' \sim \mathrm{N}_p(\mathbf{0}, \mathbf{I}_p)$. Therefore, by Example 1.3.2, $\mathbf{Y}'\mathbf{A}\mathbf{Y} = \mathbf{Y}'\mathbf{P}\mathbf{P}'\mathbf{Y} = \mathbf{X}'\mathbf{X} = \sum_{i=1}^{p} X_i^2 \sim \chi_p^2(\lambda = 0)$. Next assume that $\mathbf{Y}'\mathbf{A}\mathbf{Y} \sim \chi_p^2(\lambda = 0)$. Therefore, the moment generating function of $\mathbf{Y}'\mathbf{A}\mathbf{Y}$ is $(1-2t)^{-p/2}$. But the moment generating function of $\mathbf{Y}'\mathbf{A}\mathbf{Y}$ is also defined as

$$\begin{aligned} m_{\mathbf{Y}'\mathbf{A}\mathbf{Y}}(t) &= \mathrm{E}\left[e^{t\mathbf{Y}'\mathbf{A}\mathbf{Y}}\right] \\ &= \int \cdots \int (2\pi)^{-n/2} e^{[t\mathbf{y}'\mathbf{A}\mathbf{y} - \mathbf{y}'\mathbf{y}/2]} dy_1 \ldots dy_n \\ &= \int \cdots \int (2\pi)^{-n/2} e^{[\mathbf{y}'(\mathbf{I}_n - 2t\mathbf{A})\mathbf{y}/2]} dy_1 \ldots dy_n \\ &= |\mathbf{I}_n - 2t\mathbf{A}|^{-1/2}. \end{aligned}$$

The final equality holds since the last integral equation is the integral of a multivariate normal distribution (without the Jacobian $|\mathbf{I}_n - 2t\mathbf{A}|^{1/2}$) with mean vector $\mathbf{0}_{n\times 1}$ and covariance matrix $(\mathbf{I}_n - 2t\mathbf{A})^{-1}$. The two forms of the moment generating function must be equal for all t in some neighborhood of zero. Therefore,

$$(1-2t)^{-p/2} = |\mathbf{I}_n - 2t\mathbf{A}|^{-1/2}$$

or

$$(1-2t)^{p} = |\mathbf{I}_n - 2t\mathbf{A}|.$$

Let $\mathbf{Q}$ be the $n \times n$ matrix of eigenvectors and $\mathbf{D}$ be the $n \times n$ diagonal matrix of eigenvalues of $\mathbf{A}$ where the eigenvalues are given by $\lambda_1, \ldots, \lambda_n$. By Theorem 1.1.3, $\mathbf{Q}'\mathbf{A}\mathbf{Q} = \mathbf{D}$, $\mathbf{Q}'\mathbf{Q} = \mathbf{I}_n$, and

$$\begin{aligned} |\mathbf{I}_n - 2t\mathbf{A}| &= (|\mathbf{Q}'\mathbf{Q}|)(|\mathbf{I}_n - 2t\mathbf{A}|) \\ &= |\mathbf{Q}'(\mathbf{I}_n - 2t\mathbf{A})\mathbf{Q}| \\ &= |\mathbf{I}_n - 2t\mathbf{Q}'\mathbf{A}\mathbf{Q}| \\ &= |\mathbf{I}_n - 2t\mathbf{D}] \\ &= \prod_{i=1}^{n}(1 - 2t\lambda_i). \end{aligned}$$

Therefore, $(1-2t)^p = \Pi_{i=1}^{n}(1-2t\lambda_i)$. The left side of the equation is a polynomial in $2t$ with highest power p. The right side of the equation therefore must have highest power p in $2t$ also, implying that $(n-p)$ of the λ_i's are zero. Thus, the equation becomes $(1-2t)^p = \Pi_{i=1}^{p}(1-2t\lambda_i)$. Taking logarithms of each side and equating coefficients produces $1 - 2t = 1 - 2t\lambda_i$ for $i = 1, \ldots, p$. The solution to these equations is $\lambda_1 = \cdots = \lambda_p = 1$. Therefore, by Theorem 1.1.8, $\mathbf{A}$ is an idempotent matrix. ■

Thus far we have concentrated on central chi-square random variables (i.e., $\lambda = 0$). However, in general, if the $n \times 1$ random vector $\mathbf{Y} \sim \mathrm{N}_n(\boldsymbol{\mu}, \mathbf{I}_n)$ then $\mathbf{Y}'\mathbf{AY} \sim \chi^2_p(\lambda)$ where $\boldsymbol{\mu}$ is any $n \times 1$ mean vector and the noncentrality parameter is given by $\lambda = \boldsymbol{\mu}'\mathbf{A}\boldsymbol{\mu}/2$.

The next theorem considers the distribution of $\mathbf{Y}'\mathbf{AY}$ when $\mathbf{Y} \sim \mathrm{N}_n(\boldsymbol{\mu}, \boldsymbol{\Sigma})$ and $\boldsymbol{\Sigma}$ is a positive definite matrix of rank n.

Theorem 3.1.2 *Let the $n \times 1$ random vector $\mathbf{Y} \sim \mathrm{N}_n(\boldsymbol{\mu}, \boldsymbol{\Sigma})$ where $\boldsymbol{\Sigma}$ is an $n \times n$ positive definite matrix of rank n. Then $\mathbf{Y}'\mathbf{AY} \sim \chi^2_p(\lambda = \boldsymbol{\mu}'\mathbf{A}\boldsymbol{\mu}/2)$ if and only if any of the following conditions are satisfied: (1)$\mathbf{A\Sigma}$ (or $\mathbf{\Sigma A}$) is an idempotent matrix of rank p or (2)$\mathbf{A\Sigma A} = \mathbf{A}$ and $\mathbf{A}$ has rank p.*

Proof: Let $\mathbf{Z} = \mathbf{T}^{-1}(\mathbf{Y} - \boldsymbol{\mu})$ where $\boldsymbol{\Sigma} = \mathbf{TT}'$. By Theorem 2.1.2 with $n \times n$ matrix $\mathbf{B} = \mathbf{T}^{-1}$ and $n \times 1$ vector $\mathbf{b} = -\mathbf{T}^{-1}\boldsymbol{\mu}$, $\mathbf{Z} \sim \mathrm{N}_n(\mathbf{0}, \mathbf{I}_n)$. Furthermore, $\mathbf{Y}'\mathbf{AY} = (\mathbf{TZ} + \boldsymbol{\mu})'\mathbf{A}(\mathbf{TZ} + \boldsymbol{\mu}) = (\mathbf{Z} + \mathbf{T}^{-1}\boldsymbol{\mu})'\mathbf{T}'\mathbf{AT}(\mathbf{Z} + \mathbf{T}^{-1}\boldsymbol{\mu}) = \mathbf{V}'\mathbf{RV}$ where $\mathbf{V} = \mathbf{Z} + \mathbf{T}^{-1}\boldsymbol{\mu}$ and $\mathbf{R} = \mathbf{T}'\mathbf{AT}$. By Theorem 2.1.2 with $n \times n$ matrix $\mathbf{B} = \mathbf{I}_n$ and $n \times 1$ vector $\mathbf{b} = \mathbf{T}^{-1}\boldsymbol{\mu}$, $\mathbf{V} \sim \mathrm{N}_n(\mathbf{T}^{-1}\boldsymbol{\mu}, \mathbf{I}_n)$. By Theorem 3.1.1, $\mathbf{V}'\mathbf{RV} \sim \chi^2_p(\lambda)$ if and only if $\mathbf{R}$ is idempotent of rank p. But $\mathbf{R}$ is idempotent if and only if $(\mathbf{T}'\mathbf{AT})(\mathbf{T}'\mathbf{AT}) = \mathbf{T}'\mathbf{AT}$ or equivalently $\mathbf{A\Sigma} = \mathbf{A\Sigma A\Sigma}$. Therefore, $\mathbf{Y}'\mathbf{AY} \sim \chi^2_p(\lambda)$ if and only if $\mathbf{A\Sigma}$ is idempotent. Finally, $p = \mathrm{rank}(\mathbf{R}) = \mathrm{rank}(\mathbf{T}'\mathbf{AT}) = \mathrm{rank}(\mathbf{ATT}') = \mathrm{rank}(\mathbf{A\Sigma})$ since $\mathbf{T}$ is nonsingular. Also, $\lambda = (\mathbf{T}^{-1}\boldsymbol{\mu})'\mathbf{R}(\mathbf{T}^{-1}\boldsymbol{\mu})/2 = \boldsymbol{\mu}'\mathbf{T}^{-1\prime}\mathbf{T}'\mathbf{ATT}^{-1}\boldsymbol{\mu} = \boldsymbol{\mu}'\mathbf{A}\boldsymbol{\mu}/2$. The proofs for $\mathbf{\Sigma A}$ idempotent of rank p or $\mathbf{A\Sigma A} = \mathbf{A}$ of rank p are left to the reader. ■

It is convenient to make an observation at this point. In most applications it is more natural to show that $\mathbf{A\Sigma}$ is a multiple of an idempotent matrix. We therefore state the following two corollaries which are direct consequences of Theorem 3.1.2.

Corollary 3.1.2(a) *Let the $n \times 1$ random vector $\mathbf{Y} \sim \mathrm{N}_n(\boldsymbol{\mu}, \boldsymbol{\Sigma})$ where $\boldsymbol{\Sigma}$ is an $n \times n$ positive definite matrix of rank n. Then $\mathbf{Y}'\mathbf{AY} \sim c\chi^2_p(\lambda = \boldsymbol{\mu}'\mathbf{A}\boldsymbol{\mu}/(2c))$ if and only if (1) $\mathbf{A\Sigma}$(or$\mathbf{\Sigma A}$) is a multiple of an idempotent matrix of rank p where the multiple is c or (2) $\mathbf{A\Sigma A} = c\mathbf{A}$ and $\mathbf{A}$ has rank p.*

Corollary 3.1.2(b) *If the $n \times 1$ random vector $\mathbf{Y} \sim \mathrm{N}_n(\mathbf{0}, \sigma^2\mathbf{V})$ where $\mathbf{V}$ is an $n \times n$ positive definite matrix of known constants then $\mathbf{Y}'\mathbf{V}^{-1}\mathbf{Y}/\sigma^2 \sim \chi^2_n(\lambda = 0)$.*

There are cases where $\mathbf{Y} \sim \mathrm{N}_n(\boldsymbol{\mu}, \boldsymbol{\Sigma})$ but $\mathbf{A\Sigma}$ is not a multiple of an idempotent matrix. Such situations are covered by the next theorem.

Theorem 3.1.3 *If the $n \times 1$ random vector $\mathbf{Y} \sim \mathrm{N}_n(\boldsymbol{\mu}, \boldsymbol{\Sigma})$ where $\boldsymbol{\Sigma}$ is a positive definite matrix of rank n, then $\mathbf{Y}'\mathbf{AY} \sim \sum_{i=1}^{p} \lambda_i W_i^2$ where p = rank $(\mathbf{A\Sigma})$; W_i^2*

are independent $\chi_1^2(\delta_i)$ random variables for $i = 1, \ldots, p$; and $\lambda_1, \ldots, \lambda_p$ are the nonzero eigenvalues of $\mathbf{A\Sigma}$.

Proof: Let $\mathbf{Z} = \mathbf{T}^{-1}(\mathbf{Y} - \boldsymbol{\mu})$ where $\boldsymbol{\Sigma} = \mathbf{TT}'$. By Theorem 2.1.2 with $n \times n$ matrix $\mathbf{B} = \mathbf{T}^{-1}$ and $n \times 1$ vector $\mathbf{b} = -\mathbf{T}^{-1}\boldsymbol{\mu}$, $\mathbf{Z} \sim \mathrm{N}_n(\mathbf{0}, \mathbf{I}_n)$. Furthermore, $\mathbf{Y}'\mathbf{AY} = (\mathbf{TZ} + \boldsymbol{\mu})'\mathbf{A}(\mathbf{TZ} + \boldsymbol{\mu}) = (\mathbf{Z} + \mathbf{T}^{-1}\boldsymbol{\mu})'\mathbf{T}'\mathbf{AT}(\mathbf{Z} + \mathbf{T}^{-1}\boldsymbol{\mu}) = (\mathbf{Z} + \mathbf{T}^{-1}\boldsymbol{\mu})'\boldsymbol{\Gamma}\mathbf{D}\boldsymbol{\Gamma}'(\mathbf{Z} + \mathbf{T}^{-1}\boldsymbol{\mu})$ where $\mathbf{T}'\mathbf{AT}$ is an $n \times n$ symmetric matrix, $\boldsymbol{\Gamma}$ is the $n \times n$ orthogonal matrix of eigenvectors of $\mathbf{T}'\mathbf{AT}$, and $\mathbf{D}$ is the $n \times n$ diagonal matrix of eigenvalues of $\mathbf{T}'\mathbf{AT}$ such that $\mathbf{T}'\mathbf{AT} = \boldsymbol{\Gamma}\mathbf{D}\boldsymbol{\Gamma}'$. The eigenvalues of $\mathbf{T}'\mathbf{AT}$ are $\lambda_1, \ldots, \lambda_p, 0, \ldots, 0$ and rank $(\mathbf{T}'\mathbf{AT}) = p$. Let $\mathbf{W} = (W_1, \ldots, W_n)' = \boldsymbol{\Gamma}'(\mathbf{Z} + \mathbf{T}^{-1}\boldsymbol{\mu})$. Therefore, $\mathbf{Y}'\mathbf{AY} = \mathbf{W}'\mathbf{DW} = \sum_{i=1}^{p} \lambda_i W_i^2$. By Theorem 2.1.2 with $n \times n$ matrix $\mathbf{B} = \boldsymbol{\Gamma}'$ and $n \times 1$ vector $\mathbf{b} = \boldsymbol{\Gamma}'\mathbf{T}^{-1}\boldsymbol{\mu}$, $\mathbf{W} \sim \mathrm{N}_n(\boldsymbol{\Gamma}'\mathbf{T}^{-1}\boldsymbol{\mu}, \mathbf{I}_n)$. Therefore, W_i^2 are independent $\chi_1^2(\delta_i \geq 0)$ random variables for $i = 1, \ldots, p$. Furthermore, $p = \mathrm{rank}(\mathbf{T}'\mathbf{AT}) = \mathrm{rank}(\mathbf{ATT}') = \mathrm{rank}(\mathbf{A\Sigma})$ because $\mathbf{T}$ is nonsingular. Finally, the eigenvalues of $\mathbf{T}'\mathbf{AT}$ are found by solving the polynomial equation

$$|\mathbf{I}_n - \lambda\mathbf{T}'\mathbf{AT}| = 0.$$

Premultiplying the above expression by $|\mathbf{T}'^{-1}|$ and postmultiplying by $|\mathbf{T}'|$ we obtain

$$|\mathbf{T}'^{-1}|(|\mathbf{I}_n - \lambda\mathbf{T}'\mathbf{AT}|)|\mathbf{T}'| = 0$$

$$|\mathbf{T}'^{-1}\mathbf{T}' - \lambda\mathbf{T}'^{-1}\mathbf{T}'\mathbf{ATT}'| = 0$$

$$|\mathbf{I}_n - \lambda\mathbf{A\Sigma}| = 0$$

Thus, the eigenvalues $\mathbf{T}'\mathbf{AT}$ are the eigenvalues of $\mathbf{A\Sigma}$. ■

We now reexamine the distributions of a number of quadratic forms previously derived in Section 2.3.

Example 3.1.1 From Example 2.3.1 let the $n \times 1$ random vector $\mathbf{Y} \sim \mathrm{N}_n(\alpha\mathbf{1}_n, \sigma^2\mathbf{I}_n)$. By Corollary 3.1.2(a), $\sum_{i=1}^{n}(Y_i - \bar{Y})^2 = \mathbf{Y}'\left(\mathbf{I}_n - \frac{1}{n}\mathbf{J}_n\right)\mathbf{Y} \sim \sigma^2\chi_{n-1}^2$ $(\lambda = 0)$ since

$$\lambda = (\alpha\mathbf{1}_n)'\left(\mathbf{I}_n - \frac{1}{n}\mathbf{J}_n\right)(\alpha\mathbf{1}_n)/(2\sigma^2) = 0$$

and

$$\left(\mathbf{I}_n - \frac{1}{n}\mathbf{J}_n\right)(\sigma^2\mathbf{I}_n) = \sigma^2\left(\mathbf{I}_n - \frac{1}{n}\mathbf{J}_n\right),$$

which is a multiple of an idempotent matrix of rank $n - 1$. Furthermore, let $n(\bar{Y} - \alpha)^2 = n[(1/n)\mathbf{1}_n'(\mathbf{Y}-\alpha\mathbf{1}_n)]'[(1/n)\mathbf{1}_n'(\mathbf{Y}-\alpha\mathbf{1}_n)] = (\mathbf{Y}-\alpha\mathbf{1}_n)'(\frac{1}{n}\mathbf{J}_n)(\mathbf{Y}-\alpha\mathbf{1}_n)$. By

Theorem 2.1.2 with $n \times n$ matrix $\mathbf{B} = \mathbf{I}_n$ and $n \times 1$ vector ($\mathbf{b} = -\alpha\mathbf{1}_n$, $\mathbf{Y} - \alpha\mathbf{1}_n \sim \mathrm{N}_n(\mathbf{0}, \sigma^2\mathbf{I}_n)$. Therefore, by Corollary 3.1.2(a), $n(\bar{Y} - \alpha)^2 \sim \sigma^2\chi_1^2(\lambda = 0)$ since

$$\lambda = (\mathbf{0}_{n\times 1})'\left(\frac{1}{n}\mathbf{J}_n\right)(\mathbf{0}_{n\times 1})/(2\sigma^2) = 0$$

and

$$\left(\frac{1}{n}\mathbf{J}_n\right)(\sigma^2\mathbf{I}_n) = \sigma^2\left(\frac{1}{n}\mathbf{J}_n\right)$$

which is a multiple of an idempotent matrix of rank 1.

Example 3.1.2 Consider the two-way cross classification from Example 2.3.2 where the $st \times 1$ random vector $\mathbf{Y} = (Y_{11}, \ldots, Y_{1t}, \ldots, Y_{s1}, \ldots, Y_{st})' \sim \mathrm{N}_{st}(\boldsymbol{\mu}, \Sigma)$, $\boldsymbol{\mu} = \alpha\mathbf{1}_s \otimes \mathbf{1}_t$, and $\boldsymbol{\Sigma} = \sigma_S^2\mathbf{I}_s \otimes \mathbf{J}_t + \sigma_T^2\mathbf{J}_s \otimes \mathbf{I}_t + \sigma_{ST}^2\mathbf{I}_s \otimes \mathbf{I}_t$. The sum of squares due to the random factor S is $\mathbf{Y}'\mathbf{A}_2\mathbf{Y}$ where $\mathbf{A}_2 = (\mathbf{I}_s - \frac{1}{s}\mathbf{J}_s) \otimes \frac{1}{t}\mathbf{J}_t$. By Corollary 3.1.2(a), $\mathbf{Y}'\mathbf{A}_2\mathbf{Y} \sim (\sigma_{ST}^2 + t\sigma_S^2)\chi_{s-1}^2(\lambda_2 = 0)$ since $\mathbf{A}_2\Sigma = t\sigma_S^2[(\mathbf{I}_s - \frac{1}{s}\mathbf{J}_s) \otimes \frac{1}{t}\mathbf{J}_t] + \sigma_{ST}^2[(\mathbf{I}_s - \frac{1}{s}\mathbf{J}_s) \otimes \frac{1}{t}\mathbf{J}_t] = (\sigma_{ST}^2 + t\sigma_s^2)\mathbf{A}_2$,

$$\lambda_2 = (\alpha\mathbf{1}_s \otimes \mathbf{1}_t)'\left[\left(\mathbf{I}_s - \frac{1}{s}\mathbf{J}_s\right) \otimes \frac{1}{t}\mathbf{J}_t\right](\alpha\mathbf{1}_s \otimes \mathbf{1}_t) = 0$$

and $\mathbf{A}_2$ is an idempotent matrix of rank $s - 1$. The sum of squares due to the mean is $\mathbf{Y}'\mathbf{A}_1\mathbf{Y}$ where $\mathbf{A}_1 = \frac{1}{s}\mathbf{J}_s \otimes \frac{1}{t}\mathbf{J}_t$. By Corollary 3.1.2(a), $\mathbf{Y}'\mathbf{A}_1\mathbf{Y} \sim (t\sigma_S^2 + s\sigma_T^2 + \sigma_{ST}^2)\chi_1^2(\lambda_1)$ since $A_1\boldsymbol{\Sigma} = (t\sigma_S^2 + s\sigma_T^2 + \sigma_{ST}^2)\mathbf{A}_1$, $\mathbf{A}_1$ is an idempotent matrix of rank 1, and

$$\lambda_1 = \{1/[2(t\sigma_s^2 + s\sigma_T^2 + \sigma_{ST}^2)]\}\ (\alpha\mathbf{1}_s \otimes \mathbf{1}_t)'\left[\frac{1}{s}\mathbf{J}_s \otimes \frac{1}{t}\mathbf{J}_t\right](\alpha\mathbf{1}_s \otimes \mathbf{1}_t)$$

$$= st\alpha^2/[2(t\sigma_S^2 + s\sigma_T^2 + \sigma_{ST}^2)]$$

The distributions of the other quadratic forms in Example 2.3.2 can be derived in a similar manner and are left to the reader.

3.2 INDEPENDENCE

The independence of two quadratic forms is examined in the next theorem.

Theorem 3.2.1 *Let* $\mathbf{A}$ *and* $\mathbf{B}$ *be* $n \times n$ *constant matrices. Let the* $n \times 1$ *random vector* $\mathbf{Y} \sim \mathrm{N}_n(\boldsymbol{\mu}, \Sigma)$. *The quadratic forms* $\mathbf{Y}'\mathbf{A}\mathbf{Y}$ *and* $\mathbf{Y}'\mathbf{B}\mathbf{Y}$ *are independent if and only if* $\mathbf{A}\Sigma\mathbf{B} = \mathbf{0}$ (*or* $\mathbf{B}\Sigma\mathbf{A} = \mathbf{0}$).

Proof: The matrices $\mathbf{A}$, Σ, and $\mathbf{B}$ are symmetric. Therefore, $\mathbf{A}\Sigma\mathbf{B} = \mathbf{0}$ is equivalent to $\mathbf{B}\Sigma A = \mathbf{0}$. Assume $\mathbf{A}\Sigma\mathbf{B} = \mathbf{0}$. Since $\boldsymbol{\Sigma}$ is positive definite, by

Theorem 1.1.5, there exists an $n \times n$ nonsingular matrix $\mathbf{S}$ such that $\mathbf{S\Sigma S'} = \mathbf{I}_n$. Then $\mathbf{Z} = \mathbf{SY} \sim \mathrm{N}_n(\mathbf{S}\boldsymbol{\mu}, \mathbf{I}_n)$. Let $\mathbf{G} = (\mathbf{S}^{-1})'\mathbf{AS}^{-1}$ and $\mathbf{H} = (\mathbf{S}^{-1})'\mathbf{BS}^{-1}$. Therefore, $\mathbf{Y'AY} = \mathbf{Z'GZ}$, $\mathbf{Y'BY} = \mathbf{Z'HZ}$, and $\mathbf{A\Sigma B} = \mathbf{S'GS\Sigma S'HS} = \mathbf{S'GHS}$. Thus, the statement $\mathbf{A\Sigma B} = \mathbf{0}$ implies $\mathbf{Y'AY}$ and $\mathbf{Y'BY}$ are independent and the statement $\mathbf{GH} = \mathbf{0}$, implies $\mathbf{Z'GZ}$ and $\mathbf{Z'HZ}$ are independent, are equivalent. Since $\mathbf{G}$ is symmetric, there exists an orthogonal matrix $\mathbf{P}$ such that $\mathbf{G} = \mathbf{P'DP}$, where $a = \mathrm{rank}(\mathbf{A}) = \mathrm{rank}(\mathbf{G})$ and $\mathbf{D}$ is a diagonal matrix with a nonzero diagonal elements. Without loss of generality, assume that

$$\mathbf{D} = \begin{bmatrix} \mathbf{D}_a & \mathbf{0} \\ \mathbf{0} & \mathbf{0} \end{bmatrix}$$

where $\mathbf{D}_a$ is the $a \times a$ diagonal matrix containing the nonzero elements of $\mathbf{D}$. Let $\mathbf{X} = \mathbf{PZ} \sim \mathrm{N}_n(\mathbf{PS}\boldsymbol{\mu}, \mathbf{I}_n)$ and partition $\mathbf{X}$ as $[\mathbf{X}_1', \mathbf{X}_2']'$, where $\mathbf{X}_1$ is $a \times 1$. Note that $\mathbf{X}_1$ and $\mathbf{X}_2$ are independent. Then $\mathbf{Z'GZ} = \mathbf{Z'P'DPZ} = \mathbf{X'DX} = \mathbf{X}_1'\mathbf{D}_a\mathbf{X}_1$ and $\mathbf{Z'HZ} = \mathbf{X'PHP'X} = \mathbf{X'CX}$ where the symmetric matrix $\mathbf{C} = \mathbf{PHP'}$. If $\mathbf{GH} = \mathbf{0}$ then $\mathbf{P'DPP'CP} = \mathbf{0}$, which implies $\mathbf{DC} = \mathbf{0}$. Partitioning $\mathbf{C}$ to conform with $\mathbf{D}$,

$$\mathbf{0} = \mathbf{DC} = \begin{bmatrix} \mathbf{D}_a & \mathbf{0} \\ \mathbf{0} & \mathbf{0} \end{bmatrix} \begin{bmatrix} \mathbf{C}_{11} & \mathbf{C}_{12} \\ \mathbf{C}_{12}' & \mathbf{C}_{22} \end{bmatrix} = \begin{bmatrix} \mathbf{D}_a\mathbf{C}_{11} & \mathbf{D}_a\mathbf{C}_{12} \\ \mathbf{0} & \mathbf{0} \end{bmatrix},$$

which implies $\mathbf{C}_{11} = \mathbf{0}$ and $\mathbf{C}_{12} = \mathbf{0}$. Therefore,

$$\mathbf{C} = \begin{bmatrix} \mathbf{0} & \mathbf{0} \\ \mathbf{0} & \mathbf{C}_{22} \end{bmatrix},$$

which implies $\mathbf{X'CX} = \mathbf{X}_2'\mathbf{C}_{22}\mathbf{X}_2$, which is independent of $\mathbf{X}_1'\mathbf{D}_a\mathbf{X}_1$. Therefore, $\mathbf{Z'GZ}$ and $\mathbf{Z'HZ}$ are independent. The proof of the converse statement is supplied by Searle (1971). ■

The following theorem considers the independence of a quadratic form and linear combinations of a normally distributed random vector.

Theorem 3.2.2 *Let* $\mathbf{A}$ *and* $\mathbf{B}$ *be* $n \times n$ *and* $m \times n$ *constant matrices, respectively. Let the* $n \times 1$ *random vector* $\mathbf{Y} \sim \mathrm{N}_n(\boldsymbol{\mu}, \Sigma)$. *The quadratic form* $\mathbf{Y'AY}$ *and the set of linear combinations* $\mathbf{BY}$ *are independent if and only if* $\mathbf{B\Sigma A} = \mathbf{0}$ *(or* $\mathbf{A\Sigma B'} = \mathbf{0}$*).*

Proof: The "if" portion can be proven by the same method used in the proof of Theorem 3.2.1. The proof of the converse statement is supplied by Searle (1971). ■

In the following examples the independence of certain quadratic forms and linear combinations is examined.

Example 3.2.1 Consider the one-way classification described in Examples 1.2.10 and 2.1.4. The sum of squares due to the fixed factor is $\mathbf{Y}'\mathbf{A}_2\mathbf{Y}$ where $\mathbf{A}_2 = (\mathbf{I}_t - \frac{1}{t}\mathbf{J}_t) \otimes \frac{1}{r}\mathbf{J}_r$ is an idempotent matrix of rank $t-1$. Furthermore, $\mathbf{A}_2\Sigma = [(\mathbf{I}_t - \frac{1}{t}\mathbf{J}_t) \otimes \frac{1}{r}\mathbf{J}_r]\,[\sigma^2\mathbf{I}_t \otimes \mathbf{I}_r] = \sigma^2\mathbf{A}_2$. The sum of squares due to the nested replicates is $\mathbf{Y}'\mathbf{A}_3\mathbf{Y}$ where $\mathbf{A}_3 = \mathbf{I}_t \otimes (\mathbf{I}_r - \frac{1}{r}\mathbf{J}_r)$ is an idempotent matrix of rank $t(r-1)$. Likewise, $\mathbf{A}_3\Sigma = [\mathbf{I}_t \otimes (\mathbf{I}_r - \frac{1}{r}\mathbf{J}_r)]\,[\sigma^2\mathbf{I}_t \otimes \mathbf{I}_r] = \sigma^2\mathbf{A}_3$. Therefore, by Corollary 3.1.2(a), $\mathbf{Y}'\mathbf{A}_2\mathbf{Y} \sim \sigma^2\chi^2_{t-1}(\lambda_2)$ and $\mathbf{Y}'\mathbf{A}_3\mathbf{Y} \sim \sigma^2\chi^2_{t(r-1)}(\lambda_3)$ where $\lambda_2 = [(\mu_1, \ldots, \mu_t)' \otimes \mathbf{1}_r]'\,[(\mathbf{I}_t - \frac{1}{t}\mathbf{J}_t) \otimes \frac{1}{r}\mathbf{J}_r]\,[(\mu_1, \ldots, \mu_t)' \otimes \mathbf{1}_r]/(2\sigma^2) = r\sum_{i=1}^{t}(\mu_i - \bar{\mu}.)^2/(2\sigma^2)$ with $\bar{\mu}. = \sum_{i=1}^{t}\mu_i/t$ and $\lambda_3 = [(\mu_1, \ldots, \mu_t)' \otimes \mathbf{1}_r]'\,[\mathbf{I}_t \otimes (\mathbf{I}_r - \frac{1}{r}\mathbf{J}_r)]\,[(\mu_1, \ldots, \mu_t)' \otimes \mathbf{1}_r]/(2\sigma^2) = 0$. Finally, by Theorem 3.2.1, $\mathbf{Y}'\mathbf{A}_2\mathbf{Y}$ and $\mathbf{Y}'\mathbf{A}_3\mathbf{Y}$ are independent since $\mathbf{A}_2\Sigma\mathbf{A}_3 = \sigma^2\mathbf{A}_2\mathbf{A}_3 = \sigma^2[(\mathbf{I}_t - \frac{1}{t}\mathbf{J}_t) \otimes \frac{1}{r}\mathbf{J}_r]\,[\mathbf{I}_t \otimes (\mathbf{I}_r - \frac{1}{r}\mathbf{J}_r)] = \mathbf{0}_{tr \times tr}$.

Example 3.2.2 Reconsider Example 2.3.1 where $\mathbf{Y} = (Y_1, \ldots, Y_n)' \sim \mathrm{N}_n(\alpha\mathbf{1}_n, \sigma^2\mathbf{I}_n)$, $U = \Sigma_{i=1}^{n}(Y_i - \bar{Y})^2/\sigma^2 = \mathbf{y}'[(1/\sigma^2)(\mathbf{I}_n - \frac{1}{n}\mathbf{J}_n)]\mathbf{Y}$ and $\bar{Y} = (1/n)\mathbf{1}_n'\mathbf{Y}$. By Theorem 3.2.2, $\bar{Y}$ and U are independent since $(1/n)\mathbf{1}_n'[\sigma^2\mathbf{I}_n]\,[(1/\sigma^2)(\mathbf{I}_n - \frac{1}{n}\mathbf{J}_n)] = \mathbf{0}_{1\times n}$.

3.3 THE t AND F DISTRIBUTIONS

The normal and chi-square distributions were discussed at length in the previous sections. We now examine the distributions of certain functions of chi-square and normal random variables.

Definition 3.3.1 *Noncentral t Random Variable*: Let the random variable $Y \sim \mathrm{N}_1(\alpha, \sigma^2)$ and the random variable $U \sim \chi^2_n(0)$. If Y and U are independent, then the random variable $T = (Y/\sigma)/\sqrt{U/n}$ is distributed as a noncentral t random variable with n degrees of freedom and noncentrality parameter $\lambda = \alpha^2/2$. Denote this noncentral t random variable as $t_n(\lambda)$.

Definition 3.3.2 *Noncentral F Random Variable*: Let the random variable $U_1 \sim \chi^2_{n_1}(\lambda)$ and the random variable $U_2 \sim \chi^2_{n_2}(0)$. If U_1 and U_2 are independent, then the random variable $F = (U_1/n_1)/(U_2/n_2)$ is distributed as a noncentral F random variable with n_1 and n_2 degrees of freedom and noncentrality parameter λ. Denote this noncentral F random variable as $F_{n_1,n_2}(\lambda)$.

A t random variable with n degrees of freedom and a noncentrality parameter equal to zero [i.e., $t_n(\lambda = 0)$] has a central t distribution. Likewise, an F random variable with n_1 and n_2 degrees of freedom and a noncentrality parameter equal to zero [i.e., $F_{n_1,n_2}(\lambda = 0)$] has a central F distribution.

In recent years Smith and Lewis (1980, 1982), Pavur and Lewis (1983), Scariano, Neill, and Davenport (1984) and Scariano and Davenport (1984) have developed the theory of the corrected F random variable. The definition of the corrected F random variable is given next.

Definition 3.3.3 *Noncentral Corrected F Random Variable*: Let the random variable $U_1 \sim c_1\chi^2_{n_1}(\lambda)$ and the random variable $U_2 \sim c_2\chi^2_{n_2}(0)$. If U_1 and U_2 are independent, then the random variable $F_c = (c_2/c_1)[(U_1/n_1)/(U_2/n_2)] \sim F_{n_1,n_2}(\lambda)$ is called a corrected F random variable where the ratio c_2/c_1 is the correction factor.

In practice, we often encounter independent random variables U_1 and U_2, which are distributed as multiples of chi-square random variables (U_2 being a multiple of a central chi square). The random variable $F = (U_1/n_1)/(U_2/n_2)$ in this case will be distributed as a noncentral F random variable if and only if $c_1 = c_2$ (i.e., $c_2/c_1 = 1$). Generally, c_1 and c_2 will be linear combinations of unknown variance parameters.

In the following examples a number of central and noncentral t and F random variables are derived.

Example 3.3.1 Let the $n \times 1$ random vector $\mathbf{Y} = (Y_1, \ldots, Y_n)' \sim \mathrm{N}_n(\alpha\mathbf{1}_n, \sigma^2\mathbf{I}_n)$. By Example 2.1.1, $\bar{Y} \sim \mathrm{N}_1(\alpha, \sigma^2/n)$. By Example 3.1.1, $\sum_{i=1}^n (Y_i - \bar{Y})^2 = \mathbf{Y}'[(\mathbf{I}_n - \frac{1}{n}\mathbf{J}_n)]\mathbf{Y} \sim \sigma^2\chi^2_{n-1}(0)$. By Example 3.2.2. $\bar{Y}$ and $\sum_{i=1}^n (Y_i - \bar{Y})^2$ are independent. Therefore,

$$T = \sqrt{n}\,\bar{Y}/S = [\bar{Y}/(\sigma/\sqrt{n})]/\sqrt{S^2/\sigma^2} \sim t_{n-1}(\lambda)$$

where $S^2 = \sum_{i=1}^n (Y_i - \bar{Y})^2/(n-1)$ and $\lambda = \alpha^2/2$.

Example 3.3.2 Consider the one-way classification described in Example 3.2.1. It was shown that the sum of squares due to the fixed factor $\mathbf{Y}'\mathbf{A}_2\mathbf{Y} = \mathbf{Y}'[(\mathbf{I}_t - \frac{1}{t}\mathbf{J}_t) \otimes \frac{1}{r}\mathbf{J}_r]\mathbf{Y} \sim \sigma^2\chi^2_{t-1}(\lambda_2)$ and the sum of squares due to the nested replicates $\mathbf{Y}'\mathbf{A}_3\mathbf{Y} = \mathbf{Y}'[\mathbf{I}_t \otimes (\mathbf{I}_r - \frac{1}{r}\mathbf{J}_r)]\mathbf{Y} \sim \sigma^2\chi^2_{t(r-1)}(0)$. Furthermore, $\mathbf{Y}'\mathbf{A}_2\mathbf{Y}$ and $\mathbf{Y}'\mathbf{A}_3\mathbf{Y}$ are independent. Therefore, the statistic

$$F^* = \frac{\mathbf{Y}'\mathbf{A}_2\mathbf{Y}/(t-1)}{\mathbf{Y}'\mathbf{A}_3\mathbf{Y}/[t(r-1)]} \sim F_{t-1,t(r-1)}(\lambda_2)$$

where $\lambda_2 = r\sum_{i=1}^t (\mu_i - \bar{\mu}_\cdot)^2/(2\sigma^2)$. The hypothesis $\mathrm{H}_0 : \lambda_2 = 0$ versus $\mathrm{H}_1 : \lambda > 0$ is equivalent to the hypothesis $\mathrm{H}_0 : \mu_1 = \mu_2 = \cdots = \mu_t$ versus H_1 : the μ_i's are not all equal. Thus, under H_0, the statistic F^* has a central F distribution with $t-1$ and $t(r-1)$ degrees of freedom. A γ level rejection region for the hypothesis H_0 versus H_1 is as follows: Reject H_0 if $F^* > F^{\gamma}_{t-1,t(r-1)}$ where

$F^{\gamma}_{t-1,t(r-1)}$ is the 100 $(1-\gamma)^{\text{th}}$ percentile point of a central F distribution with $t-1$ and $t(r-1)$ degrees of freedom.

Example 3.3.3 Consider the two-way cross classification described in Example 3.1.2. The sums of squares due to the random factor S and due to the random interaction ST are given by $\mathbf{Y}'\mathbf{A}_2\mathbf{Y}$ and $\mathbf{Y}'\mathbf{A}_4\mathbf{Y}$, respectively. It was shown that $\mathbf{Y}'\mathbf{A}_2\mathbf{Y} = \mathbf{Y}'[(\mathbf{I}_s - \frac{1}{s}\mathbf{J}_s) \otimes \frac{1}{t}\mathbf{J}_t]\mathbf{Y} \sim (\sigma^2_{ST} + t\sigma^2_S)\chi^2_{s-1}(0)$. Furthermore, $\mathbf{A}_4\boldsymbol{\Sigma} = [(\mathbf{I}_s - \frac{1}{s}\mathbf{J}_s) \otimes (\mathbf{I}_t - \frac{1}{t}\mathbf{J}_t)]\ [\sigma^2_S\mathbf{I}_s \otimes \mathbf{J}_t + \sigma^2_T\mathbf{J}_s \otimes \mathbf{I}_t + \sigma^2_{ST}\mathbf{I}_s \otimes \mathbf{I}_t] = \sigma^2_{ST}\mathbf{A}_4$ where $\mathbf{A}_4$ is an idempotent matrix of rank $(s-1)(t-1)$. Therefore, by Corollary 3.1.2(a), $\mathbf{Y}'\mathbf{A}_4\mathbf{Y} = \mathbf{Y}'[(\mathbf{I}_s - \frac{1}{s}\mathbf{J}_s) \otimes (\mathbf{I}_t - \frac{1}{t}\mathbf{J}_t)]\mathbf{Y} \sim \sigma^2_{ST}\chi^2_{(s-1)(t-1)}(\lambda_4)$ where $\lambda_4 = (\alpha\mathbf{1}_s \otimes \mathbf{1}_t)'[(\mathbf{I}_s - \frac{1}{s}\mathbf{J}_s) \otimes (\mathbf{I}_t - \frac{1}{t}\mathbf{J}_t)](\alpha\mathbf{1}_s \otimes \mathbf{1}_t)/(2\sigma^2_{ST}) = 0$. Finally, $\mathbf{A}_2\boldsymbol{\Sigma}\mathbf{A}_4 = (\sigma^2_{ST} + t\sigma^2_S)\mathbf{A}_2\mathbf{A}_4 = \mathbf{0}_{st\times st}$. Therefore, by Theorem 3.2.1, $\mathbf{Y}'\mathbf{A}_2\mathbf{Y}$ and $\mathbf{Y}'\mathbf{A}_4\mathbf{Y}$ are independent. By Definition 3.3.3, the statistic

$$F^* = \frac{\mathbf{Y}'\mathbf{A}_2\mathbf{Y}/(s-1)}{\mathbf{Y}'\mathbf{A}_4\mathbf{Y}/[(s-1)(t-1)]} \sim \frac{(\sigma^2_{ST} + t\sigma^2_S)}{\sigma^2_{ST}} F_{s-1,(s-1)(t-1)}(0).$$

Under the hypothesis $\text{H}_0 : \sigma^2_S = 0$, the statistic F^* has a central F distribution with $s-1$ and $(s-1)(t-1)$ degrees of freedom. A γ level rejection region for the hypothesis $\text{H}_0 : \sigma^2_S = 0$ versus $\text{H}_1 : \sigma^2_S > 0$ follows; Reject H_0 if $F^* > F^{\gamma}_{s-1,(s-1)(t-1)}$.

3.4 BHAT'S LEMMA

The following lemma by Bhat (1962) is applicable in many ANOVA and regression problems. The lemma provides necessary and sufficient conditions for sums of squares to be distributed as multiples of independent chi-square random variables.

Lemma 3.4.1 *Let k and n denote fixed positive integers such that $1 \le k \le n$. Suppose $\mathbf{I}_n = \sum_{i=1}^k \mathbf{A}_i$, where each $\mathbf{A}_i$ is an $n \times n$ symmetric matrix of rank n_i with $\sum_{i=1}^k n_i = n$. If the $n \times 1$ random vector $\mathbf{Y} \sim \text{N}_n(\boldsymbol{\mu}, \boldsymbol{\Sigma})$ and the sum of squares $S_i^2 = \mathbf{Y}'\mathbf{A}_i\mathbf{Y}$ for $i = 1, \ldots, k$, then*

(a) $S_i^2 \sim c_i\chi^2_{n_i}(\lambda_i = \boldsymbol{\mu}'\mathbf{A}_i\boldsymbol{\mu}/(2c_i))$ *and*

(b) $\{S_i^2, i = 1, \ldots, k\}$ *are mutually independent*

if and only if $\boldsymbol{\Sigma} = \sum_{i=1}^k c_i\mathbf{A}_i$ *where* $c_i > 0$.

Proof: This proof is due to Scariano *et al.* (1984). Assume that the quadratic forms S_i^2 satisfy (a) and (b) given in Lemma 3.4.1. By Theorems 3.1.2 and 3.2.1, (i) the matrices $(1/c_i)\ \mathbf{A}_i\boldsymbol{\Sigma}$ are idempotent for $i = 1, \ldots, k$ and (ii) $\mathbf{A}_i\boldsymbol{\Sigma}\mathbf{A}_j = \mathbf{0}_{n\times n}$ for $i \ne j, i, j = 1, \ldots, k$. Furthermore, by Theorem 1.1.7, $\mathbf{A}_i = \mathbf{A}_i^2$ and

$\mathbf{A}_i\mathbf{A}_j = \mathbf{0}_{n\times n}$ for $i \neq j, i, j = 1, \ldots, k$. But (i) and (ii) imply that $\sum_{i=1}^{k}(1/c_i)\mathbf{A}_i\mathbf{\Sigma}$ is idempotent of rank n and thus equal to $\mathbf{I}_n$. Hence, $\Sigma = [\sum_{i=1}^{k}(1/c_i)\mathbf{A}_i]^{-1} = \sum_{i=1}^{k} c_i\mathbf{A}_i$. Conversely, assume $\Sigma = \sum_{i=1}^{k} c_i\mathbf{A}_i$. But $\mathbf{A}_i = \mathbf{A}_i^2$ and $\mathbf{A}_i\mathbf{A}_j = \mathbf{0}_{n\times n}$, so (i) and (ii) hold. Therefore, by Theorems 3.1.2 and 3.2.1, (a) and (b) hold. ■

In the next example, Bhat's lemma is applied to the two-way cross classification described in Example 2.3.2

Example 3.4.1 From Example 2.3.2, $\Sigma = [\sigma_S^2\mathbf{I}_s \otimes \mathbf{J}_t + \sigma_T^2\mathbf{J}_s \otimes \mathbf{I}_t + \sigma_{ST}^2\mathbf{I}_s \otimes \mathbf{I}_t]$, $\mathbf{A}_1 = \frac{1}{s}\mathbf{J}_s \otimes \frac{1}{t}\mathbf{J}_t$, $\mathbf{A}_2 = (\mathbf{I}_s - \frac{1}{s}\mathbf{J}_s) \otimes \frac{1}{t}\mathbf{J}_t$, $\mathbf{A}_3 = \frac{1}{s}\mathbf{J}_s \otimes (\mathbf{I}_t - \frac{1}{t}\mathbf{J}_t)$, $\mathbf{A}_4 = (\mathbf{I}_s - \frac{1}{s}\mathbf{J}_s)\otimes(\mathbf{I}_t - \frac{1}{t}\mathbf{J}_t)$, $st = \sum_{m=1}^{4}\text{rank}(\mathbf{A}_m) = 1+(s-1)+(t-1)+(s-1)(t-1)$, and $\mathbf{I}_s \otimes \mathbf{I}_t = \sum_{m=1}^{4}\mathbf{A}_m$. Furthermore, $\mathbf{A}_m\mathbf{\Sigma} = c_m\mathbf{A}_m$ for $m = 1, \ldots, 4$ where $c_1 = \sigma_{ST}^2 + t\sigma_S^2 + s\sigma_T^2$, $c_2 = \sigma_{ST}^2 + t\sigma_S^2$, $c_3 = \sigma_{ST}^2 + s\sigma_T^2$, and $c_4 = \sigma_{ST}^2$. Therefore, $\mathbf{\Sigma} = (\sum_{m=1}^{4}\mathbf{A}_m)\mathbf{\Sigma} = \sum_{m=1}^{4}(\mathbf{A}_m\mathbf{\Sigma}) = \sum_{m=1}^{4}c_m\mathbf{A}_m$. Thus, by Bhat's lemma, the quadratic forms $\mathbf{Y}'\mathbf{A}_m\mathbf{Y}$ are distributed as independent $c_m\chi^2_{\text{rank}(\mathbf{A}_m)}\{\lambda_m = (\alpha\mathbf{1}_s \otimes \mathbf{1}_t)'\mathbf{A}_m(\alpha\mathbf{1}_s \otimes \mathbf{1}_t)/(2c_m)\}$ for $m = 1, \ldots, 4$.

EXERCISES

1. Use Corollary 3.1.2(a) to find the distribution of $\sum_{i=1}^{n} w_i Y_i^2$ from Exercise 3b in Chapter 2.
2. Use Corollary 3.1.2(a) to find the distribution of $\mathbf{Y}'\mathbf{A}\mathbf{Y}$ from Exercise 4c in Chapter 2.
3. Consider the model presented in Exercise 5 of Chapter 2.
   (a) Find the distribution of $V_1 = \sum_{i=1}^{a}\sum_{j=1}^{s}\sum_{k=1}^{t}(\bar{Y}_{ij\cdot} - \bar{Y}_{i\cdot\cdot})^2$.
   (b) Find the distribution of $V_2 = \sum_{i=1}^{a}\sum_{j=1}^{s}\sum_{k=1}^{t}(Y_{ijk} - \bar{Y}_{ij\cdot})^2$.
   (c) Find the distribution of $\{V_1/[a(s-1)]\}/\{V_2/[as(t-1)]\}$.
4. Consider Exercise 6 of Chapter 2.
   (a) Use Corollary 3.1.2(a) to find the distribution of U.
   (b) Use Theorem 3.2.2 to show that U and $\bar{Y}_{\cdot}$ are independent.
5. Prove Theorem 3.1.2, part (2).
6. Use Corollary 3.1.2(a) to find the distribution of $(\bar{Y}_{1\cdot} - \bar{Y}_{2\cdot})^2$ from Exercise 9b in Chapter 2.
7. Consider Exercise 11 of Chapter 2.

(a) Use Theorem 3.2.2 to show that $Y_1 + 2Y_2 - Y_3$ is independent of $2Y_1^2 + Y_2^2 + 2Y_3^2 - 2Y_1Y_2 + 2Y_2Y_3$.

(b) Use Corollary 3.1.2(a) to find a constant c such that $c[5Y_1^2 + 2Y_2^2 + 5Y_3^2 - 4Y_1Y_2 + 2Y_1Y_3 + 4Y_2Y_3]$ has a central chi-square distribution.

8. Derive the distributions of $\mathbf{Y}'\mathbf{A}_3\mathbf{Y}$ and $\mathbf{Y}'\mathbf{A}_4\mathbf{Y}$ from Example 2.3.2.

9. Calculate the noncentrality parameters $\lambda_1, \ldots, \lambda_4$ in Example 3.4.1.

10. Let the 3×1 random vector $\mathbf{Y} = (Y_1, Y_2, Y_3)' \sim \mathrm{N}_3(\alpha\mathbf{1}_3, \sigma^2\mathbf{I}_3)$ and define $Z_1 = (Y_1^2 + Y_3^2 - 2Y_1Y_3$ and $Z_2 = Y_1^2 + Y_2^2 + Y_3^2 - Y_1Y_2 - Y_1Y_3 - Y_2Y_3$.

    (a) Find the distributions of Z_1 and Z_2.

    (b) Find the $\mathrm{E}(Z_i^k)$ for $i = 1, 2$ and any positive integer k.

11. Let the $n \times 1$ random vector $\mathbf{Y} = (Y_1, \ldots, Y_n)' \sim \mathrm{N}_n(\alpha\mathbf{1}_n, \Sigma)$ where $\Sigma = (a - b)\mathbf{I}_n + b\mathbf{J}_n$.

    (a) Find the distribution of $V = \sum_{i=1}^{n-1}(Y_i - \bar{Y}^*)^2$ where $\bar{Y}^* = \sum_{i=1}^{n-1} Y_i/(n-1)$.

    (b) Find the distribution of $(\bar{Y}^* - Y_n)/V^{1/2}$.

12. Let the $n \times 1$ random vector $\mathbf{Y} \sim \mathrm{N}_n(\boldsymbol{\mu}, \mathbf{I}_n)$. Let $\mathbf{X} = \mathbf{AY}$ where $\mathbf{A}$ is an $n \times n$ orthogonal matrix whose first row is $\boldsymbol{\mu}'/\sqrt{\boldsymbol{\mu}'\boldsymbol{\mu}}$. Let $V = X_1^2(X_1$ is the first element of vector $\mathbf{X})$ and $U = (\mathbf{X}'\mathbf{X} - V)$.

    (a) Find the distributions of U and V.

    (b) Are U and V independent? Prove your answer.

13. Let $U_i \sim \chi^2(\lambda_i)$ for $i = 1, 2$ where U_1 and U_2 are independent. Let a and b be two positive constants. Under what conditions is $aU_1 + bU_2 \sim c\chi^2(\lambda)$? Provide the values of c and λ.

14. Consider the model $Y_{ij} = \mu_i + R(T)_{(i)j}$ where $R(T)_{(i)j} \sim iid\ \mathrm{N}_1(0, \sigma^2_{R(T)})$ for $i = 1, \ldots, 3$, $j = 1, \ldots, n_i$ with $n_1 = 3$, $n_2 = 4$, and $n_3 = 2$.

    (a) Find the distribution of $U = \sum_{i=1}^{3}\sum_{j=1}^{n_i}(Y_{ij} - \bar{Y}_{i.})^2$ where $\bar{Y}_{i.} = \sum_{j=1}^{n_i} Y_{ij}/n_i$

    (b) Find the expected value of $V = \sum_{i=1}^{3}(\bar{Y}_{i.} - \bar{Y}^*)^2$ where $\bar{Y}^* = \sum_{i=1}^{3}\bar{Y}_{i.}/3$. [*Hint*: Write $V = \bar{\mathbf{Y}}'(\mathbf{I}_3 - \frac{1}{3}\mathbf{J}_3)\bar{\mathbf{Y}}$ where $\bar{\mathbf{Y}} = (\bar{Y}_{1.}, \bar{Y}_{2.}, \bar{Y}_{3.})'$.]

15. (Paired t-Test Problem) Consider an experiment with n experimental units. Suppose two observations are made on each unit. The first observation corresponds to the first level of a fixed factor, the second observation to the second level of the fixed factor. Let Y_{ij} be a random variable representing the

j^{th} observation on the ith experimental unit for $i = 1, \ldots, n$ and $j = 1, 2$. Let the $2n \times 1$ random vector $\mathbf{Y} = (Y_{11}, Y_{12}, Y_{21}, Y_{22}, \ldots, Y_{n1}, Y_{n2})'$. Let $\mathrm{E}(\mathbf{Y}_{ij}) = \mu_j$ and $\mathrm{var}(Y_{ij}) = \sigma^2$ for $i = 1, \ldots, n$ and $j = 1, 2$; and let $\mathrm{cov}(Y_{i1}, Y_{i2}) = \sigma^2\rho$ for all $i = 1, \ldots, n$. Assume $\mathbf{Y} \sim \mathrm{N}_{2n}(\boldsymbol{\mu}, \boldsymbol{\Sigma})$.

(a) Define $\boldsymbol{\mu}$ and $\boldsymbol{\Sigma}$ in terms of $\boldsymbol{\mu}_j, \sigma^2$, and ρ. (*Hint*: Use Kronecker products.)

(b) Let $T = \bar{D}/(S_D/\sqrt{n})$ where $D_i = Y_{i1} - Y_{i2}$ for $i = 1, \ldots, n$; $\bar{D} = \sum_{i=1}^n D_i/n$; and $S_D^2 = \Sigma_{i=1}^n (D_i - \bar{D})^2/(n-1)$. Find the distribution of T. [*Hint*: Start by finding the distribution of $D = (D_1, \ldots, D_n)'$.]

16. Let the $6n \times 1$ random vector $\mathbf{Y} = (Y_{111}, \ldots, Y_{11n}, Y_{121}, \ldots, Y_{12n}, \ldots, Y_{131}, \ldots, Y_{13n}, Y_{211}, \ldots, Y_{21n}, \ldots, Y_{221}, \ldots, Y_{22n}, Y_{231}, \ldots, Y_{23n})' \sim \mathrm{N}_{6n}(\mathbf{1}_2 \otimes (\mu_1, \mu_2, \mu_3)' \otimes \mathbf{1}_n, \boldsymbol{\Sigma})$ where

$$\begin{aligned}\boldsymbol{\Sigma} = \sigma_1^2&[\mathbf{I}_2 \otimes \mathbf{J}_3 \otimes \mathbf{J}_n] \\ &+ \sigma_2^2\left[\mathbf{I}_2 \otimes \left(\mathbf{I}_3 - \frac{1}{3}\mathbf{J}_3\right) \otimes \mathbf{J}_n\right] \\ &+ \sigma_3^2[\mathbf{I}_2 \otimes \mathbf{I}_3 \otimes \mathbf{I}_n].\end{aligned}$$

(a) Show $\bar{Y}_{.1.} - \bar{Y}_{.2.} = [(1/2)\mathbf{1}_2' \otimes (1, -1, 0) \otimes (1/n)\mathbf{1}_n']\mathbf{Y}$ where $\bar{Y}_{.j.} = \sum_{i=1}^2 \sum_{k=1}^n Y_{ijk}/(2n)$.

(b) Find the distribution of $\bar{Y}_{.1.} - \bar{Y}_{.2.}$.

(c) Find the distribution of $\mathbf{Y}'[(\mathbf{I}_2 - \frac{1}{2}\mathbf{J}_2) \otimes (\mathbf{I}_3 - \frac{1}{3}\mathbf{J}_3) \otimes \frac{1}{n}\mathbf{J}_n]\mathbf{Y}$.

17. Let the $(n_1 + n_2) \times 1$ random vector $\mathbf{Y} = (Y_{11}, \ldots, Y_{1n_1}, Y_{21}, \ldots, Y_{2n_2})' \sim \mathrm{N}_{n_1+n_2}(\boldsymbol{\mu}, \boldsymbol{\Sigma})$ where $\boldsymbol{\mu} = (\mu_1\mathbf{1}_{n_1}', \mu_2\mathbf{1}_{n_2}')'$ and

$$\boldsymbol{\Sigma} = \begin{bmatrix} \sigma_1^2\mathbf{I}_{n_1} & \mathbf{0} \\ \mathbf{0} & \sigma_2^2\mathbf{I}_{n_2} \end{bmatrix}.$$

If $\sigma_1^2 \neq \sigma_2^2$, this problem is called the Behrens–Fisher problem.

(a) Find the distribution of

$$V = \frac{(\bar{Y}_{1.} - \bar{Y}_{2.})^2/(1/n_1 + 1/n_2)}{\sum_{i=1}^2 \sum_{j=1}^{n_i} (Y_{ij} - Y_{i.})^2/(n_1 + n_2 - 2)}$$

when $\sigma_1^2 = \sigma_2^2$.

(b) Describe the distribution of V when $\sigma_1^2 \neq \sigma_2^2$.

4 Complete, Balanced Factorial Experiments

The main objective of this chapter is to provide sum of squares and covariance matrix algorithms for complete, balanced factorial experiments. The algoorithm rules are dependent on the model used in the analysis and on the model assumptions. Therefore, before the algorithms are presented we will discuss two different model formulations, models that admit restrictions on the random variables and models that do not admit restrictions.

4.1 MODELS THAT ADMIT RESTRICTIONS (FINITE MODELS)

We begin our model discussion with an example. Consider a group of btr experimental units. Separate the units into b homogeneous groups with tr units per group. In each group (or random block) randomly assign r replicate units to each of the t fixed treatment levels. The observed data for this two-way mixed experiment with replication are given in Figure 4.1.1.

				Random blocks B_i 1	2		b
Fixed treatments T_j	1	Reps $R(BT)_{(ij)k}$	1	y_{111}	y_{211}	$\cdots$	y_{b11}
			$\vdots$	$\vdots$	$\vdots$	$\ddots$	$\vdots$
			r	y_{11r}	y_{21r}	$\cdots$	y_{b1r}
	$\vdots$	$\vdots$		$\vdots$	$\vdots$	$\vdots$	$\vdots$
	t	Reps $R(BT)_{(ij)k}$	1	y_{1t1}	y_{2t1}	$\cdots$	y_{bt1}
			$\vdots$	$\vdots$	$\vdots$	$\ddots$	$\vdots$
			r	y_{1tr}	y_{2tr}	$\cdots$	y_{btr}

Figure 4.1.1 Two-Way Mixed Experimental Layout with Replication.

A model for this experiment is

$$Y_{ijk} = \mu_j + B_i + BT_{ij} + R(BT)_{(ij)k}$$

for $i = 1, \ldots, b$, $j = 1, \ldots, t$, and $k = 1, \ldots, r$ where Y_{ijk} is a random variable representing the kth replicate value in the ij^{th} block treatment combination; μ_j is a constant representing the mean effect of the jth fixed treatment; B_i is a random variable representing the effect of the ith random block; BT_{ij} is a random variable representing the interaction of the ith random block and the jth fixed treatment; and $R(BT)_{(ij)k}$ is a random variable representing the effect of the kth replicate unit nested in the ij^{th} block treatment combination.

We now attempt to develop a reasonable set of distributional assumptions for the random variables B_i, BT_{ij}, and $R(BT)_{(ij)k}$. Start by considering the btr observed data points in the experiment as a collection of values sampled from an entire population of possible values. The population for this experiment can be viewed as a rectangular grid with an infinite number of columns, exactly t rows, and an infinite number of possible observed values in each row–column combination (see Figure 4.1.2). The infinite number of columns represents the infinite number of blocks in the population. Each block (or column) contains exactly t rows, one for each level of the fixed treatments. Then the population contains an infinite number of replicate observed values nested in each block treatment combination. The btr observed data points for the experiment are then sampled from this infinite population of values in the following way. Exactly b blocks are selected at random from the infinite number of blocks in the population. For each block selected, all t of the treatment rows are then included in the sample. Finally, within the selected block treatment combinations, r replicate observations are randomly sampled from the infinite number of nested population replicates.

Since the r blocks are selected at random from an infinite population of possible blocks, assume that the block variables B_i for $i = 1, \ldots, b$ are independent. If the

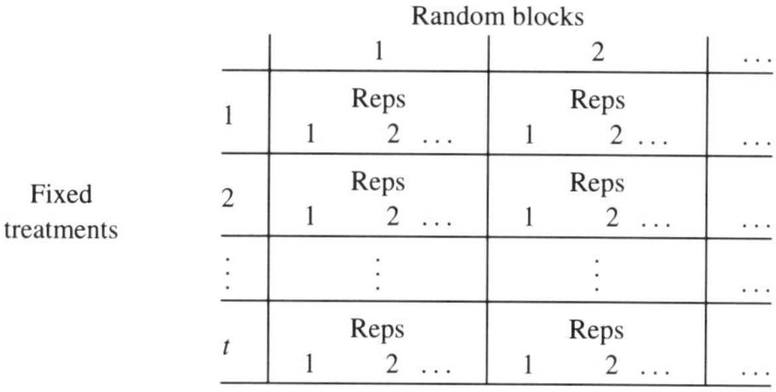

		Random blocks		
		1	2	...
	1	Reps 1 2 ...	Reps 1 2 ...	...
Fixed treatments	2	Reps 1 2 ...	Reps 1 2 ...	...
	⋮	⋮	⋮	...
	t	Reps 1 2 ...	Reps 1 2 ...	...

Figure 4.1.2 Finite Model Population Grid.

r blocks have been sampled from the same single population of blocks, then the variables B_i are identically distributed. Furthermore, assume that across the entire population of blocks the average influence of B_i is zero, that is, $\mathrm{E}(B_i) = 0$ for all $i = 1, \ldots, b$. If the random variables B_i are assumed to be normally distributed, then the assumptions above are satisfied when the b variables $B_i \sim iid\ \mathrm{N}_1(0, \sigma_B^2)$.

Now consider the random variables BT_{ij} that represent the block by treatment interaction. Recall that the population contains exactly t treatment levels for each block. Therefore, in the ith block the population contains exactly t possible values for the random variable BT_{ij}. If the average influence of the block by treatment interaction is assumed to be zero for each block, then $\mathrm{E}[BT_{ij}] = 0$ for each i. But for each i, $\mathrm{E}[BT_{ij}] = \sum_{j=1}^{t} BT_{ij}/t$ since the population contains exactly t values of BT_{ij} for each block. Therefore, $\sum_{j=1}^{t} BT_{ij} = 0$ for each i, implying that the variables $BT_{i1}, \ldots, BT_{it}$ are dependent, because the value of any one of these variables is determined by the values of the other $t - 1$ variables. Although the dependence between the BT_{ij} variables occurs within each block, the dependence does not occur across blocks. Therefore, assume that the b vectors $(BT_{11}, \ldots, BT_{1t})', \ldots, (BT_{b1}, \ldots, BT_{bt})'$ are mutually independent. If the random variables BT_{ij} are assumed to be normally distributed, then the assumptions above are satisfied when the $b(t-1) \times 1$ random vector $(\mathbf{I}_b \otimes \mathbf{P}'_t)(BT_{11}, \ldots, BT_{1t}, \ldots, BT_{b1}, \ldots, BT_{bt})' \sim \mathrm{N}_{b(t-1)}[\mathbf{0}, \sigma_{BT}^2 \mathbf{I}_b \otimes \mathbf{I}_{t-1}]$ where P'_t is the $(t-1) \times t$ lower portion of a t-dimensional Helmert matrix.

Finally, consider the nested replicate variables $R(BT)_{(ij)k}$. Within each block treatment combination, the r replicate observations are selected at random from the infinite population of nested replicates. If each block treatment combination of the population has the same distribution, then the random variables $R(BT)_{(ij)k}$ are independent, identically distributed random variables. Furthermore, within each block treatment combination, assume that the average influence of $R(BT)_{(ij)k}$ is zero, that is, $\mathrm{E}[R(BT)_{(ij)k}] = 0$ for each ij pair. If the random variables $R(BT)_{(ij)k}$ are also assumed to be normally distributed, then the assumptions above are satisfied when the btr random variables $R(BT)_{(ij)k} \sim iid\, \mathrm{N}_1(0, \sigma^2_{R(BT)})$. Furthermore, assume that random variables B_i, the $t \times 1$ random vectors $(BT_{i1}, \ldots, BT_{it})'$, and the random variables $R(BT)_{(ij)k}$ are mutually independent.

In the previous model formulation, the random variables BT_{ij} contain a finite population of possible values for each i. If the variables are assumed to have zero expectation, then the finite population induces restrictions and distributional dependencies. Note that variables representing interactions of random and fixed factors are the only types of variables that assume these restrictions. Furthermore, the dependencies occurred because of the assumed population structure of possible observed values. Kempthorne (1952) called such models *finite models*, because the fixed by random interaction components were restricted to a finite population of possible values.

In the next section we discuss models where the population is assumed to have a structure where no variable dependencies occur.

4.2 MODELS THAT DO NOT ADMIT RESTRICTIONS (INFINITE MODELS)

Consider the same experiment discussed in Section 4.1. Use a model with the same variables

$$Y_{ijk} = \mu_j + B_i + BT_{ij} + R(BT)_{(ij)k}$$

where all variables and constants represent the same effects as previously stated. In this model formulation, the population has an infinite number of random blocks. For each block, an infinite number of replicates of each of the t treatment levels exists. Each of these treatment level replicates contains an infinite number of experimental units (see Figure 4.2.1).

The btr observed values for the experiment are sampled from the population by first choosing b blocks at random from the infinite number of blocks in the population. For each selected block, one replicate of each of the t treatment levels is selected. Finally, within the selected block treatment combinations, r replicate observations are randomly sampled. Since the blocks are randomly selected from one infinite population of blocks, assume the random variables B_i are independent, identically distributed. With a normality and zero expectation assumption, let the b block random variables $B_i \sim iid\, \mathrm{N}_1(0, \sigma^2_B)$. Since the t observed treatment levels are randomly chosen from an infinite population of treatment replicates, an infinite number of possible values are available for the random variables BT_{ij}. Assume that the average influence of BT_{ij} is zero for each block. But now $\mathrm{E}[BT_{ij}] = 0$ does not imply $\sum_{j=1}^{t} BT_{ij} = 0$ for each i since the variables BT_{ij} have an infinite population. Therefore, a zero expectation does not imply dependence. With a normality assumption, let the bt random variables $BT_{ij} \sim iid\, \mathrm{N}_1(0, \sigma^2_{BT})$. Finally,

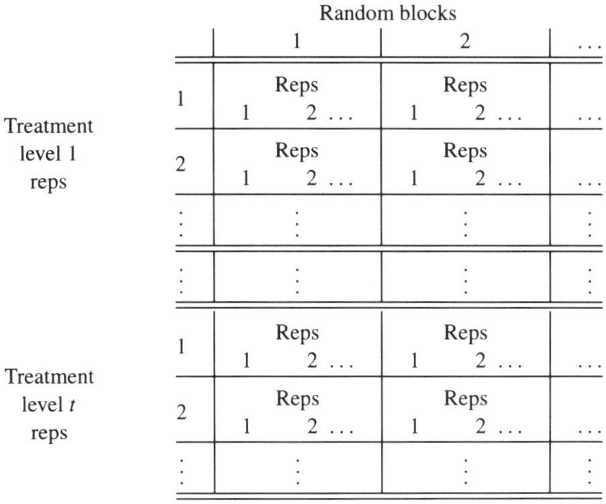

		Random blocks 1	Random blocks 2	...
Treatment level 1 reps	1	Reps 1 2 ...	Reps 1 2 ...	...
	2	Reps 1 2 ...	Reps 1 2 ...	...
	⋮	⋮	⋮	⋮
	⋮	⋮	⋮	⋮
Treatment level t reps	1	Reps 1 2 ...	Reps 1 2 ...	...
	2	Reps 1 2 ...	Reps 1 2 ...	...
	⋮	⋮	⋮	⋮

Figure 4.2.1 Infinite Model Population Grid.

within each block treatment combination, the nested replicates are assumed to be sampled from an infinite population. With a normality and zero expectation assumption, let the *btr* random variables $R(BT)_{(ij)k} \sim iid\ \mathrm{N}_1(0, \sigma^2_{R(BT)})$. Furthermore, assume that random variables B_i, the random variables BT_{ij}, and the random variables $R(BT)_{(ij)k}$ are mutually independent. Hence, in models that do not admit restrictions, all variables on the right side of the model are assumed to be independent. Kempthorne (1952) called such models *infinite models*, because it is assumed that all of the random components are sampled from infinite populations.

In passing, we raise one additional topic. Consider the previous experiment, with one replicate unit within each block treatment combination ($r = 1$). Observing only one replicate unit within each block treatment combination does not change the fact that different experimental units are intrinsically different. The random variables $R(BT)_{(ij)k}$ represents this experimental unit difference. Hence, there is some motivation for leaving the random variable $R(BT)_{(ij)k}$ in the model. However, the variance $\sigma^2_{R(BT)}$ is not estimable when $r = 1$. Therefore, it also seems reasonable to drop $R(BT)_{(ij)k}$ from the model in this case. If the $R(BT)_{(ij)k}$ variable is dropped from the model, then the expected mean squares will not contain the $\sigma^2_{R(BT)}$ term. However, all F tests in the ANOVA are the same whether $R(BT)_{(ij)k}$ is dropped or not. In addition, the covariance matrix construction is slightly less complicated if variables with nonestimable variance parameters are dropped from the model. In summary, the overall analysis is unchanged in its essential characteristics when variables with nonestimable variance parameters are dropped from the model. Thus, variables with nonestimable variance parameters are generally not used in this text.

In the next section we establish the sum of squares and covariance matrix algorithm rules for models that admit restrictions. We then show how the covariance matrix algorithm can be easily modified to accommodate models that do not admit restrictions. Finally, for completeness, we discuss how the algorithms can be modified to accommodate models that contain variables with nonestimable variance parameters. Therefore, the algorithms can be applied to any complete, balanced experiment, using finite or infinite models, and using models with or without nonestimable variance parameters.

4.3 SUM OF SQUARES AND COVARIANCE MATRIX ALGORITHMS

In Section 4.1 a finite model was presented for an experiment with r replicate observations nested in bt block treatment combinations. This experiment is now used to establish the sum of squares and covariance matrix algorithm rules for finite models.

Let the $btr \times 1$ random vector $\mathbf{Y} = (Y_{111}, \ldots, Y_{11r}, \ldots, Y_{bt1}, \ldots, Y_{btr})'$. The covariance matrix $\Sigma = \text{cov}(\mathbf{Y})$ for the finite model is given by

$$\begin{aligned}\Sigma &= \sigma_B^2[\mathbf{I}_b \otimes \mathbf{J}_t \otimes \mathbf{J}_r] \\ &\quad + \sigma_{BT}^2\left[\mathbf{I}_b \otimes \left(\mathbf{I}_t - \frac{1}{t}\mathbf{J}_t\right) \otimes \mathbf{J}_r\right] \\ &\quad + \sigma_{R(BT)}^2[\mathbf{I}_b \otimes \mathbf{I}_t \otimes \mathbf{I}_r].\end{aligned}$$

The rules for constructing the matrix Σ are given in the following paragraphs. The matrix Σ is constructed in tabular form. The first rule describes the construction of the table, and the second rule describes the matrix terms that fill the table.

Rule Σ1 List the variances of all random factors and interactions, one variance in each row. Construct column headings where the first column heading designates the main factor letters and the second heading designates the number of levels of the factor. Place brackets ([ ]), Kronecker product symbols ($\otimes$), and plus signs (+) in each row, as described in Example 4.3.1.

Example 4.3.1 Rule Σ1.

Factor		B		T		R	
Levels $= l$		b		t		r	
σ_B^2	[		$\otimes$		$\otimes$		]
$+\ \sigma_{BT}^2$	[		$\otimes$		$\otimes$		]
$+\ \sigma_{R(BT)}^2$	[		$\otimes$		$\otimes$		].

Rule Σ2 Compare the factor letters in the first column heading with each subscript letter on the variance (i.e., B from σ_B^2, T from σ_{BT}^2).

Rule Σ2.1 If the factor letter does *not* match a subscript letter on the variance, place a $\mathbf{J}_l$ in the Kronecker product with the same row as the variance and the same column as the factor letter, where l is the number of levels of the factor. In the examples below, new matrix additions to the covariance matrix are underlined.

Example 4.3.1 (*continued*) Rule Σ2.1

$$
\begin{array}{lllllll}
\text{Factor} & & & B & T & R & \\
\text{Levels} = l & & & b & t & r & \\
 & \sigma_B^2 & [& & \otimes\, \underline{\mathbf{J}_t} & \otimes\, \underline{\mathbf{J}_r} &] \\
 & +\sigma_{BT}^2 & [& & \otimes & \otimes\, \underline{\mathbf{J}_r} &] \\
 & +\sigma_{R(BT)}^2 & [& & \otimes & \otimes &]
\end{array}
$$

Rule Σ2.2 If a factor letter corresponding to a fixed non-nested factor matches a subscript letter on the variance, place $\mathbf{I}_l - \frac{1}{l}\mathbf{J}_l$ in the Kronecker product.

Example 4.3.1 (*continued*) Rule Σ2.2

$$
\begin{array}{lllllll}
\text{Factor} & & & B & T & R & \\
\text{Levels} = l & & & b & t & r & \\
 & \sigma_B^2 & [& & \otimes\, \mathbf{J}_t & \otimes\, \mathbf{J}_r &] \\
 & +\sigma_{BT}^2 & \Big[& & \otimes\, \underline{\left(\mathbf{I}_t - \frac{1}{t}\mathbf{J}_t\right)} & \otimes\, \mathbf{J}_r & \Big] \\
 & +\sigma_{R(BT)}^2 & [& & \otimes & \otimes &].
\end{array}
$$

Rule Σ2.3 Place $\mathbf{I}_l$ elsewhere.

Example 4.3.1 (*continued*) Rule Σ2.3

$$
\begin{array}{lllllll}
\text{Factor} & & & B & T & R & \\
\text{Levels} = l & & & b & t & r & \\
 & \sigma_B^2 & [& \underline{\mathbf{I}_b} & \otimes\, \mathbf{J}_t & \otimes\, \mathbf{J}_r &] \\
 & \sigma_{BT}^2 & \Big[& \underline{\mathbf{I}_b} & \otimes \left(\mathbf{I}_t - \frac{1}{t}\mathbf{J}_t\right) & \otimes\, \mathbf{J}_r & \Big] \\
 & \sigma_{R(BT)}^2 & [& \underline{\mathbf{I}_b} & \otimes\, \underline{\mathbf{I}_t} & \otimes\, \underline{\mathbf{I}_r} &].
\end{array}
$$

The covariance algorithm can be applied to complete, balanced finite models with any number of fixed and random main effects, interactions, or nested factors.

For infinite models, the same covariance matrix algorithm can be used, except that Rule $\Sigma 2.2$ is omitted. That is, for infinite models that do not contain restrictions on variables representing interactions of random and fixed factors, the covariance matrix is constructed following Rules $\Sigma 1$, $\Sigma 2$, $\Sigma 2.1$, and $\Sigma 2.3$. The next example illustrates the construction of covariance matrices in such models.

Example 4.3.2 Consider the experiment in Example 4.3.1, but now use an infinite model that does not admit restrictions. Using Rules $\Sigma 1$, $\Sigma 2$, $\Sigma 2.1$, and $\Sigma 2.3$, the covariance matrix is given by

$$\begin{aligned}\Sigma = \sigma_B^2 & \quad [\mathbf{I}_b \otimes \mathbf{J}_t \otimes \mathbf{J}_r] \\ + \sigma_{BT}^2 & \quad [\mathbf{I}_b \otimes \mathbf{I}_t \otimes \mathbf{J}_r] \\ + \sigma_{R(BT)}^2 & \quad [\mathbf{I}_b \otimes \mathbf{I}_t \otimes \mathbf{I}_r].\end{aligned}$$

Although finite and infinite models are motivated in different ways and produce different covariance structures, algebraically the two covariance structures are simply reparameterizations of each other. To illustrate this point, consider the previous experiment with b random blocks, t fixed treatments, and r random replicates per treatment. First, rewrite the covariance matrix for the finite model in the following equivalent form.

$$\begin{aligned}\Sigma = \left(\sigma_B^2 - \frac{1}{t}\sigma_{BT}^2\right) & \quad [\mathbf{I}_b \otimes \mathbf{J}_t \otimes \mathbf{J}_r] \\ + \sigma_{BT}^2 & \quad [\mathbf{I}_b \otimes \mathbf{I}_t \otimes \mathbf{J}_r] \\ + \sigma_{R(BT)}^2 & \quad [\mathbf{I}_b \otimes \mathbf{I}_t \otimes \mathbf{I}_r].\end{aligned}$$

Now to distinguish the parameters in the finite and infinite models, rename the variance parameters σ_B^2, σ_{BT}^2, and $\sigma_{R(BT)}^2$ in the infinite model as $\sigma_{B^*}^2$, $\sigma_{BT^*}^2$, and $\sigma_{R(BT)^*}^2$, respectively. Then the covariance matrix for the infinite model becomes

$$\begin{aligned}\Sigma = \sigma_{B^*}^2 & \quad [\mathbf{I}_b \otimes \mathbf{J}_t \otimes \mathbf{J}_r] \\ + \sigma_{BT^*}^2 & \quad [\mathbf{I}_b \otimes \mathbf{I}_t \otimes \mathbf{J}_r] \\ + \sigma_{R(BT)^*}^2 & \quad [\mathbf{I}_b \otimes \mathbf{I}_t \otimes \mathbf{I}_r].\end{aligned}$$

Note that the finite and infinite model covariance matrices are equal with $\sigma_{B^*}^2 = \sigma_B^2 - \frac{1}{t}\sigma_{BT}^2$, $\sigma_{BT^*}^2 = \sigma_{BT}^2$ and $\sigma_{R(BT)^*}^2 = \sigma_{R(BT)}^2$. Therefore, the two covariance structures are simply reparameterizations of each other.

An algorithm is now given for constructing the matrices $\mathbf{A}_s$ in the sums of squares $\mathbf{Y}'\mathbf{A}_s\mathbf{Y}$. This sum of squares algorithm applies to finite and infinite models.

The subscripts s on the matrices $\mathbf{A}_s$ are numbered sequentially with $\mathbf{A}_1$ being associated with the sum of squares for the overall mean μ_0, $\mathbf{A}_2$ being associated with the sum of squares for the next factor, etc. In the example considered in this section, $\mathbf{A}_2$ is associated with the blocks B, $\mathbf{A}_3$ with the treatments T, $\mathbf{A}_4$ with interaction of B and T, and $\mathbf{A}_5$ with the nested replicates $R(BT)$. The sum of squares matrices $\mathbf{A}_s$ are constructed in tabular form. The first rule describes the table; the second rule describes the matrix terms in the table.

Rule A1 Construct two row headings where the first designates the letters of the factors and interactions, while the second designates the matrices $\mathbf{A}_s$ associated with those factors and interactions. Construct two column headings where the first is the factor letter and the second is the number of levels, l, of the factor. Place brackets ([ ]), Kronecker product symbols ($\otimes$), and equal signs ($=$) in each row, as described in Example 4.3.3.

Example 4.3.3 Rule A1.

	Factor			B		T		R	
	Levels $= l$			b		t		r	
μ_0	$\mathbf{A}_1$	$=$	[		$\otimes$		$\otimes$		]
B	$\mathbf{A}_2$	$=$	[		$\otimes$		$\otimes$		]
T	$\mathbf{A}_3$	$=$	[		$\otimes$		$\otimes$		]
BT	$\mathbf{A}_4$	$=$	[		$\otimes$		$\otimes$		]
$R(BT)$	$\mathbf{A}_5$	$=$	[		$\otimes$		$\otimes$		].

Rule A2 Compare the factor letters in the first row heading with the factor letters in the first column heading.

Rule A2.1 If a row factor letter does *not* match the column factor heading, place a $\frac{1}{l}\mathbf{J}_l$ in the Kronecker product. Note that μ_0 does not match any column factor heading; hence, the Kronecker product for $\mathbf{A}_1$ will be comprised of terms of the form $\frac{1}{l}\mathbf{J}_l$.

Example 4.3.3 (continued) Rule A2.1

	Factor		B		T		R	
	Levels $= l$		b		t		r	
μ_0	$\mathbf{A}_1$	$=$	$\left[\ \underline{\frac{1}{b}\mathbf{J}_b}\right.$	$\otimes$	$\underline{\frac{1}{t}\mathbf{J}_t}$	$\otimes$	$\underline{\frac{1}{r}\mathbf{J}_r}$	$\left.\right]$

$$
\begin{array}{rl}
B\ \mathbf{A}_2 = & \left[\qquad\quad \otimes \underline{\frac{1}{t}\mathbf{J}_t} \quad \otimes \underline{\frac{1}{r}\mathbf{J}_r} \quad \right] \\
T\ \mathbf{A}_3 = & \left[\underline{\frac{1}{b}\mathbf{J}_b} \quad \otimes \qquad\quad \otimes \underline{\frac{1}{r}\mathbf{J}_r} \quad \right] \\
BT\ \mathbf{A}_4 = & \left[\qquad\quad \otimes \qquad\quad \otimes \underline{\frac{1}{r}\mathbf{J}_r} \quad \right] \\
R(BT)\ \mathbf{A}_5 = & [\qquad\quad \otimes \qquad\quad \otimes \qquad\quad].
\end{array}
$$

Rule A2.2 If a non-nested factor in the row heading matches the column factor heading, place an $\mathbf{I}_l - \frac{1}{l}\mathbf{J}_l$ in the Kronecker product.

Example 4.3.3 (continued) Rule A2.2

$$
\begin{array}{llllll}
 & \text{Factor} & B & T & R & \\
 & \text{Levels} = l & b & t & r & \\
\mu_0 & \mathbf{A}_1 = & \left[\underline{\frac{1}{b}\mathbf{J}_b} \right. & \otimes \underline{\frac{1}{t}\mathbf{J}_t} & \otimes \underline{\frac{1}{r}\mathbf{J}_r} & \left. \right] \\
B & \mathbf{A}_2 = & \left[\left(\mathbf{I}_b - \frac{1}{b}\mathbf{J}_b\right) \right. & \otimes \frac{1}{t}\mathbf{J}_t & \otimes \frac{1}{r}\mathbf{J}_r & \left. \right] \\
T & \mathbf{A}_3 = & \left[\frac{1}{b}\mathbf{J}_b \right. & \otimes \left(\mathbf{I}_t - \frac{1}{t}\mathbf{J}_t\right) & \otimes \frac{1}{r}\mathbf{J}_r & \left. \right] \\
BT & \mathbf{A}_4 = & \left[\left(\mathbf{I}_b - \frac{1}{b}\mathbf{J}_b\right) \right. & \otimes \left(\mathbf{I}_t - \frac{1}{t}\mathbf{J}_t\right) & \otimes \frac{1}{r}\mathbf{J}_r & \left. \right] \\
R(BT) & \mathbf{A}_5 = & \left[\underline{\mathbf{I}_b} \right. & \otimes \underline{\mathbf{I}_t} & \otimes \left(\mathbf{I}_r - \frac{1}{r}\mathbf{J}_r\right) & \left. \right].
\end{array}
$$

Rule A2.3 Place $\mathbf{I}_l$ elsewhere.

Example 4.3.3 (continued) Rule A2.3

$$
\begin{array}{llllll}
 & \text{Factor} & B & T & R & \\
 & \text{Levels} = l & b & t & r & \\
\mu_0 & \mathbf{A}_1 = & \left[\underline{\frac{1}{b}\mathbf{J}_b} \right. & \otimes \underline{\frac{1}{t}\mathbf{J}_t} & \otimes \underline{\frac{1}{r}\mathbf{J}_r} & \left. \right]
\end{array}
$$

$$
\begin{aligned}
B\ \mathbf{A}_2 &= \left[\left(\mathbf{I}_b - \tfrac{1}{b}\mathbf{J}_b\right) \otimes \tfrac{1}{t}\mathbf{J}_t \otimes \tfrac{1}{r}\mathbf{J}_r\right]\\
T\ \mathbf{A}_3 &= \left[\tfrac{1}{b}\mathbf{J}_b \otimes \left(\mathbf{I}_t - \tfrac{1}{t}\mathbf{J}_t\right) \otimes \tfrac{1}{r}\mathbf{J}_r\right]\\
BT\ \mathbf{A}_4 &= \left[\left(\mathbf{I}_b - \tfrac{1}{b}\mathbf{J}_b\right) \otimes \left(\mathbf{I}_t - \tfrac{1}{t}\mathbf{J}_t\right) \otimes \tfrac{1}{r}\mathbf{J}_r\right]\\
R(BT)\ \mathbf{A}_5 &= \left[\underline{\mathbf{I}_b} \otimes \underline{\mathbf{I}_t} \otimes \left(\mathbf{I}_r - \tfrac{1}{r}\mathbf{J}_r\right)\right].
\end{aligned}
$$

The sums of squares for each factor and interaction can be written as quadratic forms, $\mathbf{Y}'\mathbf{A}_s\mathbf{Y}$, using these Kronecker product matrices $\mathbf{A}_s$. For example, the sum of squares for the factor B is $\mathbf{Y}'\mathbf{A}_2\mathbf{Y} = \mathbf{Y}'[(\mathbf{I}_b - \frac{1}{b}\mathbf{J}_b) \otimes \frac{1}{t}\mathbf{J}_t \otimes \frac{1}{r}\mathbf{J}_r]\mathbf{Y}$.

In Examples 4.3.1, 4.3.2, and 4.3.3 all the variance parameters in the model were estimable. In the next example we apply the algorithms to a model that contains a nonestimable variance parameter.

Example 4.3.4 Consider an experiment with b random blocks, t fixed treatments, and $r = 1$ replicate observations per block treatment combination. Assume the model

$$Y_{ijk} = \mu_j + B_i + BT_{ij} + R(BT)_{(ij)k}$$

for $i = 1, \ldots, b$, $j = 1, \ldots, t$, and $k = 1$. Thus, $\sigma^2_{R(BT)}$ is nonestimable and $R(BT)_{(ij)k}$ could be dropped from the model. However, if $R(BT)_{(ij)k}$ is left in the model, then the sum of squares and covariance matrix algorithms are applied just as they were in Examples 4.3.1, 4.3.2, and 4.3.3 with r replaced by 1. Therefore, for the finite model, the covariance algorithm follows Rules $\Sigma1$, $\Sigma2$, $\Sigma2.1$, $\Sigma2.2$, and $\Sigma2.3$. With $r = 1$ the covariance matrix Σ is given by

$$
\begin{aligned}
\Sigma &= \sigma^2_B\ [\mathbf{I}_b \otimes \mathbf{J}_t \otimes 1] + \sigma^2_{BT}\left[\mathbf{I}_b \otimes \left(\mathbf{I}_t - \frac{1}{t}\mathbf{J}_t\right) \otimes 1\right] + \sigma^2_{R(BT)}\ [\mathbf{I}_b \otimes \mathbf{I}_t \otimes 1]\\
&= \sigma^2_B\ [\mathbf{I}_b \otimes \mathbf{J}_t] + \sigma^2_{BT}\left[\mathbf{I}_b \otimes \left(\mathbf{I}_t - \frac{1}{t}\mathbf{J}_t\right)\right] + \sigma^2_{R(BT)}\ [\mathbf{I}_b \otimes \mathbf{I}_t].
\end{aligned}
$$

For the infinite model, the covariance algorithm follows Rules $\Sigma1$, $\Sigma2$, $\Sigma2.1$, and $\Sigma2.3$. With $r = 1$ the covariance matrix $\boldsymbol{\Sigma}$ is

$$\Sigma = \sigma^2_B\ [\mathbf{I}_b \otimes \mathbf{J}_t] + \sigma^2_{BT}\ [\mathbf{I}_b \otimes \mathbf{I}_t] + \sigma^2_{R(BT)}\ [\mathbf{I}_b \otimes \mathbf{I}_t].$$

For both finite and infinite models, the sum of squares algorithm follows Rules $A1$, $A2$, $A2.1$, $A2.2$, and $A2.3$. With $r = 1$ the sum of squares matrices

$\mathbf{A}_1, \ldots, \mathbf{A}_5$ are given by

$$\mathbf{A}_1 = \frac{1}{b}\mathbf{J}_b \otimes \frac{1}{t}\mathbf{J}_t$$

$$\mathbf{A}_2 = \left(\mathbf{I}_b - \frac{1}{b}\mathbf{J}_b\right) \otimes \frac{1}{t}\mathbf{J}_t$$

$$\mathbf{A}_3 = \frac{1}{b}\mathbf{J}_b \otimes \left(\mathbf{I}_t - \frac{1}{t}\mathbf{J}_t\right)$$

$$\mathbf{A}_4 = \left(\mathbf{I}_b - \frac{1}{b}\mathbf{J}_b\right) \otimes \left(\mathbf{I}_t - \frac{1}{t}\mathbf{J}_t\right)$$

$$\mathbf{A}_5 = \mathbf{I}_b \otimes \mathbf{I}_t \otimes \left(1 - \frac{1}{1}1\right) = \mathbf{0}_{bt \times bt}.$$

Since $\mathbf{A}_5$ is a $bt \times bt$ matrix of zeros, the sum of squares due to effect $R(BT)$ equals zero and no estimate of $\sigma^2_{R(BT)}$ exists.

The covariance structures described in this section apply to a broad class of finite and infinite models. However, these covariance structures are based on certain distributional assumptions on the random components in the model. If other distributional assumptions are made, then the covariance algorithms are not applicable. Exercise 12 in Chapter 5 provides an example of a covariance structure that does not conform to the distributional assumptions made for the finite or infinite models.

We do not wish to confine our discussion solely to the finite and infinite covariance structures covered in this chapter. Therefore, a more general class of covariance structures is developed in Chapter 10. The finite and infinite covariance structures are a subset of this more general class.

Finally, the sums of squares and covariance algorithms presented in this chapter can be applied directly to complete, balanced factorial models. However, in Chapters 7 and 8 we show that these algorithms can also be used to construct sums of squares and covariance matrices for unbalanced data sets and data sets with missing values.

In the next section we introduce expected mean squares in complete, balanced experiments.

4.4 EXPECTED MEAN SQUARES

The mean square of an effect is the sum of squares of that effect divided by the corresponding degrees of freedom. For complete, balanced designs the mean square is $[1/tr(\mathbf{A}_s)]\mathbf{Y}'\mathbf{A}_s\mathbf{Y}$, since the degrees of freedom equal the $tr(\mathbf{A}_s)$. The expected value of the mean square, usually called the expected mean square (EMS), is a function of the mean vector $\boldsymbol{\mu} = \mathrm{E}(\mathbf{Y})$ and of the variance parameters in $\boldsymbol{\Sigma} = \mathrm{cov}(\mathbf{Y})$. The expected mean square indicates how the mean squares can be used to obtain unbiased estimates of functions of the variance parameters.

The expected mean square in complete, balanced designs is defined in the following theorem. The proof of the theorem is a direct result of Theorem 1.3.2.

Theorem 4.4.1 *Let* $\mathbf{Y}$ *be an* $n \times 1$ *random vector associated with the observations of a complete, balanced factorial experiment with an* $n \times 1$ *mean vector* $\boldsymbol{\mu} = E(\mathbf{Y})$ *and* $n \times n$ *covariance matrix* $\boldsymbol{\Sigma} = cov(\mathbf{Y})$. *The expected mean square associated with the sum of squares* $\mathbf{Y}'\mathbf{A}\mathbf{Y}$ *is* $\mathrm{E}\{[1/tr(\mathbf{A}_s)]\mathbf{Y}'\mathbf{A}\mathbf{Y}\} = [tr(\mathbf{A}\boldsymbol{\Sigma}) + \boldsymbol{\mu}'\mathbf{A}\boldsymbol{\mu}]/tr(\mathbf{A})$ *where* $tr(\mathbf{A})$ *equals the degrees of freedom associated with* $\mathbf{Y}'\mathbf{A}\mathbf{Y}$.

Example 4.4.1 Consider the experiment described in Examples 4.3.1 and 4.3.3 in which a finite model was assumed. The sums of squares due to the random effect B and the fixed effect T are $\mathbf{Y}'\mathbf{A}_2\mathbf{Y}$ and $\mathbf{Y}'\mathbf{A}_3\mathbf{Y}$, respectively, where $\mathbf{A}_2 = (\mathbf{1}_b - \frac{1}{b}\mathbf{J}_b) \otimes \frac{1}{t}\mathbf{J}_t \otimes \frac{1}{r}\mathbf{J}_r$, $\mathbf{A}_3 = \frac{1}{b}\mathbf{J}_b \otimes (\mathbf{I}_t - \frac{1}{t}\mathbf{J}_t) \otimes \frac{1}{r}\mathbf{J}_r$, and the $btr \times 1$ random vector $\mathbf{Y} = (Y_{111}, \ldots, Y_{11r}, \ldots, Y_{bt1}, \ldots, Y_{btr})'$. The mean vector $\boldsymbol{\mu} = \mathrm{E}(\mathbf{Y}) = \mathrm{E}(Y_{111}, \ldots, Y_{11r}, \ldots, Y_{bt1}, \ldots, Y_{btr})' = \mathbf{1}_b \otimes (\mu_1, \ldots, \mu_t)' \otimes \mathbf{1}_r$. Note that $\mathbf{A}_2\boldsymbol{\Sigma} = [tr\sigma_B^2 + \sigma_{R(BT)}^2]\mathbf{A}_2$ and $\mathbf{A}_3\boldsymbol{\Sigma} = [r\sigma_{BT}^2 + \sigma_{R(BT)}^2]\mathbf{A}_3$. Therefore, by Theorem 4.4.1, the expected mean square of the random effect B equals

$$
\begin{aligned}
\mathrm{E}[\mathbf{Y}'\mathbf{A}_2\mathbf{Y}/tr(\mathbf{A}_2)] &= [r(\mathbf{A}_2\Sigma) + \boldsymbol{\mu}'\mathbf{A}_2\boldsymbol{\mu}]/tr(\mathbf{A}_2) = [tr\sigma_B^2 + \sigma_{R(BT)}^2] \\
&\quad + \{[\mathbf{1}_b \otimes (\mu_1, \ldots, \mu_t)' \otimes \mathbf{1}_r]'\left[\left(\mathbf{I}_b - \frac{1}{b}\mathbf{J}_b\right) \otimes \frac{1}{t}\mathbf{J}_t \otimes \frac{1}{r}\mathbf{J}_r\right] \\
&\quad \times [\mathbf{1}_b \otimes (\mu_1, \ldots, \mu_t)' \otimes \mathbf{1}_r]/(b-1)\} \\
&= tr\sigma_B^2 + \sigma_{R(BT)}^2
\end{aligned}
$$

and the expected mean square of the fixed factor T equals

$$
\begin{aligned}
\mathrm{E}[\mathbf{Y}'\mathbf{A}_3\mathbf{Y}/tr(\mathbf{A}_3)] &= [tr(\mathbf{A}_3\boldsymbol{\Sigma}) + \boldsymbol{\mu}'\mathbf{A}_3\boldsymbol{\mu}]/tr(\mathbf{A}_3) = [tr\sigma_B^2 + \sigma_{R(BT)}^2] \\
&\quad + \{[\mathbf{1}_b \otimes (\mu_1, \ldots, \mu_t)' \otimes \mathbf{1}_r]'\left[\frac{1}{b}\mathbf{J}_b \otimes \left(\mathbf{I}_t - \frac{1}{t}\mathbf{J}_t\right) \otimes \frac{1}{r}\mathbf{J}_r\right] \\
&\quad \times [\mathbf{1}_b \otimes (\mu_1, \ldots, \mu_t)' \otimes \mathbf{1}_r]/(t-1)\} \\
&= [r\sigma_{BT}^2 + \sigma_{R(BT)}^2] + br\sum_{j=1}^{t}(\mu_j - \bar{\mu}.)^2/(t-1).
\end{aligned}
$$

Thus, the expected mean square of the random effect B provides an unbiased estimate of $tr\sigma_B^2 + \sigma_{R(BT)}^2$. Likewise, the expected mean square of the fixed factor T provides an unbiased estimate of $r\sigma_B^2 + \sigma_{R(BT)}^2 + br\Sigma_{j=1}^t(\mu_j - \bar{\mu}\cdot)^2/(t-1)$. The EMSs of the other effects can be calculated in a similar manner and are left to the reader.

4.5 ALGORITHM APPLICATIONS

The sum of squares and covariance matrix algorithms are now applied to a series of complete, balanced factorial experiments. As mentioned in Section 4.3, finite and infinite model covariance structures are reparameterizations of each other. Therefore, the choice between the finite and infinite model is somewhat arbitrary. For the remainder of this text, the finite model is assumed. Therefore, unless specifically stated, subsequent covariance matrices for complete, balanced factorial experiments will always be constructed with Rules $\Sigma 1$, $\Sigma 2$, $\Sigma 2.1$, $\Sigma 2.2$, and $\Sigma 2.3$.

Example 4.5.1 Consider a two-way cross classification where the first two factors S and T are fixed with $i = 1, \ldots, s$ and $j = 1, \ldots, t$ levels and the third factor is a set of $k = 1, \ldots, r$ random replicates nested in the st combinations of the first two factors. Let Y_{ijk} be a random variable representing the kth replicate observation in the ij^{th} combination of the two fixed factors. Let the $str \times 1$ random vector $\mathbf{Y} = (Y_{111}, \ldots, Y_{11r}, \ldots, Y_{st1}, \ldots, Y_{str})'$. The model is

$$Y_{ijk} = \mu_{ij} + R(ST)_{(ij)k}$$

where μ_{ij} are constants representing the mean effect of the ij^{th} combination of the two fixed factors and $R(ST)_{(ij)k}$ are str random variables representing the effect of the nested replicates. Assume that the str random variables $R(ST)_{(ij)k} \sim iid\ \mathrm{N}_1(0, \sigma_{R(ST)}^2)$. Therefore, the $str \times 1$ random vector $\mathbf{Y} \sim \mathrm{N}_{str}(\boldsymbol{\mu}, \boldsymbol{\Sigma})$ where the $str \times 1$ mean vector

$$\begin{aligned}\boldsymbol{\mu} &= \mathrm{E}(Y_{111}, \ldots, Y_{11r}, \ldots, Y_{st1}, \ldots, Y_{str})' \\ &= (\mu_{11}\mathbf{1}_r, \ldots, \mu_{st}\mathbf{1}_r)' \\ &= (\mu_{11}, \ldots, \mu_{st})' \otimes \mathbf{1}_r\end{aligned}$$

and the $str \times str$ covariance matrix

$$\boldsymbol{\Sigma} = \sigma_{R(ST)}^2\ [\mathbf{I}_s \otimes \mathbf{I}_t \otimes \mathbf{I}_r].$$

In addition, by the sum of squares algorithm is Section 4.3, $\mathbf{Y}'\mathbf{A}_1\mathbf{Y}, \ldots, \mathbf{Y}'\mathbf{A}_5\mathbf{Y}$ are the sums of squares due to the overall mean, the fixed factor S, the fixed factor T, the fixed interaction of S and T, and the nested replicates, respectively, where

$$\mathbf{A}_1 = \frac{1}{s}\mathbf{J}_s \otimes \frac{1}{t}\mathbf{J}_t \otimes \frac{1}{r}\mathbf{J}_r$$

$$\mathbf{A}_2 = \left(\mathbf{I}_s - \frac{1}{s}\mathbf{J}_s\right) \otimes \frac{1}{t}\mathbf{J}_t \otimes \frac{1}{r}\mathbf{J}_r$$

$$\mathbf{A}_3 = \frac{1}{s}\mathbf{J}_s \otimes \left(\mathbf{I}_t - \frac{1}{t}\mathbf{J}_t\right) \otimes \frac{1}{r}\mathbf{J}_r$$

$$\mathbf{A}_4 = \left(\mathbf{I}_s - \frac{1}{s}\mathbf{J}_s\right) \otimes \left(\mathbf{I}_t - \frac{1}{t}\mathbf{J}_t\right) \otimes \frac{1}{r}\mathbf{J}_r$$

$$\mathbf{A}_5 = \mathbf{I}_s \otimes \mathbf{I}_t \otimes \left(\mathbf{I}_r - \frac{1}{r}\mathbf{J}_r\right).$$

Furthermore, $\mathbf{A}_m\Sigma = \sigma^2_{R(ST)}\mathbf{A}_m$ for $m = 1, \ldots, 5$ where each $\mathbf{A}_m$ is idempotent with $\text{rank}(\mathbf{A}_1) = 1$, $\text{rank}(\mathbf{A}_2) = s - 1$, $\text{rank}(\mathbf{A}_3) = t - 1$, $\text{rank}(\mathbf{A}_4) = (s-1)(t-1)$, and $\text{rank}(\mathbf{A}_5) = st(r-1)$. Thus, by Corollary 3.1.2(a), $\mathbf{Y}'\mathbf{A}_m\mathbf{Y} \sim \sigma^2_{R(ST)}\chi^2_{\text{rank}(\mathbf{A}_m)}(\lambda_m)$ for $m = 1, \ldots, 5$ where

$$\lambda_1 = \frac{[(\mu_{11}, \ldots, \mu_{st})' \otimes \mathbf{1}_r]'\left[\frac{1}{s}\mathbf{J}_s \otimes \frac{1}{t}\mathbf{J}_t \otimes \frac{1}{r}\mathbf{J}_r\right][(\mu_{11}, \ldots, \mu_{st})' \otimes \mathbf{1}_r]}{2\sigma^2_{R(ST)}}$$

$$= str\bar{\mu}^2_{..}/(2\sigma^2_{R(ST)})$$

$$\lambda_2 = \frac{[(\mu_{11}, \ldots, \mu_{st})' \otimes \mathbf{1}_r]'\left[\left(\mathbf{I}_s - \frac{1}{s}\mathbf{J}_s\right) \otimes \frac{1}{t}\mathbf{J}_t \otimes \frac{1}{r}\mathbf{J}_r\right][(\mu_{11}, \ldots, \mu_{st})' \otimes \mathbf{1}_r]}{2\sigma^2_{R(ST)}}$$

$$= rt\sum_{i=1}^{s}(\bar{\mu}_{i.} - \bar{\mu}_{..})^2/(2\sigma^2_{R(ST)})$$

$$\lambda_3 = \frac{[(\mu_{11}, \ldots, \mu_{st})' \otimes \mathbf{1}_r]'\left[\frac{1}{s}\mathbf{J}_s \otimes \left(\mathbf{I}_t - \frac{1}{t}\mathbf{J}_t\right) \otimes \frac{1}{r}\mathbf{J}_r\right][(\mu_{11}, \ldots, \mu_{st})' \otimes \mathbf{1}_r]}{2\sigma^2_{R(ST)}}$$

$$= rs\sum_{j=1}^{t}(\bar{\mu}_{.j} - \bar{\mu}_{..})^2/(2\sigma^2_{R(ST)})$$

$$\lambda_4 = \frac{[(\mu_{11},\ldots,\mu_{st})'\otimes\mathbf{1}_r]'\left[(\mathbf{I}_s - \frac{1}{s}\mathbf{J}_s)\otimes(\mathbf{I}_t - \frac{1}{t}\mathbf{J}_t)\otimes\frac{1}{r}\mathbf{J}_r\right][(\mu_{11},\ldots,\mu_{st})'\otimes\mathbf{1}_r]}{2\sigma^2_{R(ST)}}$$

$$= r\sum_{i=1}^{s}\sum_{j=1}^{t}(\bar{\mu}_{ij} - \bar{\mu}_{i.} - \bar{\mu}_{.j} + \bar{\mu}_{..})^2/(2\sigma^2_{R(ST)})$$

$$\lambda_5 = \frac{[(\mu_{11},\ldots,\mu_{st})'\otimes\mathbf{1}_r]'\left[\mathbf{I}_s\otimes\mathbf{I}_t\otimes\left(\mathbf{I}_r - \frac{1}{r}\mathbf{J}_r\right)\right][(\mu_{11},\ldots,\mu_{st})'\otimes\mathbf{1}_r]}{2\sigma^2_{R(ST)}}$$

$$= 0.$$

By Theorem 3.2.1, the random variables $\mathbf{Y}'\mathbf{A}_m\mathbf{Y}$ are mutually independent since $\mathbf{A}_m\mathbf{\Sigma}\mathbf{A}_n = \sigma^2_{R(ST)}\mathbf{A}_m\mathbf{A}_n = \mathbf{0}_{str\times str}$ for $m, n = 1,\ldots,5$ and $m \neq n$. The EMSs associated with each sum of squares are calculated using Theorem 4.4.1:

$$\text{EMS (mean)} = [\text{tr}(\mathbf{A}_1\mathbf{\Sigma}) + \boldsymbol{\mu}'\mathbf{A}_1\boldsymbol{\mu}]/\text{tr}(\mathbf{A}_1) = \sigma^2_{R(ST)} + rst\bar{\mu}^2_{..}$$

$$\text{EMS (fixed factor } S) = [\text{tr}(\mathbf{A}_2\mathbf{\Sigma}) + \boldsymbol{\mu}'\mathbf{A}_2\boldsymbol{\mu}]/\text{tr}(\mathbf{A}_2)$$

$$= \sigma^2_{R(ST)} + rt\sum_{i=1}^{s}(\bar{\mu}_{i.} - \bar{\mu}_{..})^2/(s-1)$$

$$\text{EMS (fixed factor } T) = [\text{tr}(\mathbf{A}_3\mathbf{\Sigma}) + \boldsymbol{\mu}'\mathbf{A}_3\boldsymbol{\mu}]/\text{tr}(\mathbf{A}_3)$$

$$= \sigma^2_{R(ST)} + rs\sum_{j=1}^{t}(\bar{\mu}_{.j} - \bar{\mu}_{..})^2/(t-1)$$

$$\text{EMS (fixed inter } ST) = [\text{tr}(\mathbf{A}_4\mathbf{\Sigma}) + \boldsymbol{\mu}'\mathbf{A}_4\boldsymbol{\mu}]/\text{tr}(\mathbf{A}_4) = \sigma^2_{R(ST)}$$

$$+ r\sum_{i=1}^{s}\sum_{j=1}^{t}(\mu_{ij} - \bar{\mu}_{i.} - \bar{\mu}_{.j} + \bar{\mu}_{..})^2/(s-1)(t-1)$$

$$\text{EMS (random } R(ST)) = [\text{tr}(\mathbf{A}_5\mathbf{\Sigma}) + \boldsymbol{\mu}'\mathbf{A}_5\boldsymbol{\mu}]/\text{tr}(\mathbf{A}_5)$$

$$= \sigma^2_{R(ST)}.$$

Bhat's lemma 3.4.1 also applies since $\Sigma^5_{m=1}\text{rank}(\mathbf{A}_m) = 1 + (s-1) + (t-1) + (s-1)(t-1) + r(s-1)(t-1) = str$, $\mathbf{I}_s\otimes\mathbf{I}_t\otimes\mathbf{I}_r = \Sigma^5_{m=1}\mathbf{A}_m$, and $\mathbf{\Sigma} = (\Sigma^5_{m=1}\mathbf{A}_m)\mathbf{\Sigma} = \Sigma^5_{m=1}(\mathbf{A}_m\mathbf{\Sigma}) = \Sigma^5_{m=1}\sigma^2_{R(ST)}\mathbf{A}_m$. Therefore, (i) $\mathbf{Y}'\mathbf{A}_m\mathbf{Y} \sim \sigma^2_{R(ST)}\chi^2_{\text{rank }(\mathbf{A}_m)}(\lambda_m)$ and (ii) $\mathbf{Y}'\mathbf{A}_m\mathbf{Y}$ are mutually independent for $m = 1,\ldots,5$.

Example 4.5.2 Consider the same two-way layout as in Example 4.5.1 except now let S and T both be random factors. The model is

$$Y_{ijk} = \alpha + S_i + T_j + ST_{ij} + R(ST)_{(ij)k}$$

where α is a constant representing the overall mean effect; s_i are random variables representing the effect of the first random factor; T_j are the random variables representing the second random factor; ST_{ij} are random variables representing the interaction between S and T; and $R(ST)_{(ij)k}$ are random variable defined as in Example 4.5.1. Assume the s random variables $S_i \sim iid\ \mathrm{N}_1(0, \sigma_S^2)$; the t random variables $T_j \sim iid\ \mathrm{N}_1(0, \sigma_T^2)$; the st random variables $ST_{ij} \sim iid\ \mathrm{N}_1(0, \sigma_{ST}^2)$; and the str random variables $R(ST)_{(ij)k} \sim iid\ \mathrm{N}_1(0, \sigma_{R(ST)})$. Furthermore, assume that $\{S_i, i = 1, \ldots, s\}$, $\{T_j, j = 1, \ldots, t\}$, $\{ST_{ij}, i = 1, \ldots, s, j = 1 \ldots, t\}$, and $\{R(ST)_{(ij)k}, i = 1, \ldots, s, j = 1, \ldots, t, k = 1, \ldots, r\}$ are mutually independent sets of random variables. Therefore, the $str \times 1$ random vector $\mathbf{Y} \sim \mathrm{N}_{str}(\boldsymbol{\mu}, \boldsymbol{\Sigma})$ where the $str \times 1$ mean vector

$$\boldsymbol{\mu} = \mathrm{E}(Y_{111}, \ldots, Y_{11r}, \ldots, Y_{st1}, \ldots, Y_{str})' = \alpha \mathbf{1}_s \otimes \mathbf{1}_t \otimes \mathbf{1}_r$$

and, by the covariance algorithm, the $str \times str$ covariance matrix

$$\begin{aligned}\Sigma = {} & \sigma_S^2\ [\mathbf{I}_s \otimes \mathbf{J}_t \otimes \mathbf{J}_r] + \sigma_T^2\ [\mathbf{J}_s \otimes \mathbf{I}_t \otimes \mathbf{J}_r] \\ & + \sigma_{ST}^2\ [\mathbf{I}_s \otimes \mathbf{I}_t \otimes \mathbf{J}_r] + \sigma_{R(ST)}^2\ [\mathbf{I}_s \otimes \mathbf{I}_t \otimes \mathbf{I}_r].\end{aligned}$$

The sum of squares matrices are not dependent on whether the factors are fixed or random. Therefore, the sum of squares $\mathbf{Y}'\mathbf{A}_m\mathbf{Y}$ for $m = 1, \ldots, 5$ are the same as those given in Example 4.5.1. Furthermore, $\mathbf{A}_m\boldsymbol{\Sigma} = c_m\mathbf{A}_m$ for $m = 1, \ldots, 5$ where

$$\begin{aligned} c_1 &= tr\sigma_S^2 + sr\sigma_T^2 + r\sigma_{ST}^2 + \sigma_{R(ST)}^2 \\ c_2 &= tr\sigma_S^2 + r\sigma_{ST}^2 + \sigma_{R(ST)}^2 \\ c_3 &= sr\sigma_T^2 + r\sigma_{ST}^2 + \sigma_{R(ST)}^2 \\ c_4 &= r\sigma_{ST}^2 + \sigma_{R(ST)}^2 \\ c_5 &= \sigma_{R(ST)}^2. \end{aligned}$$

It is left to the reader to show $\lambda_1 = str\alpha^2/(2c_1)$ and $\lambda_2 = \lambda_3 = \lambda_4 = \lambda_5 = 0$. By Bhat's lemma, or by Theorem 3.2.1 and Corollary 3.1.2(a), $\mathbf{Y}'\mathbf{A}_m\mathbf{Y} \sim c_m\chi^2_{\mathrm{rank}(\mathbf{A}_m)}(\lambda_m)$ where the $\mathrm{rank}(\mathbf{A}_m)$ are given in Example 4.5.1. The EMS associated with each sum of squares can be calculated using Theorem 4.4.1 and are given by EMS(mean) $= c_1 + str\alpha^2$, EMS(random factor S) $= c_2$, EMS(random factor T) $= c_3$, EMS (random inter ST) $= c_4$ and EMS(random reps $R(ST)$) $= c_5$.

Examples 4.4.1, 4.5.1, and 4.5.2 provide the EMSs for three factorial experiments. Snedecor and Cochran (1978, p. 367) provide EMSs for the same three experiments in their Table 12.11.1. Snedecor and Cochran's EMSs are the same as the EMSs presented in this text, although they use different notation. Snedecor and Cochran's A, B, AB, error, $\sigma^2, \sigma_A^2, \sigma_B^2, \sigma_{AB}^2, \kappa_A^2, \kappa_B^2, \kappa_{AB}^2, a, b$, and n are equivalent to our $T, S, ST, R(ST), \sigma_{R(ST)}^2, \sigma_T^2, \sigma_S^2, \sigma_{ST}^2, \sum_{j=1}^{t}(\bar{\mu}_{.j} - \bar{\mu}_{..})^2/(t-1)$, $\sum_{i=1}^{s}(\bar{\mu}_{i.} - \bar{\mu}_{..})^2/(s-1)$, $\sum_{i=1}^{s}\sum_{j=1}^{t}(\mu_{ij} - \bar{\mu}_{i.} - \bar{\mu}_{.j} + \bar{\mu}_{..})^2/[(s-1)(t-1)]$, t, s, and r, respectively.

Example 4.5.3 Consider the split plot experiment discussed by Kempthorne (1952, sixth printing 1967, pp. 374–375). The experiment has a random replicate factor, R, with $i = 1, \ldots, r$ levels; a set of fixed whole plot treatments, T, with $j = 1, \ldots, t$ levels; and a set of fixed split plot treatments, S, with $k = 1, \ldots, s$ levels. In Kempthorne's Table 19.1 he lists the following sources of variation: replicates R, whole plot treatments T, replicate by whole plot interaction RT, split plot treatments S, split by whole plot interaction ST, and remainder. The remainder is equal to the interaction RS plus the interaction RST, or equivalently, the replicate by split plot treatment interaction nested in whole plot treatments $RS(T)$. The fixed portions of the experiment are designated by T, S, and ST with subscripts j and k. The random portions of the experiment are designated by R, RT, and $RS(T)$. A model that identifies these sources of variation is

$$Y_{ijk} = \mu_{jk} + R_i + RT_{ij} + RS(T)_{i(j)k}$$

where the r random variables $R_i \sim \mathrm{N}_1(0, \sigma_R^2)$; the $r(t-1) \times 1$ random vector $(\mathbf{I}_r \otimes \mathbf{P}_t')(RT_{11}, \ldots, RT_{rt})' \sim \mathrm{N}_{r(t-1)}(\mathbf{0}, \sigma_{RT}^2 \mathbf{I}_r \otimes \mathbf{I}_{t-1})$; and the $rt(s-1) \times 1$ random vector $(\mathbf{I}_r \otimes \mathbf{I}_t \mathbf{P}_s')(RS(T)_{1(1)1}, \ldots, RS(T)_{r(t)s})' \sim \mathrm{N}_{rt(s-1)}(\mathbf{0}, \sigma_{RS(T)}^2 \mathbf{I}_r \otimes \mathbf{I}_t \otimes \mathbf{I}_{s-1})$ where $\mathbf{P}_t'$ and $\mathbf{P}_s'$ are $(t-1) \times t$ and $(s-1) \times s$ lower portions of t- and s-dimensional Helmert matrices, respectively. Furthermore, these three sets of variables are mutually independent. Thus, the $rts \times 1$ random vector $\mathbf{Y} = (Y_{111}, \ldots, Y_{11s}, \ldots, Y_{rt1}, \ldots, Y_{rts})' \sim \mathrm{N}_{rts}(\boldsymbol{\mu}, \Sigma)$ where the $rts \times 1$ mean vector

$$\begin{aligned}\boldsymbol{\mu} &= \mathrm{E}(Y_{111}, \ldots, Y_{11s}, \ldots, Y_{rt1}, \ldots, Y_{rts})' \\ &= \mathbf{1}_r \otimes (\mu_{11}, \ldots, \mu_{1s}, \ldots, \mu_{t1}, \ldots, \mu_{ts})'\end{aligned}$$

and, by the covariance matrix algorithm, the $rts \times rts$ covariance matrix is

$$\begin{aligned}\mathbf{\Sigma} &= \sigma_R^2 \left[\mathbf{I}_r \otimes \mathbf{J}_t \otimes \mathbf{J}_s\right] + \sigma_{RT}^2 \left[\mathbf{I}_r \otimes \left(\mathbf{I}_t - \frac{1}{t}\mathbf{J}_t\right) \otimes \mathbf{J}_s\right] \\ &\quad + \sigma_{RS(T)}^2 \left[\mathbf{I}_r \otimes \mathbf{I}_t \otimes \left(\mathbf{I}_s - \frac{1}{s}\mathbf{J}_s\right)\right].\end{aligned}$$

By the sum of squares algorithm, the matrices $\mathbf{A}_1, \ldots, \mathbf{A}_7$ are

$$\text{Overall mean} \quad \mathbf{A}_1 = \frac{1}{r}\mathbf{J}_r \otimes \frac{1}{t}\mathbf{J}_t \otimes \frac{1}{s}\mathbf{J}_s$$

$$\text{Replicates } R \quad \mathbf{A}_2 = \left(\mathbf{I}_r - \frac{1}{r}\mathbf{J}_r\right) \otimes \frac{1}{t}\mathbf{J}_t \otimes \frac{1}{s}\mathbf{J}_s$$

$$\text{Whole plot treatments } T \quad \mathbf{A}_3 = \frac{1}{r}\mathbf{J}_r \otimes \left(\mathbf{I}_t - \frac{1}{t}\mathbf{J}_t\right) \otimes \frac{1}{s}\mathbf{J}_s$$

$$RT \quad \mathbf{A}_4 = \left(\mathbf{I}_r - \frac{1}{r}\mathbf{J}_r\right) \otimes \left(\mathbf{I}_t - \frac{1}{t}\mathbf{J}_t\right) \otimes \frac{1}{s}\mathbf{J}_s$$

$$\text{Split plot treatments } S \quad \mathbf{A}_5 = \frac{1}{r}\mathbf{J}_r \otimes \frac{1}{t}\mathbf{J}_t \otimes \left(\mathbf{I}_s - \frac{1}{s}\mathbf{J}_s\right)$$

$$ST \quad \mathbf{A}_6 = \frac{1}{r}\mathbf{J}_r \otimes \left(\mathbf{I}_t - \frac{1}{t}\mathbf{J}_t\right) \otimes \left(\mathbf{I}_s - \frac{1}{s}\mathbf{J}_s\right)$$

$$RS(T) \quad \mathbf{A}_7 = \left(\mathbf{I}_r - \frac{1}{r}\mathbf{J}_r\right) \otimes \mathbf{I}_t \otimes \left(\mathbf{I}_s - \frac{1}{s}\mathbf{J}_s\right).$$

Furthermore, $\mathbf{A}_m \Sigma = c_m \mathbf{A}_m$ for $m = 1, \ldots, 7$ where $c_1 = c_2 = st\sigma_R^2$, $c_3 = c_4 = s\sigma_{RT}^2$, and $c_5 = c_6 = c_7 = \sigma_{RS(T)}^2$; $\mathbf{I}_r \otimes \mathbf{I}_t \otimes \mathbf{I}_s = \Sigma_{m=1}^7 \mathbf{A}_m$; $\Sigma_{m=1}^7 \text{rank}\ (\mathbf{A}_m) = 1 + (r-1) + (t-1) + (r-1)(t-1) + (s-1) + (t-1)(s-1) + (r-1)t(s-1) = rts$; and $\mathbf{\Sigma} = (\sum_{m=1}^7 \mathbf{A}_m)\Sigma = (\sum_{m=1}^7 \mathbf{A}_m \Sigma) = \sum_{m=1}^7 \mathbf{c}_m \mathbf{A}_m$. Therefore, by Bhat's Lemma 3.4.1, (i) $\mathbf{Y}'\mathbf{A}_m\mathbf{Y} \sim c_m \chi^2_{\text{rank}(\mathbf{A}_m)}(\lambda_m)$ and (ii) $\mathbf{Y}'\mathbf{A}_m\mathbf{Y}$ are mutually independent for $m = 1, \ldots, 7$ where

$$\lambda_1 = \frac{[\mathbf{1}_r \otimes (\mu_{11}, \ldots, \mu_{ts})']' \left[\frac{1}{r}\mathbf{J}_r \otimes \frac{1}{t}\mathbf{J}_t \otimes \frac{1}{s}\mathbf{J}_s\right] [\mathbf{1}_r \otimes (\mu_{11}, \ldots, \mu_{ts})']}{(2st\sigma_R^2)}$$

$$= r\bar{\mu}^2../(2\sigma_R^2),$$

$$\lambda_2 = \frac{[\mathbf{1}_r \otimes (\mu_{11}, \ldots, \mu_{ts})']' \left[\left(\mathbf{I}_r - \frac{1}{r}\mathbf{J}_r\right) \otimes \frac{1}{t}\mathbf{J}_t \otimes \frac{1}{s}\mathbf{J}_s\right] [\mathbf{1}_r \otimes (\mu_{11}, \ldots, \mu_{ts})']}{2st\sigma_R^2}$$

$$= 0$$

$$\lambda_3 = \frac{[\mathbf{1}_r \otimes (\mu_{11}, \ldots, \mu_{ts})']' \left[\frac{1}{r}\mathbf{J}_r \otimes \left(\mathbf{I}_t - \frac{1}{t}\mathbf{J}_t\right) \otimes \frac{1}{s}\mathbf{J}_s\right] [\mathbf{1}_r \otimes (\mu_{11}, \ldots, \mu_{ts})']}{2s\sigma^2_{RT}}$$

$$= r\sum_{j=1}^{t} (\bar{\mu}_{j.} - \bar{\mu}_{..})^2/(2\sigma^2_{RT})$$

$$\lambda_4 = \frac{[\mathbf{1}_r \otimes (\mu_{11}, \ldots, \mu_{ts})']' \left[\left(\mathbf{I}_r - \frac{1}{r}\mathbf{J}_r\right) \otimes \left(\mathbf{I}_t - \frac{1}{t}\mathbf{J}_t\right) \otimes \frac{1}{s}\mathbf{J}_s\right] [\mathbf{1}_r \otimes (\mu_{11}, \ldots, \mu_{ts})']}{2s\sigma^2_{RT}}$$

$$= 0$$

$$\lambda_5 = \frac{[\mathbf{1}_r \otimes (\mu_{11}, \ldots, \mu_{ts})']' \left[\frac{1}{r}\mathbf{J}_r \otimes \frac{1}{t}\mathbf{J}_t \otimes \left(\mathbf{I}_s - \frac{1}{s}\mathbf{J}_s\right)\right] [\mathbf{1}_r \otimes (\mu_{11}, \ldots, \mu_{ts})']}{2\sigma^2_{RS(T)}}$$

$$= rt\sum_{k=1}^{s} (\bar{\mu}_{.k} - \bar{\mu}_{..})^2/(2\sigma^2_{RS(T)})$$

$$\lambda_6 = \frac{[\mathbf{1}_r \otimes (\mu_{11}, \ldots, \mu_{ts})']' \left[\frac{1}{r}\mathbf{J}_r \otimes \left(\mathbf{I}_t - \frac{1}{t}\mathbf{J}_t\right) \otimes \left(\mathbf{I}_s - \frac{1}{s}\mathbf{J}_s\right)\right] [\mathbf{1}_r \otimes (\mu_{11}, \ldots, \mu_{ts})']}{2\sigma^2_{RS(T)}}$$

$$= r\sum_{j=1}^{t}\sum_{k=1}^{s} (\mu_{jk} - \bar{\mu}_{j.} - \bar{\mu}_{.k} + \bar{\mu}_{..})^2/(2\sigma^2_{RS(T)})$$

$$\lambda_7 = \frac{[\mathbf{1}_r \otimes (\mu_{11}, \ldots, \mu_{ts})']' \left[\left(\mathbf{I}_r - \frac{1}{r}\mathbf{J}_r\right) \otimes \mathbf{I}_t \otimes \left(\mathbf{I}_s - \frac{1}{s}\mathbf{J}_s\right)\right] [\mathbf{1}_r \otimes (\mu_{11}, \ldots, \mu_{ts})']}{2\sigma^2_{RS(T)}}$$

$$= 0.$$

By Theorem 4.4.1 the expected mean squares are

$$\text{EMS (overall mean)} = st\sigma^2_R + rst\bar{\mu}_{..}^2$$
$$\text{EMS (replicate } R) = st\sigma^2_R$$
$$\text{EMS (whole plot } T) = s\sigma^2_{RT} + rs\sum_{j=1}^{t} (\bar{\mu}_{j.} - \bar{\mu}_{..})^2/(t-1)$$
$$\text{EMS } (RT) = s\sigma^2_{RT}$$

$$\text{EMS (split plot } S) = \sigma^2_{RS(T)} + rt\sum_{k=1}^{s}(\bar{\mu}_{.k} - \bar{\mu}_{..})^2/(s-1)$$

$$\text{EMS } (ST) \quad = \sigma^2_{RS(T)} + r\sum_{j=1}^{t}\sum_{k=1}^{s}(\mu_{jk} - \bar{\mu}_{j.} - \bar{\mu}_{.k} + \bar{\mu}_{..})^2/ [(t-1)(s-1)]$$

$$\text{EMS } (RS(T)) \quad = \sigma^2_{RS(T)}.$$

Kempthorne (1952) provides the EMSs for T, RT, S, ST, and $RS(T)$ in his Table 19.2. Kempthorne's EMSs are the same as the EMSs given here, although Kempthorne uses different notation. Kempthorne's $\sigma^2, t_j, \sigma_s^2, s_k$, and $(ts)_{jk}$ are equivalent to our $s\sigma^2_{RT}$, $(\bar{\mu}_{j.} - \bar{\mu}_{..})$, $\sigma^2_{RS(T)}$, $(\bar{\mu}_{.k} - \bar{\mu}_{..})$, and $(\bar{\mu}_{ij} - \bar{\mu}_{j.} - \bar{\mu}_{.k} + \bar{\mu}_{..})$, respectively.

Example 4.5.4 Consider an experiment where bst experimental units are divided into b homogeneous sets of st units. Within each of the b sets (or, equivalently, within each of the b random blocks B) the st units are randomly assigned to the st combinations of the two fixed factors S and T where factor S has s levels and T has t levels. This is a classic random block design where the fixed treatments are identified by two fixed factors. Let Y_{ijk} be the random variable representing the observation in the kth level of factor T, the jth level of factor S, and the ith block for $i = 1, \ldots, b$, $j = 1, \ldots, s$, and $k = 1, \ldots, t$. This experiment is characterized by the model

$$Y_{ijk} = \mu_{jk} + B_i + R_{ijk}$$

where μ_{jk} are st constants representing the mean effect of the jk^{th} combination of factors S and T; the B_i are random variables representing the random effect of blocks; and the R_{ijk} are random variables representing the random residual or remainder. It is our intention to write a covariance matrix that contains two variance components, one associated with the variance of the random variables B_i and one associated with the variance of the random variables R_{ijk}. However, to use the covariance algorithm, we must first rewrite the variables R_{ijk} in terms of the factor letters B, S, and T. Note that R_{ijk} can be equivalently written as

$$R_{ijk} = BS_{ij} + BT_{ik} + BST_{ijk}$$

where BS_{ij}, BT_{ik}, and BST_{ijk} are random variables representing the interaction of B with S, B with T, and B with ST, respectively. We could proceed from here by putting the last two equations together to produce a model

$$Y_{ijk} = \mu_{jk} + B_i + BS_{ij} + BT_{ik} + BST_{ijk}.$$

However, the last model has four sets of random variables and would thus require a definition with four, not two, random components. The solution to the problem is to view the st combinations of fixed factors S and T as one fixed factor, say, V, with st levels. If we let $v = 1, \ldots, st$ designate the levels of factor V then the residual R_{ijk} can be rewritten as

$$R_{ijk} = BS_{ij} + BT_{ik} + BST_{ijk} = BV_{iv}.$$

Now assume the random variables $B_i \sim \mathrm{N}_1(0, \sigma_B^2)$ and the $b(st-1) \times 1$ random vector $(\mathbf{I}_b \otimes \mathbf{P}'_{st})\ (BV_{11}, \ldots, BV_{b,st})' \sim \mathrm{N}_{b(st-1)}(\mathbf{0}, \sigma_{BV}^2 \mathbf{I}_b \otimes \mathbf{I}_{st-1})$ where $\mathbf{P}'_{st}$ is the $(st-1) \times st$ lower portion of an st-dimensional Helmert matrix. Thus, the $bst \times 1$ random vector $\mathbf{Y} = (Y_{111}, \ldots, Y_{11t}, \ldots, Y_{bs1}, \ldots, Y_{bst})' \sim \mathrm{N}_{bst}(\boldsymbol{\mu}, \Sigma)$ where the $bst \times 1$ mean vector

$$\begin{aligned}\boldsymbol{\mu} &= \mathrm{E}(Y_{111}, \ldots, Y_{11t}, \ldots, Y_{bs1}, \ldots, Y_{bst})' \\ &= \mathbf{1}_b \otimes (\mu_{11}, \ldots, \mu_{1t}, \ldots, \mu_{s1}, \ldots, \mu_{st})'.\end{aligned}$$

The covariance matrix algorithm can be applied using the two factor letters B and V where the number of levels is b and st, respectively. The $bst \times bst$ covariance matrix, $\boldsymbol{\Sigma}$, is constructed here:

Factor Letter	B	V
Levels	b	st

$$\begin{aligned}\boldsymbol{\Sigma} = \ & \sigma_B^2 \ [\mathbf{I}_b \otimes \mathbf{J}_{st}] \\ & + \sigma_{BV}^2 \left[\mathbf{I}_b \otimes \left(\mathbf{I}_{st} - \frac{1}{st}\mathbf{J}_{st}\right)\right].\end{aligned}$$

That is,

$$\begin{aligned}\boldsymbol{\Sigma} = \ & \sigma_B^2 \ [\mathbf{I}_b \otimes \mathbf{J}_s \otimes \mathbf{J}_t] \\ & + \sigma_{BV}^2 \ [\mathbf{I}_b \otimes \mathbf{I}_s \otimes \mathbf{I}_t] \\ & - \sigma_{BV}^2 \left[\mathbf{I}_b \otimes \frac{1}{s}\mathbf{J}_s \otimes \frac{1}{t}\mathbf{J}_t\right].\end{aligned}$$

The sum of squares algorithm is now used to calculate matrices $\mathbf{A}_1, \ldots, \mathbf{A}_6$. Recall that the sum of squares remainder equals the sum of the sum of squares due to the interactions BS, BT, and BST. Thus, $\mathbf{A}_6$ equals the sum of the matrices associated with the sum of squares due to BS, BT, and BST.

Overall mean $\quad \mathbf{A}_1 = \frac{1}{b}\mathbf{J}_b \otimes \frac{1}{s}\mathbf{J}_s \otimes \frac{1}{t}\mathbf{J}_t$

Blocks $B \quad \mathbf{A}_2 = \left(\mathbf{I}_b - \frac{1}{b}\mathbf{J}_b\right) \otimes \frac{1}{s}\mathbf{J}_s \otimes \frac{1}{t}\mathbf{J}_t$

$$\text{Factor } S \quad \mathbf{A}_3 = \frac{1}{b}\mathbf{J}_b \otimes \left(\mathbf{I}_s - \frac{1}{s}\mathbf{J}_s\right) \otimes \frac{1}{t}\mathbf{J}_t$$

$$\text{Factor } T \quad \mathbf{A}_4 = \frac{1}{b}\mathbf{J}_b \otimes \frac{1}{s}\mathbf{J}_s \otimes \left(\mathbf{I}_t - \frac{1}{t}\mathbf{J}_t\right)$$

$$ST \quad \mathbf{A}_5 = \frac{1}{b}\mathbf{J}_b \otimes \left(\mathbf{I}_s - \frac{1}{s}\mathbf{J}_s\right) \otimes \left(\mathbf{I}_t - \frac{1}{t}\mathbf{J}_t\right)$$

$$\begin{aligned}\text{Remainder} \quad \mathbf{A}_6 = &\left[\left(\mathbf{I}_b - \frac{1}{b}\mathbf{J}_b\right) \otimes \left(\mathbf{I}_s - \frac{1}{s}\mathbf{J}_s\right) \otimes \frac{1}{t}\mathbf{J}_t\right] \\ &+ \left[\left(\mathbf{I}_b - \frac{1}{b}\mathbf{J}_b\right) \otimes \frac{1}{s}\mathbf{J}_s \otimes \left(\mathbf{I}_t - \frac{1}{t}\mathbf{J}_t\right)\right] \\ &+ \left[\left(\mathbf{I}_b - \frac{1}{b}\mathbf{J}_b\right) \otimes \left(\mathbf{I}_s - \frac{1}{s}\mathbf{J}_s\right) \otimes \left(\mathbf{I}_t - \frac{1}{t}\mathbf{J}_t\right)\right].\end{aligned}$$

Again $\mathbf{A}_m\boldsymbol{\Sigma} = c_m\mathbf{A}_m$ for $m = 1, \ldots, 6$ where $c_1 = c_2 = st\sigma_B^2, c_3 = c_4 = c_5 = c_6 = \sigma_{BV}^2$, $\mathbf{I}_b \otimes \mathbf{I}_s \otimes \mathbf{I}_t = \Sigma_{m=1}^6 \mathbf{A}_m$, $\Sigma_{m=1}^6 \text{rank } (\mathbf{A}_m) = 1 + (b-1) + (s-1) + (t-1) + (s-1)(t-1) + (b-1)[(s-1) + (t-1) + (s-1)(t-1)] = bst$, $\boldsymbol{\Sigma} = (\sum_{m=1}^6 \mathbf{A}_m)\Sigma = (\sum_{m=1}^6 \mathbf{A}_m\Sigma) = \sum_{m=1}^6 c_m\mathbf{A}_m$. Therefore, by Bhat's lemma 3.4.1, (i) $\mathbf{Y}'\mathbf{A}_m\mathbf{Y} \sim c_m\chi^2_{\text{rank}(\mathbf{A}_m)}(\lambda_m)$ and (ii) $\mathbf{Y}'\mathbf{A}_m\mathbf{Y}$ are mutually independent for $m = 1, \ldots, 6$ where

$$\begin{aligned}\lambda_1 &= [\mathbf{1}_b \otimes (\mu_{11}, \ldots, \mu_{1t}, \ldots, \mu_{s1}, \ldots, \mu_{st})']'\left[\frac{1}{b}\mathbf{J}_b \otimes \frac{1}{s}\mathbf{J}_s \otimes \frac{1}{t}\mathbf{J}_t\right] \\ &\quad \times [\mathbf{1}_b \otimes (\mu_{11}, \ldots, \mu_{1t}, \ldots, \mu_{s1}, \ldots, \mu_{st})']/(2st\sigma_B^2) \\ &= b\bar{\mu}^2../(2\sigma_B^2)\end{aligned}$$

$$\begin{aligned}\lambda_2 &= [\mathbf{1}_b \otimes (\mu_{11}, \ldots, \mu_{1t}, \ldots, \mu_{s1}, \ldots, \mu_{st})']'\left[\left(\mathbf{I}_b - \frac{1}{b}\mathbf{J}_b\right) \otimes \frac{1}{s}\mathbf{J}_s \otimes \frac{1}{t}\mathbf{J}_t\right] \\ &\quad \times [\mathbf{1}_b \otimes (\mu_{11}, \ldots, \mu_{1t}, \ldots, \mu_{s1}, \ldots, \mu_{st})']/(2st\sigma_B^2) \\ &= 0\end{aligned}$$

$$\begin{aligned}\lambda_3 &= [\mathbf{1}_b \otimes (\mu_{11}, \ldots, \mu_{1t}, \ldots, \mu_{s1}, \ldots, \mu_{st})']'\left[\frac{1}{b}\mathbf{J}_b \otimes \left(\mathbf{I}_s - \frac{1}{s}\mathbf{J}_s\right) \otimes \frac{1}{t}\mathbf{J}_t\right] \\ &\quad \times [\mathbf{1}_b \otimes (\mu_{11}, \ldots, \mu_{1t}, \ldots, \mu_{s1}, \ldots, \mu_{st})']/(2\sigma_{BV}^2) \\ &= bt\sum_{j=1}^{s}(\bar{\mu}_{j.} - \bar{\mu}..)^2/(2\sigma_{BV}^2)\end{aligned}$$

$$\lambda_4 = [\mathbf{1}_b \otimes (\mu_{11}, \ldots, \mu_{1t}, \ldots, \mu_{s1}, \ldots, \mu_{st})']' \left[\frac{1}{b}\mathbf{J}_b \otimes \frac{1}{s}\mathbf{J}_s \otimes \left(\mathbf{I}_t - \frac{1}{t}\mathbf{J}_t\right)\right]$$
$$\times [\mathbf{1}_b \otimes (\mu_{11}, \ldots, \mu_{1t}, \ldots, \mu_{s1}, \ldots, \mu_{st})']/(2\sigma^2_{BV})$$
$$= bs \sum_{j=1}^{t} (\bar{\mu}_{.k} - \bar{\mu}_{..})^2/(2\sigma^2_{BV})$$

$$\lambda_5 = [\mathbf{1}_b \otimes (\mu_{11}, \ldots, \mu_{1t}, \ldots, \mu_{s1}, \ldots, \mu_{st})']'$$
$$\times \left[\frac{1}{b}\mathbf{J}_b \otimes \left(\mathbf{I}_s - \frac{1}{s}\mathbf{J}_s\right) \otimes \left(\mathbf{I}_t - \frac{1}{t}\mathbf{J}_t\right)\right]$$
$$\times [\mathbf{1}_b \otimes (\mu_{11}, \ldots, \mu_{1t}, \ldots, \mu_{s1}, \ldots, \mu_{st})']/(2\sigma^2_{BV})$$
$$= b \sum_{j=1}^{s} \sum_{k=1}^{t} (\mu_{jk} - \bar{\mu}_{j.} - \bar{\mu}_{.k} - \bar{\mu}_{..})^2/(2\sigma^2_{BV})$$

$$\lambda_6 = [\mathbf{1}_b \otimes (\mu_{11}, \ldots, \mu_{1t}, \ldots, \mu_{s1}, \ldots, \mu_{st})']'$$
$$\times \left\{\left[\left(\mathbf{I}_b - \frac{1}{b}\mathbf{J}_b\right) \otimes \left(\mathbf{I}_s - \frac{1}{s}\mathbf{J}_s\right) \otimes \frac{1}{t}\mathbf{J}_t\right]\right.$$
$$+ \left[\left(\mathbf{I}_b - \frac{1}{b}\mathbf{J}_b\right) \otimes \frac{1}{s}\mathbf{J}_s \otimes \left(\mathbf{I}_t - \frac{1}{t}\mathbf{J}_t\right)\right]$$
$$\left. + \left[\left(\mathbf{I}_b - \frac{1}{b}\mathbf{J}_b\right) \otimes \left(\mathbf{I}_s - \frac{1}{s}\mathbf{J}_s\right) \otimes \left(\mathbf{I}_t - \frac{1}{t}\mathbf{J}_t\right)\right]\right\}$$
$$\times [\mathbf{1}_b \otimes (\mu_{11}, \ldots, \mu_{1t}, \ldots, \mu_{s1}, \ldots, \mu_{st})']'/(2\sigma^2_{BV})$$
$$= 0.$$

By Theorem 4.4.1, the expected mean squares are

$$\text{EMS (overall mean)} = st\sigma^2_B + bst\bar{\mu}^2_{..}$$
$$\text{EMS (random block } B) = st\sigma^2_B$$
$$\text{EMS (fixed factor } S) = \sigma^2_{BV} + bt \sum_{j=1}^{s} (\bar{\mu}_{j.} - \bar{\mu}_{..})^2/(s-1)$$
$$\text{EMS (fixed factor } T) = \sigma^2_{BV} + bs \sum_{k=1}^{t} (\bar{\mu}_{.k} - \bar{\mu}_{..})^2/(t-1)$$

$$\text{EMS }(ST) = \sigma^2_{BV} + \frac{b\sum_{j=1}^{s}\sum_{k=1}^{t}(\bar{\mu}_{jk} - \bar{\mu}_{j.} - \bar{\mu}_{.k} + \bar{\mu}_{..})^2}{[(s-1)(t-1)]}$$

$$\text{EMS (remainder } R) = \sigma^2_{BV}.$$

EXERCISES

1. Consider an experiment where r replicate units are nested in each of the s levels of a fixed factor s. The t levels of a second fixed factor T are then applied to each of the sr experimental units. A model for this experiment is

$$Y_{ijk} = \mu_{ik} + R(S)_{(i)j} + RT(S)_{(i)jk}$$

for $i = 1, \ldots, s$, $j = 1, \ldots, r$, and $k = 1, \ldots, t$ where R represents the replicate units nested in the s levels of factor S. Assume the $str \times 1$ random vector $\mathbf{Y} = (Y_{111}, \ldots, Y_{11t}, \ldots, Y_{1r1}, \ldots, Y_{1rt}, \ldots, Y_{s11}, \ldots, Y_{s1t}, \ldots, Y_{sr1}, \ldots, Y_{srt})' \sim \text{N}_{srt}(\boldsymbol{\mu}, \boldsymbol{\Sigma})$.

   (a) Derive $\boldsymbol{\mu}$ and $\boldsymbol{\Sigma}$.

   (b) Write out the ANOVA table and define the matrices $\mathbf{A}_m$ used in the sums of squares $\mathbf{Y}'\mathbf{A}_m\mathbf{Y}$ for $m = 1, \ldots, 7$ where $\mathbf{A}_7$ is the matrix corresponding to the sum of squares total.

   (c) Derive the distributions of $\mathbf{Y}'\mathbf{A}_m\mathbf{Y}$ for $m = 1, \ldots, 6$.

   (d) Calculate all the expected mean squares.

   (e) Construct all "appropriate" F statistics and explicitly define the hypothesis being tested in each case. Prove that all the statistics constructed above have F distributions.

2. Consider a factorial experiment with three fixed factors S, T, and U where the factors have s, t, and u levels, respectively. Within the stu combinations of the three main factors, r random replicate observations are observed. Let R represent the random nested replicate factor. Assume the $stur \times 1$ random vector $\mathbf{Y} = (Y_{1111}, \ldots, Y_{111r}, \ldots, Y_{stu1}, \ldots, Y_{stur})' \sim \text{N}_{stur}(\boldsymbol{\mu}, \boldsymbol{\Sigma})$.

   (a) Write a model for this experiment and derive the mean vector $\boldsymbol{\mu}$ and the covariance matrix Σ.

   (b) Write out the ANOVA table and define the matrices $\mathbf{A}_m$ used in the sums of squares $\mathbf{Y}'\mathbf{A}_m\mathbf{Y}$ for $m = 1, \ldots, 10$ where $\mathbf{A}_{10}$ is the matrix corresponding to the sum of squares total.

   (c) Derive the distributions of $\mathbf{Y}'\mathbf{A}_m\mathbf{Y}$ for $m = 1, \ldots, 9$.

(d) Calculate all the expected mean squares.

(e) Construct all "appropriate" F statistics and explicitly define the hypothesis being tested in each case. Prove that all the statistics constructed above have F distributions.

3. Redo Exercise 2 above when S is a fixed factor and T and U are random factors.

4. Redo Exercise 2 when S, T, and U are all random factors.

5. Consider the paired t-test problem introduced in Exercise 15 of Chapter 3. However, now view the problem as an experiment where two levels of a fixed factor T are applied to each of the n levels of a random factor B. Thus, the n levels of B are the n experimental units. The model for this problem is

$$Y_{ij} = \mu_j + B_i + BT_{ij}$$

where μ_j are constants representing the average effect of the jth level of fixed factors T, B_i are random variables representing the effect of the ith experimental unit, and BT_{ij} are random variables representing the interaction effect of factors B and T. Assume the $2n \times 1$ random vector $\mathbf{Y} = (Y_{11}, Y_{12}, Y_{21}, Y_{22}, \ldots, Y_{n1}, Y_{n2})' \sim \mathrm{N}_{2n}(\boldsymbol{\mu}, \boldsymbol{\Sigma})$.

(a) Derive $\boldsymbol{\mu}$ and use the covariance algorithm to define the covariance matrix Σ in terms of σ_B^2 and σ_{BT}^2.

(b) Write out the ANOVA table and define the matrices $\mathbf{A}_m$ used in the sums of squares $\mathbf{Y}'\mathbf{A}_m\mathbf{Y}$ for $m = 1, \ldots, 5$ where $\mathbf{A}_5$ is the matrix corresponding to the sum of squares total.

(c) Derive the distributions of $\mathbf{Y}'\mathbf{A}_m\mathbf{Y}$ for $m = 1, \ldots, 4$.

(d) Calculate all the expected mean squares.

(e) Construct a statistic to test $\mathrm{H}_0 : \mu_1 = \mu_2$ versus $\mathrm{H}_1 : \mu_1 \neq \mu_2$. Does the statistic you constructed have an F distribution? Prove your answer.

(f) Prove that the mean square due to factor T divided by the mean square due to the BT interaction is equal to T^2 where T is defined in Exercise 15b from Chapter 3.

6. An animal scientist was interested in evaluating the effect of five different types of feed on the weight gain of cattle. An experiment was run where 60 cattle were randomly assigned to 15 pens with 4 cattle in each pen. Each pen then received a certain type of feed. The assignment of feeds to pens was done randomly with 3 pens in each of the five feed types. Weight observations on each head of cattle were then taken at three different times. Let Y_{ijkl} be a random variable representing the weight gain observed at the lth time period, on the kth head of cattle, in the jth pen, and being fed the ith feed type for

$i = 1, 2, 3, 4, 5$, $j = 1, 2, 3$, $k = 1, 2, 3, 4$, and $l = 1, 2, 3$. Assume the 180×1 random vector $\mathbf{Y} = (Y_{1111}, \ldots, Y_{1113}, \ldots, Y_{5341}, \ldots, Y_{5343})' \sim \mathrm{N}_{180}(\boldsymbol{\mu}, \Sigma)$. A model for this problem is

$$Y_{ijkl} = \mu_{il} + P(F)_{(i)j} + C(FP)_{(ij)k} + PT(F)_{(i)jl} + CT(FP)_{(ij)kl}$$

where μ_{il} are constants representing the average effect of the ith feed type at the lth time and $P(F)_{(i)j}$, $C(FP)_{(ij)k}$, $PT(F)_{(i)jl}$, and $CT(FP)_{(ij)kl}$ are random variables representing the effects of pens nested in feed types, cattle nested in feed pens, the interaction of time by pens nested in feeds, and the interaction of time and cattle nested in feed pens, respectively.

(a) Derive $\boldsymbol{\mu}$ and $\boldsymbol{\Sigma}$.

(b) Write out the ANOVA table and define the matrices $\mathbf{A}_m$ used in the sums of squares $\mathbf{Y}'\mathbf{A}_m\mathbf{Y}$ for $m = 1, \ldots, 9$ where $\mathbf{A}_9$ is the matrix corresponding to the sum of squares total.

(c) Derive the distributions of $\mathbf{Y}'\mathbf{A}_m\mathbf{Y}$ for $m = 1, \ldots, 8$.

(d) Calculate all the expected mean squares.

(e) Construct all "appropriate" F statistics and explicitly define the hypothesis being tested in each case. Prove that all the statistics constructed above have F distributions.

7. A pump manufacturer wanted to evaluate how well his assembled pumps performed. He ran the following experiment. He randomly selected 10 people to assemble his pumps, randomly dividing them into two groups of 5. He then trained both groups to assemble pumps, but one group received more rigorous instruction. The two groups were therefore identified to be of two skill levels. Each person then assembled two pumps, one pump by one method of assembly and a second pump by a second method of assembly. Each assembled pump then pumped water for a fixed amount of time and then repeated the operation later for the same length of time. The amount of water pumped in each time period was recorded. The order of the operation (first time period or second) was also recorded. Let Y_{ijkl} be a random variable representing the amount of water pumped during the lth time period or order, on a pump assembled by the kth method and the jth person in the ith skill level for $i = 1, 2$, $j = 1, 2, 3, 4, 5$, $k - 1, 2$, and $l = 1, 2$. Assume the 40×1 random vector $\mathbf{Y} = (Y_{1111}, Y_{1112}, Y_{1121}, Y_{1122}, \ldots, Y_{2511}, Y_{2512}, Y_{2521}, Y_{2522})' \sim \mathrm{N}_{40}(\boldsymbol{\mu}, \Sigma)$. A model for this problem is

$$Y_{ijkl} = \mu_{ikl} + P(S)_{(i)j} + PM(S)_{(i)jk} + PO(SM)_{(i)j(k)l}$$

where μ_{ikl} are constants representing the average effect of the lth order in the kth method with the ith skill level, $P(S)_{(i)j}$, $PM(S)_{(i)jk}$, and $PO(SM)_{(i)j(k)l}$ are random variables representing the effects of people nested in skill levels, the interaction of methods and people nested in skill levels, and the interaction of order and people nested in skill levels and methods.

(a) Derive $\boldsymbol{\mu}$ and $\boldsymbol{\Sigma}$.

(b) Write out the ANOVA table and define the matrices $\mathbf{A}_m$ used in the sums of squares $\mathbf{Y}'\mathbf{A}_m\mathbf{Y}$ for $m = 1, \ldots, 12$ where $\mathbf{Y}'\mathbf{A}_{12}\mathbf{Y}$ is the sum of squares total.

(c) Derive the distributions of $\mathbf{Y}'\mathbf{A}_m\mathbf{Y}$ for $m = 1, \ldots, 11$.

(d) Calculate all the expected mean squares.

(e) Construct all "appropriate" F statistics and explicitly define the hypothesis being tested in each case. Prove that all the statistics constructed above have F distributions.

8. Consider Exercise 7.

(a) Calculate the standard error of $\bar{Y}_{1\ldots} - \bar{Y}_{2\ldots}$ where $\bar{Y}_{i\ldots} = \sum_{j=1}^{5} \sum_{k=1}^{2} \sum_{l=1}^{2} Y_{ijkl}/20$ for $i = 1, 2$.

(b) Find an unbiased estimator of $\bar{\mu}_{1.12} - \bar{\mu}_{2.12}$ where $\bar{\mu}_{i.kl} = \sum_{j=1}^{5} \mu_{ijkl}/5$ and calculate the standard error of the estimator.

9. Prove that the necessary and sufficient conditions of Bhat's Lemma 3.4.1 are satisfied in the following situation. Consider any complete, balanced factorial experiment with n observations where the $n \times 1$ random vector $\mathbf{Y} \sim \mathrm{N}_n(\boldsymbol{\mu}, \boldsymbol{\Sigma})$. The covariance matrix algorithm rules $\Sigma 1$, $\Sigma 2$, $\Sigma 2.1$, $\Sigma 2.2$, and $\Sigma 2.3$ are used to derive $\boldsymbol{\Sigma}$. The sum of squares algorithm rules A1, A2, A2.1, A2.2, and A2.3 are used to derive the k sum of squares matrices $\mathbf{A}_1 \ldots, \mathbf{A}_k$ with $\sum_{i=1}^{k} \mathbf{A}_i = \mathbf{I}_n$.

5 Least-Squares Regression

In this chapter the least-squares estimation procedure is examined. The topic is introduced through a regression example. Later in the chapter the regression model format is applied to a broad class of problems, including factorial experiments.

5.1 ORDINARY LEAST-SQUARES ESTIMATION

We begin with a simple example. An engineer wants to relate the fuel consumption of a new type of automobile to the speed of the vehicle and the grade of the road traveled. He has a fleet of n vehicles. Each vehicl0e is assigned to operate at a constant speed (in miles per hour) on a specific grade (in percent grade) and the fuel consumption (in ml/sec) is recorded. The engineer believes that the expected fuel consumption is a linear function of the speed of the vehicle and the speed of the vehicle times the grade of the road. Let Y_i be a random variable that represents the observed fuel consumption of the ith vehicle, operating at a fixed speed, on a road with a constant grade. Let x_{i1} represent the speed of the ith vehicle and let x_{i2} represent the speed times the grade of the ith vehicle. The expected fuel consumption of the ith vehicle can be represented by

$$E(Y_i) = \beta_0 + \beta_1 x_{i1} + \beta_2 x_{i2}$$

where β_0, β_1, and β_2 are unknown parameters. Due to qualities intrinsic to each vehicle, the observed fuel consumptions differ somewhat from the expected fuel consumptions. Therefore, the observed fuel consumption of the ith vehicle is represented by

$$Y_i = E(Y_i) + E_i$$

or

$$Y_i = \beta_0 + \beta_1 x_{i1} + \beta_2 x_{i2} + E_i$$

where E_i is a random variable representing the difference between the observed fuel consumption and the expected fuel consumption of the ith vehicle. An example data set for this fuel, speed, grade experiment is provided in Table 5.1.1. In a more general setting consider a problem where the expected value of a random variable Y_i is assumed to be a linear combination of $p - 1$ different variables $x_{i1}, x_{i2}, \ldots, x_{i,p-1}$. That is,

$$E(Y_i) = \beta_0 + \beta_1 x_{i1} + \cdots + \beta_{p-1} x_{i,p-1}.$$

Adding a component of error, E_i, to represent the difference between the observed value of Y_i and the expected value of Y_i we obtain

$$Y_i = \beta_0 + \beta_1 x_{i1} + \cdots + \beta_{p-1} x_{i,p-1} + E_i.$$

By taking expectations on the right and left sides of the preceding two equations, we obtain $E(E_i) = 0$ for all $i = 1, \ldots, n$.

Table 5.1.1
Fuel, Speed, Grade Data Set

i	Fuel Y_i	Speed x_{i1}	Grade	Speed $\times$ Grade x_{i2}
1	1.7	20	0	0
2	2.0	20	0	0
3	1.9	20	0	0
4	1.6	20	0	0
5	3.2	20	6	120
6	2.0	50	0	0
7	2.5	50	0	0
8	5.4	50	6	300
9	5.7	50	6	300
10	5.1	50	6	300

The model just discussed can be expressed in matrix form by noting that

$$
\begin{aligned}
Y_1 &= \beta_0 + \beta_1 x_{11} + \cdots + \beta_{p-1} x_{1,p-1} + E_1 \\
Y_2 &= \beta_0 + \beta_1 x_{21} + \cdots + \beta_{p-1} x_{2,p-1} + E_2 \\
&\vdots \qquad\qquad \vdots \\
Y_n &= \beta_0 + \beta_1 x_{n1} + \cdots + \beta_{p-1} x_{n,p-1} + E_n
\end{aligned}
$$

or

$$\mathbf{Y} = \mathbf{X}\boldsymbol{\beta} + \mathbf{E}$$

where the $n \times 1$ random vector $\mathbf{Y} = (Y_1, \ldots, Y_n)'$, the $p \times 1$ vector $\boldsymbol{\beta} = (\beta_0, \beta_1 \ldots, \beta_{p-1})'$, the $n \times 1$ random vector $\mathbf{E} = (E_1, \ldots, E_n)'$ and the $n \times p$ matrix

$$\mathbf{X} = \begin{bmatrix} 1 & x_{11} & \cdots & x_{1,p-1} \\ 1 & x_{21} & \cdots & x_{2,p-1} \\ \vdots & \vdots & & \vdots \\ 1 & x_{n1} & \cdots & x_{n,p-1} \end{bmatrix}.$$

Furthermore, $\mathrm{E}(E_i) = 0$ for all $i = 1, \ldots, n$ implies $\mathrm{E}(\mathbf{E}) = \mathbf{0}_{n\times 1}$. Therefore $\mathrm{E}(\mathbf{Y}) = \mathbf{X}\boldsymbol{\beta}$. For the present, assume that the E_i's are independent, identically distributed random variables where $\mathrm{var}(E_i) = \sigma^2$ for all $i = 1, \ldots, n$. Since the E_i's are independent, $\mathrm{cov}(E_i, E_j) = 0$ for all $i \neq j$. Therefore, the covariance matrix of $\mathbf{E}$ is given by $\boldsymbol{\Sigma} = \mathrm{cov}(\mathbf{E}) = \sigma^2 \mathbf{I}_n$. In later sections of this chapter more complicated error structures are considered.

Note that $\boldsymbol{\Sigma}$ has been used to represent the covariance matrix of the $n \times 1$ random error vector $\mathbf{E}$. However, $\boldsymbol{\Sigma}$ is also the covariance matrix of the $n \times 1$ random vector $\mathbf{Y}$ since

$$
\begin{aligned}
\mathrm{cov}(\mathbf{Y}) &= \mathrm{E}[(\mathbf{Y} - \mathbf{X}\boldsymbol{\beta})\,(\mathbf{Y} - \mathbf{X}\boldsymbol{\beta})'] \\
&= \mathrm{E}[\mathbf{E}\mathbf{E}'] \\
&= \boldsymbol{\Sigma}.
\end{aligned}
$$

Since the x_{ij} values are known for $i = 1, \ldots, n$ and $j = 1, \ldots, p-1$, $\bar{x}_{.j} = \sum_{i=1}^{n} x_{ij}/n$ can be calculated for any j. Therefore, the preceding model can be equivalently written as

$$Y_i = \beta_0^* + \beta_1(x_{i1} - \bar{x}_{.1}) + \cdots + \beta_{p-1}(x_{i,p-1} - \bar{x}_{.p-1}) + E_i$$

where $\beta_0 = \beta_0^* - \beta_1 \bar{x}_{.1} - \cdots - \beta_{p-1} \bar{x}_{p-1}$. In matrix form

$$\mathbf{Y} = \mathbf{X}^* \boldsymbol{\beta}^* + \mathbf{E}$$

where

$$\boldsymbol{\beta}^* = (\beta_0^*, \beta_1, \ldots, \beta_{p-1})',$$

$$\mathbf{X}^* = \begin{bmatrix} 1 & x_{11} - \bar{x}_{.1} & \cdots & x_{1,p-1} - \bar{x}_{.p-1} \\ 1 & x_{21} - x_{.1} & \cdots & x_{2,p-1} - \bar{x}_{.p-1} \\ \vdots & \vdots & & \vdots \\ 1 & x_{n1} - \bar{x}_{.1} & \cdots & x_{n,p-1} - \bar{x}_{.p-1} \end{bmatrix} = [\mathbf{1}_n | \mathbf{X}_c]$$

and $\mathbf{X}_c$ is an $n \times (p-1)$ matrix such that $\mathbf{1}_n'\mathbf{X}_c = \mathbf{0}_{1\times(p-1)}$. This later model form is called a *centered model*. Without loss of generality, a model can always be assumed to be centered since any model $\mathbf{Y} = \mathbf{X}\boldsymbol{\beta} + \mathbf{E}$ can be written as $\mathbf{Y} = \mathbf{X}^*\boldsymbol{\beta}^* + \mathbf{E}$. The asterisks on the centered model are subsequently dropped since $\mathbf{X}$ can always be considered a centered matrix if necessary.

In the next example, the 10×3 centered matrix $\mathbf{X}$ is derived for the example data set.

Example 5.1.1 For the example data given in Table 5.1.1, the average speed is $\bar{x}_{.1} = [5(20) + 5(50)]/10 = 35$ and the average value of speed $\times$ grade is $\bar{x}_{.2} = [6(0) + (1)120 + 3(300)]/10 = 102$. Therefore, the 10×3 centered matrix $\mathbf{X} = (\mathbf{1}_{10}|\mathbf{X}_c)$ where

$$\mathbf{X}_c = \begin{bmatrix} (-15, & -102) \otimes \mathbf{1}_4 \\ (-15, & 18) \otimes 1 \\ (\ 15, & -102) \otimes \mathbf{1}_2 \\ (\ 15, & 198) \otimes \mathbf{1}_3 \end{bmatrix}.$$

The main objective to this section is to develop a procedure to estimate the p unknown parameters $\beta_1, \beta_1, \ldots, \beta_{p-1}$. One method that provides such estimators is called the *ordinary least-squares procedure*. The ordinary least-squares estimators of $\beta_0, \beta_1, \ldots, \beta_{p-1}$ are obtained by minimizing the quadratic form Q with respect to the $p \times 1$ vector $\boldsymbol{\beta}$ where

$$\begin{aligned} Q &= (\mathbf{Y} - \mathbf{X}\boldsymbol{\beta})'(\mathbf{Y} - \mathbf{X}\boldsymbol{\beta}) \\ &= \mathbf{Y}'\mathbf{Y} - 2\boldsymbol{\beta}'\mathbf{X}'\mathbf{Y} + \boldsymbol{\beta}'\mathbf{X}'\mathbf{X}\boldsymbol{\beta}. \end{aligned}$$

To derive the estimators, take the derivative of Q with respect to the vector $\boldsymbol{\beta}$, set the resulting expression equal to zero, and solve for $\boldsymbol{\beta}$. That is,

$$\partial Q/\partial \boldsymbol{\beta} = -2\mathbf{X}'\mathbf{Y} + 2\mathbf{X}'\mathbf{X}\boldsymbol{\beta} = \mathbf{0}_{p\times 1}$$

or $\mathbf{X}'\mathbf{X}\boldsymbol{\beta} = \mathbf{X}'\mathbf{Y}$. If $\mathbf{X}'\mathbf{X}$ is nonsingular [i.e., rank $(\mathbf{X}'\mathbf{X}) = p$] then the least-squares estimator of $\boldsymbol{\beta}$ is

$$\hat{\boldsymbol{\beta}} = (\mathbf{X}'\mathbf{X})^{-1}\mathbf{X}'\mathbf{Y}.$$

Thus, the ordinary least-squares estimator of the $p \times 1$ vector β is a set of linear transformations of the random vector $\mathbf{Y}$ where $(\mathbf{X}'\mathbf{X})^{-1}\mathbf{X}'$ is the $p \times n$ transformation matrix. If $\mathrm{E}(\mathbf{Y}) = \mathbf{X}\beta$, $\hat{\beta}$ is an unbiased estimator of β since

$$\begin{aligned} \mathrm{E}(\hat{\beta}) &= (\mathbf{X}'\mathbf{X})^{-1}\mathbf{X}'\mathrm{E}(\mathbf{Y}) \\ &= (\mathbf{X}'\mathbf{X})^{-1}\mathbf{X}'\mathbf{X}\beta = \beta. \end{aligned}$$

Furthermore, the $p \times p$ covariance matrix of $\hat{\beta}$ is given by

$$\begin{aligned} \mathrm{cov}(\hat{\beta}) &= (\mathbf{X}'\mathbf{X})^{-1}\mathbf{X}'(\sigma^2\mathbf{I}_n)\mathbf{X}(\mathbf{X}'\mathbf{X})^{-1} \\ &= \sigma^2(\mathbf{X}'\mathbf{X})^{-1}. \end{aligned}$$

when $\mathbf{\Sigma} = \mathrm{cov}(\mathbf{Y}) = \sigma^2\mathbf{I}_n$.

It is also generally of interest to estimate the unknown parameter σ^2. The quadratic form

$$\hat{\sigma}^2 = \mathbf{Y}'[\mathbf{I}_n - \mathbf{X}(\mathbf{X}'\mathbf{X})^{-1}\mathbf{X}']\mathbf{Y}/(n-p).$$

provides an unbiased estimator of σ^2 when $\Sigma = \sigma^2\mathbf{I}_n$ since

$$\begin{aligned} \mathrm{E}(\hat{\sigma}^2) &= \mathrm{E}\{\mathbf{Y}'[\mathbf{I}_n - \mathbf{X}(\mathbf{X}'\mathbf{X})^{-1}\mathbf{X}']\mathbf{Y}/(n-p)\} \\ &= \{\mathrm{tr}[(\mathbf{I}_n - \mathbf{X}(\mathbf{X}'\mathbf{X})^{-1}\mathbf{X}')(\sigma^2\mathbf{I}_n)] \\ &\quad + \beta'\mathbf{X}'(\mathbf{I}_n - \mathbf{X}(\mathbf{X}'\mathbf{X})^{-1}\mathbf{X}')\mathbf{X}\beta\}/(n-p) \\ &= \{\sigma^2(n-p) + (\beta'\mathbf{X}'\mathbf{X}\beta \\ &\quad - \beta'\mathbf{X}'\mathbf{X}(\mathbf{X}'\mathbf{X})^{-1}\mathbf{X}'\mathbf{X}\beta)\}/(n-p) \\ &= \sigma^2. \end{aligned}$$

In the next example the ordinary least-squares estimates of β and σ^2 are calculated for the example data set. The IML procedure in SAS has been used to generate all the example calculations in this chapter. The PROC IML programs and outputs for this chapter are presented in Appendix 1.

Example 5.1.2 For the example data given in Table 5.1.1, the least-squares estimate of β is given by

$$\hat{\beta} = (\mathbf{X}'\mathbf{X})^{-1}\mathbf{X}'\mathbf{Y} = (3.11, 0.01348, 0.01061)'.$$

Therefore, the prediction equation is

$$\hat{Y} = 3.11 + 0.01348(x_1 - 35) + 0.01061(x_2 - 102)$$

or

$$\hat{Y} = 1.556 + 0.01348x_1 + 0.01061x_2.$$

The ordinary least-squares estimator of σ^2 is given by

$$\hat{\sigma}^2 = \mathbf{Y}'[\mathbf{I}_n - \mathbf{X}(\mathbf{X}'\mathbf{X})^{-1}\mathbf{X}']\mathbf{Y}/7 = 0.05988.$$

5.2 BEST LINEAR UNBIASED ESTIMATORS

In many problems it is of interest to estimate linear combinations of $\beta_0, \ldots, \beta_{p-1}$, say, $\mathbf{t}'\boldsymbol{\beta}$, where $\mathbf{t}$ is any nonzero $p \times 1$ vector of known constants. In the next definition the "best" linear unbiased estimator of $\mathbf{t}'\boldsymbol{\beta}$ is identified.

Definition 5.2.1 *Best Linear Unbiased Estimator (BLUE) of* $\mathbf{t}'\boldsymbol{\beta}$: The best linear unbiased estimator of $\mathbf{t}'\boldsymbol{\beta}$ is

(i) a linear function of the observed vector $\mathbf{Y}$, that is, a function of the form $\mathbf{a}'\mathbf{Y} + a_0$ where $\mathbf{a}$ is an $n \times 1$ vector of constants and a_0 is a scalar and

(ii) the unbiased estimator of $\mathbf{t}'\boldsymbol{\beta}$ with the smallest variance.

In the next important theorem $\mathbf{t}'\hat{\boldsymbol{\beta}} = \mathbf{t}'(\mathbf{X}'\mathbf{X})^{-1}\mathbf{X}'\mathbf{Y}$ is shown to be the BLUE of $\mathbf{t}'\boldsymbol{\beta}$ when $\mathrm{E}(\mathbf{E}) = \mathbf{0}$ and $\mathrm{cov}(\mathbf{E}) = \sigma^2\mathbf{I}_n$. The theorem is called the Gauss–Markov theorem.

Theorem 5.2.1 *Let* $\mathbf{Y} = \mathbf{X}\beta + \mathbf{E}$ *where* $\mathrm{E}(\mathbf{E}) = \mathbf{0}$ *and cov*$(\mathbf{E}) = \sigma^2\mathbf{I}_n$*. Then the least-squares estimator of* $\mathbf{t}'\boldsymbol{\beta}$ *is given by* $\mathbf{t}'\hat{\boldsymbol{\beta}} = \mathbf{t}'(\mathbf{X}'\mathbf{X})^{-1}\mathbf{X}'\mathbf{Y}$ *and* $\mathbf{t}'\hat{\boldsymbol{\beta}}$ *is the BLUE of* $\mathbf{t}'\boldsymbol{\beta}$.

Proof: First, the least-squares estimator of $\mathbf{t}'\boldsymbol{\beta}$ is shown to be $\mathbf{t}'\hat{\boldsymbol{\beta}}$. Let $\mathbf{T}$ be a $p \times p$ nonsingular matrix such that $\mathbf{T} = (\mathbf{t}|\mathbf{T}_0)$ where $\mathbf{t}$ is a $p \times 1$ vector and $\mathbf{T}_0$ is a $p \times (p-1)$ matrix. If $\mathbf{R} = \mathbf{T}'^{-1}$ then

$$\begin{aligned}
\mathbf{Y} &= \mathbf{X}\beta + \mathbf{E} \\
&= \mathbf{XRT}'\beta + \mathbf{E} \\
&= \mathbf{U}\omega + \mathbf{E}
\end{aligned}$$

where $\mathbf{U} = \mathbf{XR}$ and

$$\omega = \mathbf{T}'\beta = \begin{bmatrix} \mathbf{t}'\boldsymbol{\beta} \\ \mathbf{T}_0'\beta \end{bmatrix}.$$

The least-squares estimate of ω is given by

$$\begin{aligned}
\hat{\omega} &= (\mathbf{U}'\mathbf{U})^{-1}\mathbf{U}'\mathbf{Y} \\
&= (\mathbf{R}'\mathbf{X}'\mathbf{XR})^{-1}\mathbf{R}'\mathbf{X}'\mathbf{Y} \\
&= \mathbf{R}^{-1}(\mathbf{X}'\mathbf{X})^{-1}\mathbf{R}'^{-1}\mathbf{R}'\mathbf{X}'\mathbf{Y}
\end{aligned}$$

$$= \mathbf{T}'(\mathbf{X}'\mathbf{X})^{-1}\mathbf{X}'\mathbf{Y} = \mathbf{T}'\hat{\boldsymbol{\beta}}$$
$$= \begin{bmatrix} \mathbf{t}'\hat{\boldsymbol{\beta}} \\ \mathbf{T}_0'\hat{\boldsymbol{\beta}} \end{bmatrix}.$$

Therefore, $\mathbf{t}'\hat{\boldsymbol{\beta}}$ is the least-squares estimator of $\mathbf{t}'\boldsymbol{\beta}$. Next, $\mathbf{t}'\hat{\boldsymbol{\beta}}$ is shown to be the BLUE of $\mathbf{t}'\boldsymbol{\beta}$. Linear estimators of $\mathbf{t}'\boldsymbol{\beta}$ take the form $\mathbf{a}'\mathbf{Y}+a_0$. Since $\mathbf{t}'(\mathbf{X}'\mathbf{X})^{-1}\mathbf{X}'$ is known, without loss of generality, let $\mathbf{a}' = \mathbf{t}'(\mathbf{X}'\mathbf{X})^{-1}\mathbf{X}' + \mathbf{b}'$. Then linear unbiased estimators of $\mathbf{t}'\boldsymbol{\beta}$ satisfy the relationship

$$\begin{aligned} \mathbf{t}'\boldsymbol{\beta} = \mathrm{E}(\mathbf{a}'\mathbf{Y} + a_0) &= \mathrm{E}(\mathbf{t}'(\mathbf{X}'\mathbf{X})^{-1}\mathbf{X}'\mathbf{Y} + \mathbf{b}'\mathbf{Y} + a_0) \\ &= \mathbf{t}'(\mathbf{X}'\mathbf{X})^{-1}\mathbf{X}'\mathbf{X}\boldsymbol{\beta} + \mathbf{b}'\mathbf{X}\boldsymbol{\beta} + a_0 \\ &= \mathbf{t}'\boldsymbol{\beta} + \mathbf{b}'\mathbf{X}\boldsymbol{\beta} + a_0. \end{aligned}$$

Therefore, in the class of linear unbiased estimators $\mathbf{b}'\mathbf{X}\boldsymbol{\beta} + a_0 = 0$ for all $\boldsymbol{\beta}$. But for this expression to hold for all $\boldsymbol{\beta}$, $\mathbf{b}'\mathbf{X} = \mathbf{0}_{1\times p}$ and $a_0 = 0$. Now calculate and minimize the variance of the estimator $\mathbf{a}'\mathbf{Y} + a_0$ within the class of unbiased estimators of $\mathbf{t}'\boldsymbol{\beta}$, (i.e., when $\mathbf{b}'\mathbf{X} = \mathbf{0}_{1\times p}$ and $a_0 = 0$).

$$\begin{aligned} \mathrm{var}(\mathbf{a}'\mathbf{Y} + a_0) &= \mathrm{var}(\mathbf{a}'\mathbf{Y}) \\ &= \mathrm{var}(\mathbf{t}'(\mathbf{X}'\mathbf{X})^{-1}\mathbf{X}'\mathbf{Y} + \mathbf{b}'\mathbf{Y}) \\ &= \sigma^2(\mathbf{t}'(\mathbf{X}'\mathbf{X})^{-1}\mathbf{t} + \mathbf{b}'\mathbf{b}). \end{aligned}$$

But σ^2 and $\mathbf{t}'(\mathbf{X}'\mathbf{X})^{-1}\mathbf{t}$ are constants. Therefore, $\mathrm{var}(\mathbf{a}'\mathbf{Y}+a_0)$ is minimized when $\mathbf{b}'\mathbf{b} = 0$ or when $\mathbf{b} = \mathbf{0}_{p\times 1}$. Therefore, the BLUE of $\mathbf{t}'\boldsymbol{\beta}$ has variance $\sigma^2\mathbf{t}'(\mathbf{X}'\mathbf{X})^{-1}\mathbf{t}$. But $\mathbf{t}'\hat{\boldsymbol{\beta}}$ is a linear unbiased estimator of $\mathbf{t}'\boldsymbol{\beta}$ with variance $\sigma^2\mathbf{t}'(\mathbf{X}'\mathbf{X})^{-1}\mathbf{t}$. Therefore, $\mathbf{t}'\hat{\boldsymbol{\beta}}$ is the BLUE of $\mathbf{t}'\boldsymbol{\beta}$. ■

Example 5.2.1 Consider the example data set given in Table 5.1.1. By the Gauss–Markov theorem, the best linear unbiased estimate of $\beta_1 - \beta_2$ is $\mathbf{t}'\hat{\boldsymbol{\beta}} = (0, 1, -1)(3.11, 0.01348, 0.01061)' = 0.00287$.

5.3 ANOVA TABLE FOR THE ORDINARY LEAST-SQUARES REGRESSION FUNCTION

An ANOVA table can be constructed that partitions the total sum of squares into the sum of squares due to the overall mean, the sum of squares due to $\beta_1, \ldots, \beta_{p-1}$, and the sum of squares due to the residual. The ANOVA table for this model is given in Table 5.3.1. The sum of squares under the column "SS" can be applied to any form of the $n \times p$ matrix $\mathbf{X}$. The sum of squares under the column "SS Centered" can be applied to centered matrices $\mathbf{X} = [\mathbf{1}_n|\mathbf{X}_c]$ where $\mathbf{1}_n'\mathbf{X}_c = \mathbf{0}_{1\times(p-1)}$.

Table 5.3.1
Ordinary Least-Squares ANOVA Table

Source	df	SS	SS Centered
Overall mean	1	$\mathbf{Y}'\frac{1}{n}\mathbf{J}_n\mathbf{Y}$	$=\mathbf{Y}'\frac{1}{n}\mathbf{J}_n\mathbf{Y}$
Regression $(\beta_1,\ldots,\beta_{p-1})$	$p-1$	$\mathbf{Y}'[\mathbf{X}(\mathbf{X}'\mathbf{X})^{-1}\mathbf{X}'-\frac{1}{n}\mathbf{J}_n]\mathbf{Y}$	$=\mathbf{Y}'\mathbf{X}_c(\mathbf{X}_c'\mathbf{X}_c)^{-1}\mathbf{X}_c'\mathbf{Y}$
Residual	$n-p$	$\mathbf{Y}'[\mathbf{I}_n-\mathbf{X}(\mathbf{X}'\mathbf{X})^{-1}\mathbf{X}']\mathbf{Y}$	$=\mathbf{Y}'[\mathbf{I}_n-\frac{1}{n}\mathbf{J}_n$ $-\mathbf{X}_c(\mathbf{X}_c'\mathbf{X})^{-1}\mathbf{X}_c']\mathbf{Y}$
Total	n	$\mathbf{Y}'\mathbf{Y}$	

The expected mean squares for each effect are calculated using Theorem 1.3.2:

$$\begin{aligned}
\text{EMS (overall mean)} &= \mathrm{E}\left(\mathbf{Y}'\frac{1}{n}\mathbf{J}_n\mathbf{Y}\right)\\
&= \mathrm{tr}[(\sigma^2/n)\mathbf{J}_n] + \boldsymbol{\beta}'\mathbf{X}'\frac{1}{n}\mathbf{J}_n\mathbf{X}\boldsymbol{\beta}\\
&= \sigma^2 + \boldsymbol{\beta}'[\mathbf{1}_n|\mathbf{X}_c]'\frac{1}{n}\mathbf{J}_n[\mathbf{1}_n|\mathbf{X}_c]\boldsymbol{\beta}\\
&= \sigma^2 + \boldsymbol{\beta}'\begin{bmatrix}\mathbf{1}_n'\mathbf{1}_n\\ \mathbf{X}_c'\mathbf{1}_n\end{bmatrix}[\mathbf{1}_n'\mathbf{1}_n|\mathbf{1}_n'\mathbf{X}_c]\boldsymbol{\beta}/n\\
&= \sigma^2 + \boldsymbol{\beta}'\begin{bmatrix}n^2 & \mathbf{0}\\ \mathbf{0} & \mathbf{0}\end{bmatrix}\boldsymbol{\beta}/n\\
&= \sigma^2 + n\beta_0^{*2}
\end{aligned}$$

$$\begin{aligned}
\text{EMS (Regression)} &= E[\mathbf{Y}'\mathbf{X}_c(\mathbf{X}_c'\mathbf{X}_c)^{-1}\mathbf{X}_c'\mathbf{Y}/(p-1)]\\
&= \{\mathrm{tr}[\sigma^2\mathbf{X}_c(\mathbf{X}_c'\mathbf{X}_c)^{-1}\mathbf{X}_c']\\
&\quad + \boldsymbol{\beta}'\mathbf{X}'\mathbf{X}_c(\mathbf{X}_c'\mathbf{X}_c)^{-1}\mathbf{X}_c'\mathbf{X}\boldsymbol{\beta}\}/(p-1)\\
&= \{\sigma^2(p-1)\\
&\quad + \boldsymbol{\beta}'\begin{bmatrix}\mathbf{1}_n'\\ \mathbf{X}_c'\end{bmatrix}\mathbf{X}_c(\mathbf{X}_c'\mathbf{X}_c)^{-1}\mathbf{X}_c'[\mathbf{1}_n|\mathbf{X}_c]\boldsymbol{\beta}\}/(p-1)\\
&= \sigma^2 + (\beta_1,\ldots,\beta_{p-1})\mathbf{X}_c'\mathbf{X}_c(\beta_1,\ldots,\beta_{p-1})'/(p-1)
\end{aligned}$$

and EMS (residual) $= \mathrm{E}(\hat{\sigma}^2) = \sigma^2$ as derived in Section 5.1. The ANOVA table for the fuel, speed, grade data set is provided in the next example.

Example 5.3.1 The ANOVA table for the data in Table 5.1.1 is given here:

Source	df	SS
Overall mean	1	96.721
Regression (β_1, β_2)	2	24.070
Residual	7	0.419
Total	10	121.210

5.4 WEIGHTED LEAST-SQUARES REGRESSION

In the first three sections of this chapter the model was confined to $\mathbf{Y} = \mathbf{X}\beta + \mathbf{E}$ where $\mathrm{E}(\mathbf{E}) = \mathbf{0}$ and $\mathrm{cov}(\mathbf{E}) = \sigma^2\mathbf{I}_n$. In this section, the model $\mathbf{Y} = \mathbf{X}\beta + \mathbf{E}$ is considered when $\mathrm{E}(\mathbf{E}) = \mathbf{0}$, $\mathrm{cov}(\mathbf{E}) = \sigma^2\mathbf{V}$, and $\mathbf{V}$ is an $n \times n$ symmetric, positive definite matrix of known constants. Because $\mathbf{V}$ is positive definite, there exists an $n \times n$ nonsingular matrix $\mathbf{T}$ such that $\mathbf{V} = \mathbf{T}\mathbf{T}'$. Premultiplying both sides of the model $\mathbf{Y} = \mathbf{X}\beta + \mathbf{E}$ by $\mathbf{T}^{-1}$ we obtain

$$\mathbf{T}^{-1}\mathbf{Y} = \mathbf{T}^{-1}\mathbf{X}\boldsymbol{\beta} + \mathbf{T}^{-1}\mathbf{E}$$

$$\mathbf{Y}_{\mathrm{w}} = \mathbf{X}_{\mathrm{w}}\boldsymbol{\beta} + \mathbf{E}_{\mathrm{w}}$$

where $\mathbf{Y}_{\mathrm{w}} = \mathbf{T}^{-1}\mathbf{Y}, \mathbf{X}_{\mathrm{w}} = \mathbf{T}^{-1}\mathbf{X}$, and $\mathbf{E}_{\mathrm{w}} = \mathbf{T}^{-1}\mathbf{E}$. Therefore, $\mathrm{E}(\mathbf{E}_{\mathrm{w}}) = \mathbf{T}^{-1}\mathrm{E}(\mathbf{E}) = \mathbf{0}_{p\times 1}$ and $\mathrm{cov}(\mathbf{E}_{\mathrm{w}}) = \mathrm{cov}(\mathbf{T}^{-1}\mathbf{E}) = \mathbf{T}^{-1}(\sigma^2\mathbf{V})\mathbf{T}^{-1\prime} = \sigma^2\mathbf{I}_n$. The weighted least-squares estimators of β and σ^2 are derived using the ordinary least-squares estimator formulas with the model $\mathbf{Y}_{\mathrm{w}} = \mathbf{X}_{\mathrm{w}}\boldsymbol{\beta} + \mathbf{E}_{\mathrm{w}}$. That is, the weighted least-squares estimators of β and σ^2 are given by

$$\begin{aligned}\hat{\beta}_{\mathrm{w}} &= (\mathbf{X}_{\mathrm{w}}'\mathbf{X}_{\mathrm{w}})^{-1}\mathbf{X}_{\mathrm{w}}'\mathbf{Y}_{\mathrm{w}} \\ &= (\mathbf{X}'\mathbf{T}^{-1\prime}\mathbf{T}^{-1}\mathbf{X})^{-1}\mathbf{X}'\mathbf{T}^{-1\prime}\mathbf{T}^{-1}\mathbf{Y} \\ &= (\mathbf{X}'\mathbf{V}^{-1}\mathbf{X})^{-1}\mathbf{X}'\mathbf{V}^{-1}\mathbf{Y}\end{aligned}$$

and

$$\begin{aligned}\hat{\sigma}_{\mathrm{w}}^2 &= (\mathbf{Y}_{\mathrm{w}} - \mathbf{X}_{\mathrm{w}}\hat{\beta}_{\mathrm{w}})'(\mathbf{Y}_{\mathrm{w}} - \mathbf{X}_{\mathrm{w}}\hat{\beta}_{\mathrm{w}})/(n-p) \\ &= [\mathbf{Y}'(\mathbf{V}^{-1} - \mathbf{V}^{-1}\mathbf{X}(\mathbf{X}'\mathbf{V}^{-1}\mathbf{X})^{-1}\mathbf{X}'\mathbf{V}^{-1})\mathbf{Y}]/(n-p).\end{aligned}$$

The Gauss–Markov theorem can also be generalized for the model $\mathbf{Y} = \mathbf{X}\boldsymbol{\beta} + \mathbf{E}$ where $\mathrm{E}(\mathbf{E}) = \mathbf{0}$ and $\mathrm{cov}(\mathbf{E}) = \sigma^2\mathbf{V}$. For this model, the weighted least-squares estimator of $\mathbf{t}'\boldsymbol{\beta}$ is given by $\mathbf{t}'\hat{\boldsymbol{\beta}}_{\mathrm{w}} = \mathbf{t}'(\mathbf{X}'\mathbf{V}^{-1}\mathbf{X})^{-1}\mathbf{X}'\mathbf{V}^{-1}\mathbf{Y}$ and $\mathbf{t}'\hat{\boldsymbol{\beta}}_{\mathrm{w}}$ is the BLUE of $\mathbf{t}'\boldsymbol{\beta}$. The proof is left to the reader.

Table 5.4.1
Weighted Least-Squares ANOVA Table

Source	df	SS
Overall mean	1	$\mathbf{Y}'\mathbf{V}^{-1}\mathbf{1}_n(\mathbf{1}_n'\mathbf{V}^{-1}\mathbf{1}_n)^{-1}\mathbf{1}_n'\mathbf{V}^{-1}\mathbf{Y}$
Regression ($\beta_1, \ldots, \beta_{p-1}$)	$p-1$	$\mathbf{Y}'\mathbf{V}^{-1}[\mathbf{X}(\mathbf{X}'\mathbf{V}^{-1}\mathbf{X})^{-1}\mathbf{X}' - \mathbf{1}_n(\mathbf{1}_n'\mathbf{V}^{-1}\mathbf{1}_n)^{-1}\mathbf{1}_n']\mathbf{V}^{-1}\mathbf{Y}$
Residual	$n-p$	$\mathbf{Y}'[\mathbf{V}^{-1} - \mathbf{V}^{-1}\mathbf{X}(\mathbf{X}'\mathbf{V}^{-1}\mathbf{X})^{-1}\mathbf{X}'\mathbf{V}^{-1}]\mathbf{Y}$
Total	n	$\mathbf{Y}'\mathbf{V}^{-1}\mathbf{Y}$

The ANOVA table for weighted least-squares regression functions can be constructed using Table 5.3.1 and substituting $\mathbf{T}^{-1}\mathbf{X}$ for $\mathbf{X}$ and $\mathbf{T}^{-1}\mathbf{Y}$ for $\mathbf{Y}$. The weighted least-squares ANOVA table is provided in Table 5.4.1.

In the next example a weighted least-squares ANOVA table is derived for the fuel, speed, grade data set.

Example 5.4.1 Consider the data in Table 5.1.1. Suppose the fuel consumption observations are independent but the variance of the observations at speed 50 mph is twice the variance of observations at speed 20 mph. Then $\text{cov}(\mathbf{E}) = \sigma^2\mathbf{V}$ where the 10×10 matrix

$$\mathbf{V} = \begin{bmatrix} 1 & 0 \\ 0 & 2 \end{bmatrix} \otimes \mathbf{I}_5.$$

The weighted least-squares estimates of β and σ^2 are

$$\hat{\beta}_{\text{w}} = (\mathbf{X}'\mathbf{V}^{-1}\mathbf{X})^{-1}\mathbf{X}'\mathbf{V}^{-1}\mathbf{Y} = (3.11, 0.013, 0.0107)'.$$

Therefore, the prediction equation is

$$\hat{Y} = 3.11 + 0.013(x_1 - 35) + 0.01072(x_2 - 102)$$

or

$$\hat{Y} = 1.563 + 0.013x_1 + 0.01072x_2.$$

The weighted least-squares estimator of σ^2 is given by

$$\hat{\sigma}_{\text{w}}^2 = [\mathbf{Y}'(\mathbf{V}^{-1} - \mathbf{V}^{-1}\mathbf{X}(\mathbf{X}'\mathbf{V}^{-1}\mathbf{X})^{-1}\mathbf{X}'\mathbf{V}^{-1})\mathbf{Y}]/7 = 0.0379.$$

The weighted least-squares ANOVA table is

Source	df	SS
Overall mean	1	57.4083
Regression (β_1, β_2)	2	14.5813
Residual	7	0.2654
Total	10	72.2550

Thus far in Chapter 5 no assumptions have been made about the functional form of the distribution of the $n \times 1$ random vector $\mathbf{E}$ (and hence about the distribution of $\mathbf{Y}$). However, if we want to test hypotheses or construct confidence bands on model parameters, then we need to make an assumption about the functional form of the distribution of $\mathbf{E}$. It is common to assume that the $n \times 1$ random vector $\mathbf{E}$ has a multivariate normal distribution where $\mathrm{E}(\mathbf{E}) = \mathbf{0}$ and $\mathrm{cov}(\mathbf{E}) = \boldsymbol{\Sigma}$. In the simplest case the E_i's are independent, identically distributed normal random variables such that $\mathbf{E} \sim \mathrm{N}_n(\mathbf{0}, \sigma^2\mathbf{I}_n)$. In more complicated problems, it is assumed that $\mathbf{E} \sim \mathrm{N}_n(\mathbf{0}, \boldsymbol{\Sigma})$ where $\boldsymbol{\Sigma}$ is an $n \times n$ matrix whose elements are functions of a series of unknown variance components. In Section 5.5 model adequacy is discussed when $\mathbf{E} \sim \mathrm{N}_n(\mathbf{0}, \sigma^2\mathbf{I}_n)$. In Section 5.6 least-squares regression is developed for complete, balanced factorial experiments where the $n \times n$ covariance matrix $\boldsymbol{\Sigma}$ is a function of a series of unknown variance components. Later in Chapter 6 a general discussion on confidence bands and hypothesis testing is provided.

5.5 LACK OF FIT TEST

In this section assume that the $n \times 1$ random error vector $\mathbf{E} \sim \mathrm{N}_n(\mathbf{0}, \sigma^2\mathbf{I}_n)$. It is of interest to check whether the proposed model adequately fits the data. This lack of fit test requires replicate observations at one or more of the combinations of the $x_1, x_2, \ldots, x_{p-1}$ values.

Since the elements of the $n \times 1$ random vector $\mathbf{Y} = (Y_1, \ldots, Y_n)'$ can be listed in any order, we adopt the convention that sets of Y_i values that share the same $x_1, \ldots, x_{p-1}$ values are listed next to each other in the $\mathbf{Y}$ vector. For example, in the data set from Table 5.1.1, the 10×1 vector $\mathbf{Y} = (1.7, 2.0, 1.9, 1.6, 3.2, 2.0, 2.5, 5.4, 5.7, 5.1)'$ with Y_1–Y_4 sharing a speed equal to 20 and a speed $\times$ grade equal to 0, Y_5 having a speed equal to 20 and a speed $\times$ grade equal to 120, Y_6–Y_7 sharing a speed equal to 50 and a speed $\times$ grade equal to 0, and Y_8–Y_{10} sharing a speed equal to 50 and a speed $\times$ grade equal to 300.

When replicate observations exist within combinations of the $x_1, \ldots, x_{p-1}$ values, the residual sum of squares can be partitioned into a sum of squares due to pure error plus a sum of squares due to lack of fit. The pure error component is a measure of the variation between Y_i observations that share the same $x_1, \ldots, x_{p-1}$ values.

In the example data set from Table 5.1.1, the pure error sum of squares is the sum of squares of Y_1–Y_4 around their mean plus the sum of squares of Y_5 around its mean (zero in this case), plus the sum of squares of Y_6–Y_7 around their mean plus the sum of squares of Y_8–Y_{10} around their mean.

In general the sum of squares pure error is given by

$$\text{SS (pure error)} = \mathbf{Y}'\mathbf{A}_{\text{pe}}\mathbf{Y}$$

where $\mathbf{A}_{\text{pe}}$ is an $n \times n$ block diagonal matrix with the jth block equal to $\mathbf{I}_{r_j} - \frac{1}{r_j}\mathbf{J}_{r_j}$ for $j = 1, \ldots, k$ where k is the number of combinations of $x_1, \ldots, x_{p-1}$ values that contain at least one observation and r_j is the number of Y_i values in the jth combination with $n = \sum_{j=1}^{k} r_j$. Note that $\mathbf{A}_{\text{pe}}$ is an idempotent matrix of rank $n - k$ and $\mathbf{J}_n\mathbf{A}_{\text{pe}} = \mathbf{0}_{n \times n}$. Furthermore, the first r_1 rows of the matrix $\mathbf{X}$ are the same, the next r_2 rows of $\mathbf{X}$ are the same, etc. Therefore, $\mathbf{A}_{\text{pe}}\mathbf{X} = \mathbf{0}_{n \times p}$. In balanced data structures $r_1 = r_2 = \cdots = r_k = r$, $n = rk$, and the $n \times n$ pure error matrix $\mathbf{A}_{\text{pe}}$ can be expressed as the Kronecker product $\mathbf{I}_k \otimes (\mathbf{I}_r - \frac{1}{r}\mathbf{J}_r)$.

For the fuel, speed, grade data set, the 10×10 pure error sum of squares matrix $\mathbf{A}_{\text{pe}}$ is derived in the next example.

Example 5.5.1 From the Table 5.1.1 data set, four groups of Y_i's share the same speed and grade values. Therefore, $k = 4, r_1 = 4, r_2 = 1, r_3 = 2, r_4 = 3$, and $\mathbf{A}_{\text{pe}}$ is given by

$$\mathbf{A}_{\text{pe}} = \begin{bmatrix} \mathbf{I}_4 - \frac{1}{4}\mathbf{J}_4 & & & \mathbf{0} \\ & 0_{1\times 1} & & \\ & & \mathbf{I}_2 - \frac{1}{2}\mathbf{J}_2 & \\ & \mathbf{0} & & \mathbf{I}_3 - \frac{1}{3}\mathbf{J}_3 \end{bmatrix}.$$

In this example, $r_2 = 1$ so $\mathbf{I}_{r_2} - \frac{1}{r_2}\mathbf{J}_{r_2} = 1 - \frac{1}{1}1 = 0$. Thus, the fifth diagonal element of $\mathbf{A}_{\text{pe}}$ equals the scalar 0, indicating that observation Y_5 does not contribute to the pure error sum of squares. The $\text{rank}(\mathbf{A}_{\text{pe}}) = 10 - 4 = 6$.

The sum of squares lack of fit is calculated by subtraction. Therefore,

$$\begin{aligned} \text{SS (lack of fit)} &= \text{SS (residual)} - \text{SS (pure error)} \\ &= \mathbf{Y}'[\mathbf{I}_n - \mathbf{X}(\mathbf{X}'\mathbf{X})^{-1}\mathbf{X}' - \mathbf{A}_{\text{pe}}]\mathbf{Y}. \end{aligned}$$

The sums of squares due to the overall mean, regression, lack of fit, pure error, and total are provided in Table 5.5.1.

Note that $[\mathbf{I}_n - \mathbf{X}(\mathbf{X}'\mathbf{X})^{-1}\mathbf{X}' - \mathbf{A}_{\text{pe}}]\sigma^2\mathbf{I}_n = \sigma^2[\mathbf{I}_n - \mathbf{X}(\mathbf{X}'\mathbf{X})^{-1}\mathbf{X}' - \mathbf{A}_{\text{pe}}]$ where $\mathbf{I}_n - \mathbf{X}(\mathbf{X}'\mathbf{X})^{-1}\mathbf{X}' - \mathbf{A}_{\text{pe}}$ is an idempotent matrix of rank $k - p$. Likewise, $[\mathbf{A}_{\text{pe}}]\sigma^2\mathbf{I}_n = \sigma^2\mathbf{A}_{\text{pe}}$ where $\mathbf{A}_{\text{pe}}$ is an idempotent matrix of rank $n - k$. Therefore, by

Table 5.5.1
ANOVA Table with Pure Error and Lack of Fit

Source	df	SS	SS Centered
Overall mean	1	$\mathbf{Y}'\frac{1}{n}\mathbf{J}_n\mathbf{Y}$	$= \mathbf{Y}'\frac{1}{n}\mathbf{J}_n\mathbf{Y}$
Regression $(\beta_1, \ldots, \beta_{p-1})$	$p-1$	$\mathbf{Y}'[\mathbf{X}(\mathbf{X}'\mathbf{X})^{-1}\mathbf{X}' - \frac{1}{n}\mathbf{J}_n]\mathbf{Y}$	$= \mathbf{Y}'\mathbf{X}_c(\mathbf{X}_c'\mathbf{X}_c)^{-1}\mathbf{X}_c'\mathbf{Y}$
Lack of fit	$k-p$	$\mathbf{Y}'[\mathbf{I}_n - \mathbf{X}(\mathbf{X}'\mathbf{X})^{-1}\mathbf{X}' - \mathbf{A}_{\text{pe}}]\mathbf{Y}$	$= \mathbf{Y}'[\mathbf{I}_n - \frac{1}{n}\mathbf{J}_n - \mathbf{A}_{\text{pe}}$ $-\mathbf{X}_c(\mathbf{X}_c'\mathbf{X})^{-1}\mathbf{X}_c']\mathbf{Y}$
Pure error	$n-k$	$\mathbf{Y}'\mathbf{A}_{\text{pe}}\mathbf{Y}$	$= \mathbf{Y}'\mathbf{A}_{\text{pe}}\mathbf{Y}$
Total	n	$\mathbf{Y}'\mathbf{Y}$	

Corollary 3.1.2(a), the sum of squares lack of fit $\mathbf{Y}'[\mathbf{I}_n - \mathbf{X}(\mathbf{X}'\mathbf{X})^{-1}\mathbf{X}' - \mathbf{A}_{\text{pe}}]\mathbf{Y} \sim \sigma^2\chi^2_{k-p}(\lambda_{\text{lof}})$ and the sum of squares pure error $\mathbf{Y}'\mathbf{A}_{\text{pe}}\mathbf{Y} \sim \sigma^2\chi^2_{n-k}(\lambda_{\text{pe}})$ where

$$\lambda_{\text{lof}} = [\mathrm{E}(\mathbf{Y})]'[\mathbf{I}_n - \mathbf{X}(\mathbf{X}'\mathbf{X})^{-1}\mathbf{X}' - \mathbf{A}_{\text{pe}}][\mathrm{E}(\mathbf{Y})]/(2\sigma^2) \geq 0$$

and

$$\lambda_{\text{pe}} = [\mathrm{E}(\mathbf{Y})]'\mathbf{A}_{\text{pe}}[\mathrm{E}(\mathbf{Y})]/(2\sigma^2) = \boldsymbol{\beta}'\mathbf{X}'\mathbf{A}_{\text{pe}}\mathbf{X}\boldsymbol{\beta}/(2\sigma^2) = 0$$

Furthermore, by Theorem 3.2.1, the lack of fit sum of squares and the pure error sum of squares are independent since

$$\begin{aligned}
[\mathbf{I}_n - \mathbf{X}(\mathbf{X}'\mathbf{X})^{-1}\mathbf{X}' - \mathbf{A}_{\text{pe}}](\sigma^2\mathbf{I}_n)(\mathbf{A}_{\text{pe}}) &= \sigma^2[\mathbf{A}_{\text{pe}} - \mathbf{X}(\mathbf{X}'\mathbf{X})^{-1}\mathbf{X}'\mathbf{A}_{\text{pe}} - \mathbf{A}_{\text{pe}}] \\
&= \sigma^2[\mathbf{A}_{\text{pe}} - \mathbf{A}_{\text{pe}}] \\
&= \mathbf{0}_{n\times n}.
\end{aligned}$$

Therefore, the statistic

$$F^* = \frac{\mathbf{Y}'[\mathbf{I}_n - \mathbf{X}(\mathbf{X}'\mathbf{X})^{-1}\mathbf{X}' - \mathbf{A}_{\text{pe}}]\mathbf{Y}/(k-p)}{\mathbf{Y}'\mathbf{A}_{\text{pe}}\mathbf{Y}/(n-k)} \sim F_{k-p,n-k}(\lambda_{\text{lof}}).$$

Note that if $\mathrm{E}(\mathbf{Y}) = \mathbf{X}\boldsymbol{\beta}$, then

$$\begin{aligned}
\lambda_{\text{lof}} &= [\mathrm{E}(\mathbf{Y})]'[\mathbf{I}_n - \mathbf{X}(\mathbf{X}'\mathbf{X})^{-1}\mathbf{X}' - \mathbf{A}_{\text{pe}}][\mathrm{E}(\mathbf{Y})]/(2\sigma^2) \\
&= \boldsymbol{\beta}'\mathbf{X}'[\mathbf{I}_n - \mathbf{X}(\mathbf{X}'\mathbf{X})^{-1}\mathbf{X}' - \mathbf{A}_{\text{pe}}]\mathbf{X}\boldsymbol{\beta}/(2\sigma^2) \\
&= \boldsymbol{\beta}'\mathbf{X}'\mathbf{X}\boldsymbol{\beta} - \boldsymbol{\beta}'\mathbf{X}'\mathbf{X}(\mathbf{X}'\mathbf{X})^{-1}\mathbf{X}'\mathbf{X}\boldsymbol{\beta} - \boldsymbol{\beta}'\mathbf{X}'\mathbf{A}_{\text{pe}}\mathbf{X}\boldsymbol{\beta} \\
&= 0.
\end{aligned}$$

If $\mathrm{E}(\mathbf{Y}) \neq \mathbf{X}\boldsymbol{\beta}$ then $\lambda_{\text{lof}} > 0$. Therefore, the hypothesis $\mathrm{H}_0 : \lambda_{\text{lof}} = 0$ versus $\mathrm{H}_1 : \lambda_{\text{lof}} > 0$ is equivalent to $\mathrm{H}_0 : \mathrm{E}(\mathbf{Y}) = \mathbf{X}\boldsymbol{\beta}$ versus $\mathrm{H}_1 : \mathrm{E}(\mathbf{Y}) \neq \mathbf{X}\boldsymbol{\beta}$. The statement $\mathrm{E}(\mathbf{Y}) = \mathbf{X}\boldsymbol{\beta}$ implies that the model being used in the estimation provides a good fit and therefore may be appropriate. Thus, a γ level rejection region for the hypothesis H_0 versus H_1 is as follows: Reject H_0 if $F^* > F^{\gamma}_{k-p,n-k}$.

In the next example, the pure error and lack of fit ANOVA table is provided for the fuel, speed, grade data set.

Example 5.5.2 The ANOVA table with lack of fit and pure error calculations is provided below for Table 5.1.1 data set.

Source	df	SS	MS	$F*$
Overall mean	1	96.796	96.796	
Regression (β_1, β_2)	2	24.070	12.035	
Lack of fit	1	0.014	0.014	0.21
Pure error	6	0.405	0.0675	
Total	10			

The lack of fit test statistic $F^* = 0.21 < F^{0.05}_{1,6} = 5.99$ indicating that the equation $\hat{Y} = 1.556 + 0.013848x_1 + 0.01601x_{i2}$ provides a good fit of the data.

5.6 PARTITIONING THE SUM OF SQUARES REGRESSION

In Table 5.3.1 the sum of squares regression was expressed with $p - 1$ degrees of freedom. This sum of squares represented the total influence of the variables $x_1, \ldots, x_{p-1}$ in the ordinary least-squares regression. It is often of interest to check the contribution of a particular variable (or variables) given that other variables are already in the model. Such contributions can be calculated by partitioning the $n \times p$ matrix $\mathbf{X}$ as

$$\mathbf{X} = (\mathbf{X}_1|\mathbf{X}_2|\cdots|\mathbf{X}_m)$$

where $\mathbf{X}_j$ is an $n \times p_j$ matrix for $j = 1, \ldots, m$, $p = \sum_{j=1}^{m} p_j$, and $\mathbf{X}_1 = \mathbf{1}_n$. If $R_1 = \mathbf{X}_1$, $R_2 = (\mathbf{X}_1|\mathbf{X}_2), \ldots, R_{m-1} = (\mathbf{X}_1|\mathbf{X}_2|\cdots|\mathbf{X}_{m-1})$ and $\mathbf{R}_m = \mathbf{X}$, then the sum of squares due to the p_j variables in $\mathbf{X}_j$ given that $\mathbf{X}_1, \mathbf{X}_2, \ldots, X_{j-1}$ are already in the model is given by

$$\begin{aligned} \mathrm{SS}\,(\mathbf{X}_j|\mathbf{X}_1, \mathbf{X}_2, \ldots, \mathbf{X}_{j-1}) = \mathbf{Y}'[&\mathbf{R}_j(\mathbf{R}_j'\mathbf{R}_j)^{-1}\mathbf{R}_j' \\ &- \mathbf{R}_{j-1}(\mathbf{R}_{j-1}'\mathbf{R}_{j-1})^{-1}\mathbf{R}_{j-1}']\mathbf{Y}. \end{aligned}$$

Such conditional sums of squares are often called Type I sums of squares. The entire ANOVA table with the Type I sums of squares is presented in Table 5.6.1.

Table 5.6.1
Type I Sum of Squares ANOVA Table

Source	df	Type I SS
Overall mean $\mathbf{X}_1$	$p_1 = 1$	$\mathbf{Y}'\mathbf{R}_1(\mathbf{R}_1'\mathbf{R}_1)^{-1}\mathbf{R}_1'\mathbf{Y} = \mathbf{Y}'\frac{1}{n}\mathbf{J}_n\mathbf{Y}$
$\mathbf{X}_2 \vert \mathbf{X}_1$	p_2	$\mathbf{Y}'[\mathbf{R}_2(\mathbf{R}_2'\mathbf{R}_2)^{-1}\mathbf{R}_2' - \mathbf{R}_1(\mathbf{R}_1'\mathbf{R}_1)^{-1}\mathbf{R}_1']\mathbf{Y}$
$\mathbf{X}_3 \vert \mathbf{X}_1, \mathbf{X}_2$	p_3	$\mathbf{Y}'[\mathbf{R}_3(\mathbf{R}_3'\mathbf{R}_3)^{-1}\mathbf{R}_3' - \mathbf{R}_2(\mathbf{R}_2'\mathbf{R}_2)^{-1}\mathbf{R}_2']\mathbf{Y}$
$\vdots$	$\vdots$	$\vdots$
$\mathbf{X}_m \vert \mathbf{X}_1, \mathbf{X}_2, \ldots, \mathbf{X}_{m-1}$	p_m	$\mathbf{Y}'[\mathbf{X}(\mathbf{X}'\mathbf{X})^{-1}\mathbf{X}' - \mathbf{R}_{m-1}(\mathbf{R}_{m-1}'\mathbf{R}_{m-1})^{-1}\mathbf{R}_{m-1}']\mathbf{Y}$
Residual	$n - p$	$\mathbf{Y}'(\mathbf{I}_n - \mathbf{X}(\mathbf{X}'\mathbf{X})^{-1}\mathbf{X}')\mathbf{Y}$
Total	n	$\mathbf{Y}'\mathbf{Y}$

Note that the sum of squares due to all sources of variations still add up to the total sum of squares $\mathbf{Y}'\mathbf{Y}$.

The Type I sums of squares for the fuel, speed, grade data set are provided below with the output provided in Appendix 1.

Example 5.6.1 Using the example data set from Table 5.1.1, the Type I sums of squares are provided for the overall mean, for the speed variable x_1 given the overall mean, and for the speed $\times$ grade variable x_2 given the overall mean and x_1.

Source	df	SS
Overall mean	1	96.721
$x_1 \vert$overall mean	1	10.609
$x_2 \vert$overall mean, x_1	1	13.461
Residual	7	0.419
Total	10	121.210

Some useful relationships are developed next. Note $\mathbf{R}_j'\mathbf{R}_j(\mathbf{R}_j'\mathbf{R}_j)^{-1}\mathbf{R}_j' = \mathbf{R}_j'$ for any $j = 1, \ldots, m$. That is,

$$\begin{bmatrix} \mathbf{X}_1' \\ \mathbf{X}_2' \\ \vdots \\ \mathbf{X}_j' \end{bmatrix} \mathbf{R}_j(\mathbf{R}_j'\mathbf{R}_j)^{-1}\mathbf{R}_j' = \begin{bmatrix} \mathbf{X}_1' \\ \mathbf{X}_2' \\ \vdots \\ \mathbf{X}_j' \end{bmatrix}.$$

Therefore, for any $1 \leq i \leq j \leq m$

$$\mathbf{X}_i'\mathbf{R}_j(\mathbf{R}_j'\mathbf{R}_j)^{-1}\mathbf{R}_j' = \mathbf{X}_i'$$

and

$$\mathbf{R}_i'\mathbf{R}_j(\mathbf{R}_j'\mathbf{R}_j)^{-1}\mathbf{R}_j' = \mathbf{R}_i'.$$

It is left to the reader to show that the above relationships imply that the matrices $\mathbf{R}_j(\mathbf{R}_j'\mathbf{R}_j)^{-1}\mathbf{R}_j' - \mathbf{R}_{j-1}(\mathbf{R}_{j-1}'\mathbf{R}_{j-1})^{-1}\mathbf{R}_{j-1}'$ are idempotent of rank p_j for any $j = 2, \ldots, m$ and to show that any two matrices $\mathbf{R}_i(\mathbf{R}_i'\mathbf{R}_i)^{-1}\mathbf{R}_i' - \mathbf{R}_{i-1}(\mathbf{R}_{i-1}'\mathbf{R}_{i-1})^{-1}$ $\mathbf{R}_{i-1}'$ and $\mathbf{R}_j(\mathbf{R}_j'\mathbf{R}_j)^{-1}\mathbf{R}_j' - \mathbf{R}_{j-1}(\mathbf{R}_{j-1}'\mathbf{R}_{j-1})^{-1}\mathbf{R}_{j-1}'$ are orthogonal for any $i \neq j$.

In certain problems the $n \times p_j$ matrix $\mathbf{X}_j$ is orthogonal to the $n \times p_i$ matrix $\mathbf{X}_i$ for all $i \neq j, i, j = 1, \ldots, m$. In such cases the Type I sums of squares due to $\mathbf{X}_j|\mathbf{X}_1, \ldots, \mathbf{X}_{j-1}$ reduce to much simpler forms for all $j = 1, \ldots, m$. That is, if $\mathbf{X}_i'\mathbf{X}_j = \mathbf{0}_{p_i \times p_j}$ for all $i \neq j$, then

$$\begin{aligned}
\mathbf{X}_j'\mathbf{R}_{j-1} &= \mathbf{X}_j'(\mathbf{X}_1|\mathbf{X}_2|\cdots|\mathbf{X}_{j-1}) \\
&= (\mathbf{X}_j'\mathbf{X}_1|\mathbf{X}_j'\mathbf{X}_2|\cdots|\mathbf{X}_j'\mathbf{X}_{j-1}) \\
&= \mathbf{0}_{p_j \times (p_1+p_2+\cdots+p_{j-1})}.
\end{aligned}$$

Therefore, the Type I sum of squares due to $\mathbf{X}_j|\mathbf{X}_1, \ldots, \mathbf{X}_{j-1}$ for any $j = 1, \ldots, m$ is given by

$$\begin{aligned}
\text{SS }(\mathbf{X}_j|\mathbf{X}_1, \ldots, \mathbf{X}_{j-1}) &= \mathbf{Y}'[\mathbf{R}_j(\mathbf{R}_j'\mathbf{R}_j)^{-1}\mathbf{R}_j' - \mathbf{R}_{j-1}(\mathbf{R}_{j-1}'\mathbf{R}_{j-1})^{-1}\mathbf{R}_{j-1}']\mathbf{Y} \\
&= \mathbf{Y}'\{(\mathbf{R}_{j-1}|\mathbf{X}_j)[(\mathbf{R}_{j-1}|\mathbf{X}_j)'(\mathbf{R}_{j-1}|\mathbf{X}_j)]^{-1}(\mathbf{R}_{j-1}|\mathbf{X}_j)' \\
&\quad - \mathbf{R}_{j-1}(\mathbf{R}_{j-1}'\mathbf{R}_{j-1})^{-1}\mathbf{R}_{j-1}'\}\mathbf{Y} \\
&= \mathbf{Y}'[(\mathbf{R}_{j-1}|\mathbf{X}_j)\begin{bmatrix} \mathbf{R}_{j-1}'\mathbf{R}_{j-1} & \mathbf{R}_{j-1}'\mathbf{X}_j \\ \mathbf{X}_j'\mathbf{R}_{j-1} & \mathbf{X}_j'\mathbf{X}_j \end{bmatrix}^{-1}\begin{bmatrix} \mathbf{R}_{j-1}' \\ \mathbf{X}_j' \end{bmatrix} \\
&\quad - \mathbf{R}_{j-1}(\mathbf{R}_{j-1}'\mathbf{R}_{j-1})^{-1}\mathbf{R}_{j-1}'\}\mathbf{Y} \\
&= \mathbf{Y}'[\mathbf{R}_{j-1}(\mathbf{R}_{j-1}'\mathbf{R}_{j-1})^{-1}\mathbf{R}_{j-1}' + \mathbf{X}_j(\mathbf{X}_j'\mathbf{X}_j)^{-1}\mathbf{X}_j' \\
&\quad - \mathbf{R}_{j-1}(\mathbf{R}_{j-1}'\mathbf{R}_{j-1})^{-1}\mathbf{R}_{j-1}'\}\mathbf{Y} \\
&= \mathbf{Y}'\mathbf{X}_j(\mathbf{X}_j'\mathbf{X}_j)^{-1}\mathbf{X}_j'\mathbf{Y}.
\end{aligned}$$

In the previous sections, the model $\mathbf{Y} = \mathbf{X}\boldsymbol{\beta} + \mathbf{E}$ was introduced within the context of a regression analysis. However, $\mathbf{Y} = \mathbf{X}\boldsymbol{\beta} + \mathbf{E}$ is a very general model form that can be used to describe a very broad class of experiments, including complete, balanced factorial experiments. In the next section the general linear model, $\mathbf{Y} = \mathbf{X}\boldsymbol{\beta}+\mathbf{E}$, is adapted to complete, balanced factorial experiments where the $n \times n$ covariance matrix, $\text{cov}(\mathbf{E}) = \boldsymbol{\Sigma}$, may be a complicated function of many variance components. In Chapters 6 and 7 more general applications of the model $\mathbf{Y} = \mathbf{X}\boldsymbol{\beta} + \mathbf{E}$ are introduced.

5.7 THE MODEL Y = Xβ + E IN COMPLETE, BALANCED FACTORIALS

The experiment presented in Section 4.1 has b random blocks, t fixed treatments, and r random replicates nested in each block treatment combination. The $btr \times 1$ random vector $\mathbf{Y} = (Y_{111}, \ldots, Y_{11r}, \ldots, Y_{bt1}, \ldots, Y_{btr})' \sim \mathrm{N}_{btr}(\boldsymbol{\mu}, \boldsymbol{\Sigma})$ where the $btr \times 1$ mean vector and the $btr \times btr$ covariance matrix are given by

$$\boldsymbol{\mu} = \mathbf{1}_b \otimes (\mu_1, \ldots, \mu_t)' \otimes \mathbf{1}_r$$

and

$$\begin{aligned}\boldsymbol{\Sigma} = \sigma_B^2[\mathbf{I}_b \otimes \mathbf{J}_t \otimes \mathbf{J}_r] + \sigma_{BT}^2\left[\mathbf{I}_b \otimes \left(\mathbf{I}_t - \frac{1}{t}\mathbf{J}_t\right) \otimes \mathbf{J}_r\right] \\ + \sigma_{R(BT)}^2[\mathbf{I}_b \otimes \mathbf{I}_t \otimes \mathbf{I}_r].\end{aligned}$$

This experiment can be characterized by the general linear model $\mathbf{Y} = \mathbf{X}\boldsymbol{\beta}+\mathbf{E}$. First, $\mathrm{cov}(\mathbf{E})$ equals the $btr \times btr$ covariance matrix Σ. Next, the $btr \times 1$ vector $\boldsymbol{\mu}$ must be reconciled with the $btr \times 1$ mean vector $\mathrm{E}(\mathbf{Y}) = \mathbf{X}\boldsymbol{\beta}$ from the general linear model. Note that the $btr \times 1$ mean vector $\boldsymbol{\mu}$ is a function of the t unknown parameters $\mu_1, \ldots, \mu_t$. Therefore, the general linear model mean vector $\mathbf{X}\boldsymbol{\beta}$ must also be written as a function of $\mu_1, \ldots, \mu_t$. One simple approach is to let the $t \times 1$ vector $\boldsymbol{\beta} = (\mu_1, \ldots, \mu_t)'$ and let the $btr \times t$ matrix $\mathbf{X} = \mathbf{1}_b \otimes \mathbf{I}_t \otimes \mathbf{1}_r$. Then the $btr \times 1$ mean vector of the general linear model is

$$\begin{aligned}\mathbf{X}\boldsymbol{\beta} &= (\mathbf{1}_b \otimes \mathbf{I}_t \otimes \mathbf{1}_r)(\mu_1, \ldots, \mu_t)' \\ &= (\mathbf{1}_b \otimes \mathbf{I}_t \otimes \mathbf{1}_r)[1 \otimes (\mu_1, \ldots, \mu_t)' \otimes 1] \\ &= \mathbf{1}_b \otimes (\mu_1, \ldots, \mu_t)' \otimes \mathbf{1}_r = \boldsymbol{\mu}.\end{aligned}$$

The preceding example suggests a general approach for writing the mean vector $\boldsymbol{\mu}$ as $\mathbf{X}\boldsymbol{\beta}$ for complete, balanced factorial experiments. First, if $\boldsymbol{\mu}$ is a function of p unknown parameters, then let $\boldsymbol{\beta}$ be a $p \times 1$ vector whose elements are the p unknown parameters in $\boldsymbol{\mu}$. In general these elements will be subscripted, such as μ_{ijk}. The elements of $\boldsymbol{\beta}$ should be ordered so the last subscript changes first, the second to the last subscript changes next, etc. The corresponding $\mathbf{X}$ matrix can then be constructed using a simple algorithm. The previous experiment is used to develop the algorithm rules.

Rule X1 Construct column headings where the first column heading designates main factor letters and the second heading designates the number of levels of the factor, ℓ. Place Kronecker product symbols $\otimes$ as described in Example 5.7.1.

Rule X2 Place $\mathbf{1}_\ell$ in the Kronecker product under the random factor columns.

Rule X3 Place $\mathbf{I}_\ell$ elsewhere.

Example 5.7.1 Rules **X**1, **X**2, and **X**3 for the example model.

$$\begin{array}{lccc} \text{Factor} & B & T & R \\ \text{Levels } \ell & b & t & r \end{array}$$

$$\mathbf{X} = \underline{\mathbf{1}_b} \otimes \underset{---}{\mathbf{I}_t} \otimes \underline{\mathbf{1}_r}$$

where Rule **X**2 is designated by —— and Rule **X**3 is designated by - - - -.

This formulation of the **X** matrix and its associated $\boldsymbol{\beta}$ vector is not unique. Another **X** matrix and $\boldsymbol{\beta}$ vector can be generated for the same experiment. This second formulation of **X** and $\boldsymbol{\beta}$ is motivated by the sum of squares matrices $\mathbf{A}_m$ from Section 4.2. In the example experiment, the sum of squares matrices for the mean, blocks, treatments, block by treatment interaction, and the nested replicates are given by

$$\mathbf{A}_1 = \frac{1}{b}\mathbf{J}_b \otimes \frac{1}{t}\mathbf{J}_t \otimes \frac{1}{r}\mathbf{J}_r$$

$$\mathbf{A}_2 = \left(\mathbf{I}_b - \frac{1}{b}\mathbf{J}_b\right) \otimes \frac{1}{t}\mathbf{J}_t \otimes \frac{1}{r}\mathbf{J}_r$$

$$\mathbf{A}_3 = \frac{1}{b}\mathbf{J}_b \otimes \left(\mathbf{I}_t - \frac{1}{t}\mathbf{J}_t\right) \otimes \frac{1}{r}\mathbf{J}_r$$

$$\mathbf{A}_4 = \left(\mathbf{I}_b - \frac{1}{b}\mathbf{J}_b\right) \otimes \left(\mathbf{I}_t - \frac{1}{t}\mathbf{J}_t\right) \otimes \frac{1}{r}\mathbf{J}_r$$

$$\mathbf{A}_5 = \mathbf{I}_b \otimes \mathbf{I}_t \otimes \left(\mathbf{I}_r - \frac{1}{r}\mathbf{J}_r\right),$$

respectively. Matrices $\mathbf{A}_1$ through $\mathbf{A}_5$ can be rewritten as $\mathbf{A}_1 = \mathbf{X}_1\mathbf{X}_1'$, $\mathbf{A}_2 = \mathbf{Z}_1\mathbf{Z}_1'$, $\mathbf{A}_3 = \mathbf{X}_2\mathbf{X}_2'$, $\mathbf{A}_4 = \mathbf{Z}_2\mathbf{Z}_2'$, and $\mathbf{A}_5 = \mathbf{Z}_3\mathbf{Z}_3'$ where

$$\mathbf{X}_1 = (1/\sqrt{b})\mathbf{1}_b \otimes (1/\sqrt{t})\mathbf{1}_t \otimes (1/\sqrt{r})\mathbf{1}_r$$

$$\mathbf{Z}_1 = \mathbf{P}_b \otimes (1/\sqrt{t})\mathbf{1}_t \otimes (1/\sqrt{r})\mathbf{1}_r$$

$$\mathbf{X}_2 = (1/\sqrt{b})\mathbf{1}_b \otimes \mathbf{P}_t \otimes (1/\sqrt{r})\mathbf{1}_r$$

$$\mathbf{Z}_2 = \mathbf{P}_b \otimes \mathbf{P}_t \otimes (1/\sqrt{r})\mathbf{1}_r$$

$$\mathbf{Z}_3 = \mathbf{I}_b \otimes \mathbf{I}_t \otimes \mathbf{P}_r$$

where the $(\ell - 1) \times \ell$ matrix $\mathbf{P}_\ell'$ is the lower portion of an ℓ-dimensional Helmert matrix. Note that $\mathbf{X}_1$ and $\mathbf{X}_2$ are associated with the fixed factor matrices and

$\mathbf{Z}_1$, $\mathbf{Z}_2$, and $\mathbf{Z}_3$ are associated with the random factor matrices. In this form $\mathbf{X}_1'\mathbf{X}_1 = 1$, $\mathbf{Z}_1'\mathbf{Z}_1 = \mathbf{I}_{b-1}$, $\mathbf{X}_2'\mathbf{X}_2 = \mathbf{I}_{t-1}$, $\mathbf{Z}_2'\mathbf{Z}_2 = \mathbf{I}_{(b-1)(t-1)}$, and $\mathbf{Z}_3'\mathbf{Z}_3 = \mathbf{I}_{bt(r-1)}$. Now let the $btr \times t$ matrix $\mathbf{X} = (\mathbf{X}_1|\mathbf{X}_2)$ where $\mathbf{X}_1$ and $\mathbf{X}_2$ are the $btr \times 1$ and $btr \times (t-1)$ matrices defined earlier. Note that $\mathbf{X}_1'\mathbf{X}_2 = \mathbf{0}_{1\times(t-1)}$. Then define the $t \times 1$ vector $\boldsymbol{\beta}$ such that $\mathbf{X}\boldsymbol{\beta} = \boldsymbol{\mu}$. Premultiplying this expression by $(\mathbf{X}'\mathbf{X})^{-1}\mathbf{X}'$ we obtain

$$\begin{aligned}\boldsymbol{\beta} &= (\mathbf{X}'\mathbf{X})^{-1}\mathbf{X}'\boldsymbol{\mu} = [(\mathbf{X}_1|\mathbf{X}_2)'(\mathbf{X}_1|\mathbf{X}_2)]^{-1}(\mathbf{X}_1|\mathbf{X}_2)'\boldsymbol{\mu} \\ &= \begin{bmatrix} (\mathbf{X}_1'\mathbf{X}_1)^{-1}\mathbf{X}_1' \\ \mathbf{X}_2'\mathbf{X}_2)^{-1}\mathbf{X}_2' \end{bmatrix} \boldsymbol{\mu}\end{aligned}$$

or

$$\begin{aligned}\beta_1 &= \sum_{j=1}^{t} \mu_j/\sqrt{br/t} \\ \beta_2 &= (\mu_1 - \mu_2)/\sqrt{br/2} \\ \beta_3 &= (\mu_1 + \mu_2 - 2\mu_3)/\sqrt{br/6} \\ &\vdots \qquad \vdots \\ \beta_t &= (\mu_1 + \mu_2 + \cdots + \mu_{t-1} - (t-1)\mu_t)/\sqrt{br/[t(t-1)]}.\end{aligned}$$

A third formulation of the matrix $\mathbf{X}$ can be constructed by writing $\mathbf{A}_1 = \mathbf{X}_1(\mathbf{X}_1'\mathbf{X}_1)^{-1}\mathbf{X}_1'$, $\mathbf{A}_2 = \mathbf{Z}_1(\mathbf{Z}_1'\mathbf{Z}_1)^{-1}\mathbf{Z}_1'$, $\mathbf{A}_3 = \mathbf{X}_2(\mathbf{X}_2'\mathbf{X}_2)^{-1}\mathbf{X}_2'$, $\mathbf{A}_4 = \mathbf{Z}_2(\mathbf{Z}_2'\mathbf{Z}_2)^{-1}\mathbf{Z}_2'$, and $\mathbf{A}_5 = \mathbf{Z}_3(\mathbf{Z}_3'\mathbf{Z}_3)^{-1}\mathbf{Z}_3'$ where

$$\begin{aligned}\mathbf{X}_1 &= \mathbf{1}_b \otimes \mathbf{1}_t \otimes \mathbf{1}_r \\ \mathbf{Z}_1 &= \mathbf{Q}_b \otimes \mathbf{1}_t \otimes \mathbf{1}_r \\ \mathbf{X}_2 &= \mathbf{1}_b \otimes \mathbf{Q}_t \otimes \mathbf{1}_r \\ \mathbf{Z}_2 &= \mathbf{Q}_b \otimes \mathbf{Q}_t \otimes \mathbf{1}_r \\ \mathbf{Z}_3 &= \mathbf{I}_b \otimes \mathbf{I}_t \otimes \mathbf{Q}_r\end{aligned}$$

and where the $\ell \times (\ell - 1)$ matrix $\mathbf{Q}_\ell$ is given by

$$\mathbf{Q}_\ell = \begin{bmatrix} 1 & 1 & 1 & \cdots & 1 \\ -1 & 1 & 1 & \cdots & 1 \\ 0 & -2 & 1 & \cdots & 1 \\ 0 & 0 & -3 & \cdots & 1 \\ 0 & 0 & 0 & \cdots & 1 \\ \vdots & \vdots & \vdots & \ddots & \vdots \\ 0 & 0 & 0 & \cdots & -(\ell-1) \end{bmatrix}.$$

Note that the columns of $\mathbf{Q}_\ell$ equal $\sqrt{j(j+1)}$ times the columns of $\mathbf{P}_\ell$ where j is the column number for $j = 1, \ldots, \ell - 1$. Now let the $btr \times t$ matrix $\mathbf{X} = (\mathbf{X}_1|\mathbf{X}_2)$ where $\mathbf{X}_1$ and $\mathbf{X}_2$ are the $btr \times 1$ and $btr \times (t-1)$ matrices defined above. Note $\mathbf{X}_1'\mathbf{X}_2 = \mathbf{0}_{1\times(t-1)}$.

It is apparent that a number of different forms of the matrices $\mathbf{X}, \mathbf{X}_1, \mathbf{X}_2, \ldots, \mathbf{Z}_1, \mathbf{Z}_2, \ldots$ can be constructed in complete balanced factorial designs.

Furthermore, in any particular problem, one form of the matrix $\mathbf{X}$ can be defined while another form of the $\mathbf{X}_1, \mathbf{X}_2, \ldots, \mathbf{Z}_1, \mathbf{Z}_2, \ldots$ matrices can be used to construct the sum of squares matrices. For example, in the previous experiment, the $btr \times t$ matrix $\mathbf{X}$ can be defined as $\mathbf{X} = \mathbf{1}_b \otimes \mathbf{I}_t \otimes \mathbf{1}_r$. Then with $\mathbf{X}_1 = \mathbf{1}_b \otimes \mathbf{1}_t \otimes \mathbf{1}_r, \mathbf{Z}_1 = \mathbf{Q}_b \otimes \mathbf{1}_t \otimes \mathbf{1}_r, \mathbf{X}_2 = \mathbf{1}_b \otimes \mathbf{Q}_t \otimes \mathbf{1}_r, \mathbf{Z}_2 = \mathbf{Q}_b \otimes \mathbf{Q}_t \otimes \mathbf{1}_r, \mathbf{Z}_3 = \mathbf{I}_b \otimes \mathbf{I}_t \otimes \mathbf{Q}_r$, the sum of squares matrices can be constructed as $\mathbf{A}_1 = \mathbf{X}_1(\mathbf{X}_1'\mathbf{X}_1)^{-1}\mathbf{X}_1'$, $\mathbf{A}_2 = \mathbf{Z}_1(\mathbf{Z}_1'\mathbf{Z}_1)^{-1}\mathbf{Z}_1'$, $\mathbf{A}_3 = \mathbf{X}_2(\mathbf{X}_2'\mathbf{X}_2)^{-1}\mathbf{X}_2'$, $\mathbf{A}_4 = \mathbf{Z}_2(\mathbf{Z}_2'\mathbf{Z}_2)^{-1}\mathbf{Z}_2'$, and $\mathbf{A}_5 = \mathbf{Z}_3(\mathbf{Z}_3'\mathbf{Z}_3)^{-1}\mathbf{Z}_3'$. In general, any acceptable form of the $\mathbf{X}$ matrix can be used with any acceptable form of the matrices $\mathbf{X}_1, \mathbf{X}_2, \ldots, \mathbf{Z}_1, \mathbf{Z}_2, \ldots$, where the later set of matrices is used to construct the sums of squares matrices.

EXERCISES

1. From Table 5.3.1, let $\mathbf{B}_1, \mathbf{B}_2$, and $\mathbf{B}_3$ represent the matrices for the sums of squares due to the overall mean, regression and residual, respectively. Prove $\mathbf{B}_r^2 = \mathbf{B}_r$ for $r = 1, 2, 3$ and $\mathbf{B}_r\mathbf{B}_s = \mathbf{0}$ for $r \neq s$.

2. Let $\mathbf{Y} = \mathbf{X}\boldsymbol{\beta} + \mathbf{E}$ where $\mathbf{X}$ is an $n \times p$ matrix and $\mathbf{E} \sim \mathrm{N}_n(\mathbf{0}, \sigma^2\mathbf{V})$ for any $n \times n$ symmetric, positive definite matrix $\mathbf{V}$.

   (a) Is $\hat{\boldsymbol{\beta}} = (\mathbf{X}'\mathbf{X})^{-1}\mathbf{X}'\mathbf{Y}$ an unbiased estimator of $\boldsymbol{\beta}$?

   (b) Prove that if there exists a $p \times p$ nonsingular matrix $\mathbf{F}$ such that $\mathbf{VX} = \mathbf{XF}$ then $\hat{\boldsymbol{\beta}} = \hat{\boldsymbol{\beta}}_{\mathrm{w}}$ where $\hat{\boldsymbol{\beta}}_{\mathrm{w}}$ is the weighted least-squares estimator of $\boldsymbol{\beta}$.

3. Let $\mathbf{Y} = \mathbf{X}\boldsymbol{\beta} + \mathbf{E}$ where $\mathbf{X}$ is an $n \times p$ matrix and $\mathbf{E} \sim \mathrm{N}_n(\mathbf{0}, \sigma^2\mathbf{V})$ for any $n \times n$ symmetric, positive definite matrix of known constants $\mathbf{V}$. Prove that $\mathbf{t}'(\mathbf{X}'\mathbf{V}^{-1}\mathbf{X})^{-1}\mathbf{X}'\mathbf{V}^{-1}\mathbf{Y}$ is the BLUE of $\mathbf{t}'\boldsymbol{\beta}$.

4. From Table 5.4.1, let $\mathbf{B}_1, \mathbf{B}_2$, and $\mathbf{B}_3$ represent the matrices for the sums of squares due to the overall mean, regression, and residual, respectively. Prove $\mathbf{B}_r\mathbf{V}$ is an idempotent matrix for $r = 1, 2, 3$ and $\mathbf{B}_r\mathbf{V}\mathbf{B}_s = \mathbf{0}$ for $r \neq s$.

5. Let R_j be defined as in Section 5.6. Let $\mathbf{B}_j = \mathbf{R}_j(\mathbf{R}_j'\mathbf{R}_j)^{-1}\mathbf{R}_j' - \mathbf{R}_{j-1}(\mathbf{R}_{j-1}'\mathbf{R}_{j-1})^{-1}\mathbf{R}_{j-1}'$.

   (a) Prove $\mathbf{B}_j$ is an idempotent matrix of rank $\mathbf{P}_j$ for $j = 2, \ldots, m$.

   (b) Prove $\mathbf{B}_i\mathbf{B}_j = \mathbf{0}$ for $i \neq j, i, j = 2, \ldots, m$.

6. Assume the model $\mathbf{Y} = \mathbf{X}\beta + \mathbf{E}$ where $\mathbf{X}$ is defined as in Section 5.6 and $\mathbf{E} \sim \mathrm{N}_n(\mathbf{0}, \sigma^2\mathbf{I}_n)$. Find the distributions of all the Type I sums of squares in Table 5.6.1.

7. Let $\mathbf{Y} = \mathbf{X}\boldsymbol{\beta} + \mathbf{E}$ where the $n \times (a+b)$ matrix $\mathbf{X} = [\mathbf{X}_1|\mathbf{X}_2]$ with $n \times a$ and $n \times b$ matrices $\mathbf{X}_1$ and $\mathbf{X}_2$, the $(a+b) \times 1$ vector $\boldsymbol{\beta} = (\beta_1, \ldots, \beta_a|\beta_{a+1}, \ldots, \beta_{a+b})'$, and $\mathbf{E} \sim \mathrm{N}_n(0, \sigma^2\mathbf{I}_n)$. Let $\mathbf{Y}'\mathbf{X}_1(\mathbf{X}_1'\mathbf{X}_1)^{-1}\mathbf{X}_1'\mathbf{Y}$, $\mathbf{Y}'[\mathbf{X}(\mathbf{X}'\mathbf{X})^{-1}\mathbf{X}' - \mathbf{X}_1(\mathbf{X}_1'\mathbf{X}_1)^{-1}\mathbf{X}_1']\mathbf{Y}$, and $\mathbf{Y}'[\mathbf{I}_n - \mathbf{X}(\mathbf{X}'\mathbf{X})^{-1}\mathbf{X}']\mathbf{Y}$ be the sum of squares due to $\beta_1, \ldots, \beta_a, \beta_{a+1}, \ldots, \beta_{a+b}|\beta_1, \ldots, \beta_a$ and the residual, respectively.

   (a) Prove $\mathbf{X}_i'\mathbf{X}(\mathbf{X}'\mathbf{X})^{-1}\mathbf{X}' = \mathbf{X}_i'$ for $i = 1, 2$. [*Hint:* Show $\mathbf{X}'\mathbf{X}(\mathbf{X}'\mathbf{X})^{-1}\mathbf{X}' = \mathbf{X}'$.]

   (b) Find the distributions of the sums of squares due to $\beta_{a+1}, \ldots, \beta_{a+b}|\beta_1, \ldots, \beta_a$ and the residual and show they are independent.

   (c) Construct a statistic to test the hypothesis $\mathrm{H}_0 : \beta_{a+1} = \cdots = \beta_{a+b} = 0$ versus H_1 : at least one $\beta_{a+1}, \ldots, \beta_{a+b} \neq 0$. What is the distribution of the statistic?

8. Consider the experiment from Exercise 7 in Chapter 4. Write the model as $\mathbf{Y} = \mathbf{X}\beta + \mathbf{E}$ defining all terms and the appropriate distributions explicitly.

9. Let $Y_{ij} = a + b_i x_j + E_{ij}$ for $i = 1, 2$ and $j = 1, \ldots, n$. Assume $\mathrm{E}(E_{ij}) = 0$, $\mathrm{var}(E_{ij}) = \sigma^2$, the E_{ij}'s are uncorrelated, and $\sum_{j=1}^{n} x_j = 0$.

   (a) Find the BLUE of $b_1 - b_2$. Write your answer in terms of the Y_{ij}'s and x_j's.

   (b) Find the variance of the BLUE of $b_1 - b_2$.

10. Let $Y_1 = \mu_1 + E_1, Y_2 = \mu_2 + E_2$, and $Y_3 = \mu_1 + \mu_2 + E_3$ where $\mathrm{E}(E_i) = 0$ and $\mathrm{E}(E_i^2) = 2$, $\mathrm{E}(E_i E_j) = 1$ for $i \neq j = 1, 2, 3$.

    (a) Find the BLUEs of μ_1 and μ_2.

    (b) Find the covariance between the BLUEs of μ_1 and μ_2.

    (c) Find the BLUE and the variance of the BLUE of $2\mu_1 + 3\mu_2$.

11. Let $Y_i = x_i\beta + U_i$ for $i = 1, \ldots, n$ and $0 < x_1 < x_2 < \cdots < x_n$ where $U_i = E_1 + E_2 + \cdots + E_i$ and the E_i's are uncorrelated with $\mathrm{E}(E_i) = 0$, $\mathrm{var}(E_i) = \sigma^2(x_i - x_{i-1})$ for $i > 1$ and $\mathrm{var}(E_1) = \sigma^2 x_1$.

    (a) Find the BLUE of β and show it depends only on (Y_n, x_n).

    (b) Find the variance of the BLUE of β. (*Hint*: Transform the Y_i values into Y_i^* values such that $Y_i^* = Y_i - Y_{i-1}$ for $i > 1$ and $Y_1^* = Y_1$.)

12. Consider the following design layout:

		Factor A (fixed) (i)			
		1	2	$\cdots$	a
Replicates (j)	1	Y_{11}	Y_{21}	$\cdots$	Y_{a1}
	2	Y_{12}	Y_{22}	$\cdots$	Y_{a2}
		$\vdots$	$\vdots$	$\ddots$	$\vdots$
	n	Y_{1n}	Y_{2n}	$\cdots$	Y_{an}

Let $\mathbf{Y} - \beta_0 \mathbf{1}_a \otimes \mathbf{1}_n + \mathbf{X}\beta + \mathbf{E}$ where $\mathbf{Y} = (Y_{11}, \ldots, Y_{1n}, \ldots, Y_{a1}, \ldots, Y_{an})'$, β_0 is an unknown scalar parameter, $\boldsymbol{\beta} = (\beta_1, \ldots, \beta_p)'$ is a $p \times 1$ vector of unknown parameters, $\mathbf{X} = \mathbf{X}^* \otimes \mathbf{1}_n$ with $\mathbf{X}^*$ an $a \times p$ matrix of known values, $p < a - 1$, $\mathbf{1}_a' \mathbf{X}^* = \mathbf{0}_{1 \times p}$, and the $an \times 1$ vector $\mathbf{E} \sim \mathrm{N}_{an}(\mathbf{0}, \boldsymbol{\Sigma})$ with $\boldsymbol{\Sigma} = \mathbf{I}_a \otimes (\sigma_1^2 \mathbf{I}_n + \sigma_2^2 \mathbf{J}_n)]$.

(a) Let $\mathbf{Y}'\mathbf{A}_1\mathbf{Y}$, $\mathbf{Y}'\mathbf{A}_2\mathbf{Y}$, $\mathbf{Y}'\mathbf{A}_3\mathbf{Y}$, and $\mathbf{Y}'\mathbf{A}_4\mathbf{Y}$ be the sum of squares due to β_0, β, lack of fit, and pure error, respectively. Find $\mathbf{A}_1$, $\mathbf{A}_2$, $\mathbf{A}_3$, and $\mathbf{A_4}$.

(b) Find the distributions of $\mathbf{Y}'\mathbf{A}_1\mathbf{Y}$, $\mathbf{Y}'\mathbf{A}_2\mathbf{Y}$, $\mathbf{Y}'\mathbf{A}_3\mathbf{Y}$, and $\mathbf{Y}'\mathbf{A}_4\mathbf{Y}$.

(c) Are $\mathbf{Y}'\mathbf{A}_1\mathbf{Y}$, $\mathbf{Y}'\mathbf{A}_2\mathbf{Y}$, $\mathbf{Y}'\mathbf{A}_3\mathbf{Y}$, and $\mathbf{Y}'\mathbf{A}_4\mathbf{Y}$ mutually independent?

(d) Assume no significant lack of fit in the model. Construct a test for the hypothesis $\mathrm{H}_0 : \boldsymbol{\beta} = \mathbf{0}$ versus $\mathrm{H}_1 : \boldsymbol{\beta} \neq \mathbf{0}$.

13. Consider this factorial layout:

		Factor A (fixed) (i)	
		1	2
Factor B (fixed) (j)	1	Y_{111}	Y_{211}
		Y_{112}	Y_{212}
	2	Y_{121}	Y_{221}
		Y_{122}	Y_{222}

Let the 8×1 vector $\mathbf{Y} = (Y_{111}, Y_{112}, Y_{121}, Y_{122}, Y_{211}, Y_{212}, Y_{221}, Y_{222})'$. Assume the model $\mathbf{Y} = \mathbf{X}\boldsymbol{\beta} + \mathbf{E}$ where $\mathbf{X} = [\mathbf{X}_1 | \mathbf{X}_2 | \mathbf{X}_3 | \mathbf{X}_4]$, $\mathbf{X}_1 = \mathbf{1}_2 \otimes \mathbf{1}_2 \otimes \mathbf{1}_2$, $\mathbf{X}_2 = \mathbf{Q}_2 \otimes \mathbf{1}_2 \otimes \mathbf{1}_2$, $\mathbf{X}_3 = \mathbf{1}_2 \otimes \mathbf{Q}_2 \otimes \mathbf{1}_2$, $\mathbf{X}_4 = \mathbf{Q}_2 \otimes \mathbf{Q}_2 \otimes \mathbf{1}_2$, $\mathbf{Q}_2 = (1, -1)'$, $\boldsymbol{\beta} = (\beta_1, \beta_2, \beta_3, \beta_4)$, and $\mathbf{E} \sim \mathrm{N}_8(\mathbf{0}, \sigma^2 \mathbf{I}_8)$. Note that the vector β_1 corresponds to the overall mean, β_2 to the fixed factor A, β_3 to the fixed factor B, and β_4 to the interaction of A and B.

(a) Define the sums of squares due to β_1, β_2, β_3, and β_4.

(b) Find $\hat{\boldsymbol{\beta}}$, the least-squares estimator of $\boldsymbol{\beta}$.

(c) Calculate the standard error of $\hat{\boldsymbol{\beta}}$.

14. Let $Y_{ij} = \mu + B_i + R(B)_{(i)j}$ where μ is an unknown constant, and the B_i's and $R(B)_{(i)j}$'s are uncorrelated random variables with 0 means and variances σ_B^2 and $\sigma_{R(B)}^2$, respectively, for $i = 1, \ldots, b$ and $j = 1, \ldots, r$.

    (a) Write the model as $\mathbf{Y} = \mathbf{X}\boldsymbol{\beta} + \mathbf{E}$ where $\mathrm{E}(\mathbf{E}) = \mathbf{0}$ and $\mathrm{cov}(\mathbf{E}) = \boldsymbol{\Sigma}$. Identify all terms explicitly.

    (b) Construct the matrices for the sums of squares for the usual ANOVA table for this model and derive the corresponding expected mean squares.

    (c) Assume the Y_{ij}'s are normally distributed. Find the distributions of the sums of squares defined in part b.

6 Maximum Likelihood Estimation and Related Topics

This chapter deals with maximum likelihood estimation of the parameters of the general linear model $\mathbf{Y} = \mathbf{X}\boldsymbol{\beta} + \mathbf{E}$ when $\mathbf{E} \sim \mathrm{N}_n(\mathbf{0}, \Sigma)$. The maximum likelihood estimators of $\boldsymbol{\beta}$ and Σ are the parameter values that maximize the likelihood function of the random vector $\mathbf{Y}$. In the first section of the chapter, the discussion is confined to the cases where $\Sigma = \sigma^2\mathbf{I}_n$ and $\Sigma = \sigma^2\mathbf{V}$ when $\mathbf{V}$ is known. In the second section, the concepts of invariance, completeness, sufficiency, and minimum variance unbiased estimation are discussed. In the third section, maximum likelihood estimation is developed for more general forms of Σ. Finally, the likelihood ratio test and related confidence bands on linear combinations of the $p \times 1$ vector $\boldsymbol{\beta}$ are examined.

6.1 MAXIMUM LIKELIHOOD ESTIMATORS OF $\boldsymbol{\beta}$ AND σ^2

For the present, assume the model $\mathbf{Y} = \mathbf{X}\boldsymbol{\beta} + \mathbf{E}$ where $\mathbf{E} \sim \mathrm{N}_n(\mathbf{0}, \sigma^2\mathbf{I}_n)$. Therefore, the likelihood function is given by

$$\ell(\boldsymbol{\beta}, \sigma^2, \mathbf{Y}) = (2\pi\sigma^2)^{-n/2} e^{-\{(\mathbf{Y}-\mathbf{X}\boldsymbol{\beta})'(\mathbf{Y}-\mathbf{X}\boldsymbol{\beta})/(2\sigma^2)\}}.$$

The logarithm of the likelihood function is

$$\log[\ell(\boldsymbol{\beta}, \sigma^2, \mathbf{Y})] = -\frac{n}{2}\log(2\pi) - \frac{n}{2}\log(\sigma^2) - (\mathbf{Y}-\mathbf{X}\boldsymbol{\beta})'(\mathbf{Y}-\mathbf{X}\boldsymbol{\beta})/(2\sigma^2).$$

The objective is to find the values of $\boldsymbol{\beta}$ and σ^2 that maximize the function $\log[\ell(\boldsymbol{\beta}, \sigma^2, \mathbf{Y})]$. Take derivatives of $\log[\ell(\boldsymbol{\beta}, \sigma^2, \mathbf{Y})]$ with respect to the $p \times 1$ vector $\boldsymbol{\beta}$ and σ^2, set the resulting expressions equal to zero, and solve for $\boldsymbol{\beta}$ and σ^2. That is,

$$\partial \log[\ell(\boldsymbol{\beta}, \sigma^2, \mathbf{Y})]/\partial\boldsymbol{\beta} = [-2\mathbf{X}'\mathbf{Y} + 2\mathbf{X}'\mathbf{X}\boldsymbol{\beta}]/(2\sigma^2) = 0$$

and

$$\partial \log[\ell(\boldsymbol{\beta}, \sigma^2, \mathbf{Y})]/\partial\sigma^2 = -n/(2\sigma^2) + \{(\mathbf{Y}-\mathbf{X}\boldsymbol{\beta})'(\mathbf{Y}-\mathbf{X}\boldsymbol{\beta})/(2\sigma^4)\} = 0.$$

Solving the first equation for $\boldsymbol{\beta}$ produces $\mathbf{X}'\mathbf{X}\boldsymbol{\beta} = \mathbf{X}'\mathbf{Y}$. If the $n \times p$ matrix $\mathbf{X}$ has full rank (i.e., if $\mathbf{X}'\mathbf{X}$ is nonsingular), then the maximum likelihood estimator (MLE) of $\boldsymbol{\beta}$ is given by

$$\tilde{\boldsymbol{\beta}} = (\mathbf{X}'\mathbf{X})^{-1}\mathbf{X}'\mathbf{Y}.$$

Solve the second equation for σ^2 with $\boldsymbol{\beta}$ replaced by $\tilde{\boldsymbol{\beta}}$. The resulting maximum likelihood estimator of σ^2 is

$$\tilde{\sigma}^2 = (\mathbf{Y}-\mathbf{X}\tilde{\boldsymbol{\beta}})'(\mathbf{Y}-\mathbf{X}\tilde{\boldsymbol{\beta}})/n = \mathbf{Y}'[\mathbf{I}_n - \mathbf{X}(\mathbf{X}'\mathbf{X})^{-1}\mathbf{X}']\mathbf{Y}/n.$$

The maximum likelihood estimator of $\boldsymbol{\beta}$ is a set of p linear transformations of the random vector $\mathbf{Y}$. Since $\mathbf{Y} \sim \mathrm{N}_n(\mathbf{X}\boldsymbol{\beta}, \sigma^2\mathbf{I}_n)$, by Theorem 2.1.2, the MLE is

$$\begin{aligned}\tilde{\boldsymbol{\beta}} &\sim \mathrm{N}_p[(\mathbf{X}'\mathbf{X})^{-1}\mathbf{X}'\mathbf{X}\boldsymbol{\beta}, (\mathbf{X}'\mathbf{X})^{-1}\mathbf{X}'(\sigma^2\mathbf{I}_n)\mathbf{X}(\mathbf{X}'\mathbf{X})^{-1}] \\ &= \mathrm{N}_p[\boldsymbol{\beta}, \sigma^2(\mathbf{X}'\mathbf{X})^{-1}].\end{aligned}$$

Therefore, $\tilde{\boldsymbol{\beta}}$ is an unbiased estimator of $\boldsymbol{\beta}$. Furthermore, $(\mathbf{I}_n - \mathbf{X}(\mathbf{X}'\mathbf{X})^{-1}\mathbf{X}')(\sigma^2\mathbf{I}_n) = \sigma^2(\mathbf{I}_n - \mathbf{X}(\mathbf{X}'\mathbf{X})^{-1}\mathbf{X}')$ is a multiple of an idempotent matrix of rank $n-p$. Therefore, by Corollary 3.1.2(a), $n\tilde{\sigma}^2 = \mathbf{Y}'[\mathbf{I}_n - \mathbf{X}(\mathbf{X}'\mathbf{X})^{-1}\mathbf{X}']\mathbf{Y} \sim \sigma^2\chi^2_{n-p}(\lambda)$ where

$$\begin{aligned}\lambda &= (\mathbf{X}\boldsymbol{\beta})'[\mathbf{I}_n - \mathbf{X}(\mathbf{X}'\mathbf{X})^{-1}\mathbf{X}']\mathbf{X}\boldsymbol{\beta}/(2\sigma^2) \\ &= [\boldsymbol{\beta}'\mathbf{X}'\mathbf{X}\boldsymbol{\beta} - \boldsymbol{\beta}'\mathbf{X}'\mathbf{X}(\mathbf{X}'\mathbf{X})^{-1}\mathbf{X}'\mathbf{X}\boldsymbol{\beta}]/(2\sigma^2) = 0\end{aligned}$$

The MLE $\tilde{\sigma}^2$ is not an unbiased estimator of σ^2 since $\mathrm{E}(\tilde{\sigma}^2) = \mathrm{E}[\sigma^2\chi^2_{n-p}(0)/n] = (n-p)\sigma^2/n$. Finally, by Theorem 3.2.2, $\tilde{\boldsymbol{\beta}}$ and $\tilde{\sigma}^2$ are independent since

$$\begin{aligned}(\mathbf{X}'\mathbf{X})^{-1}\mathbf{X}'(\sigma^2\mathbf{I}_n)[\mathbf{I}_n - \mathbf{X}(\mathbf{X}'\mathbf{X})^{-1}\mathbf{X}'] &= \sigma^2[(\mathbf{X}'\mathbf{X})^{-1}\mathbf{X}' \\ &\quad - (\mathbf{X}'\mathbf{X})^{-1}\mathbf{X}'\mathbf{X}(\mathbf{X}'\mathbf{X})^{-1}\mathbf{X}'] \\ &= \mathbf{0}_{p\times n}.\end{aligned}$$

In Chapter 5, $\hat{\boldsymbol{\beta}}$ and $\hat{\sigma}^2$ denoted the ordinary least-squares estimators (OLSEs) of $\boldsymbol{\beta}$ and σ^2, respectively. Note that the OLSE $\hat{\boldsymbol{\beta}}$ equals the MLE $\tilde{\boldsymbol{\beta}}$, and the OLSE $\hat{\sigma}^2$ is a multiple of the MLE $\tilde{\sigma}^2$ for the model $\mathbf{Y} = \mathbf{X}\boldsymbol{\beta} + \mathbf{E}$ when $\mathbf{E} \sim \mathrm{N}_n(\mathbf{0}, \sigma^2\mathbf{I}_n)$ [i.e., $\hat{\boldsymbol{\beta}} = \tilde{\boldsymbol{\beta}}$ and $\hat{\sigma}^2 = n\tilde{\sigma}^2/(n-p)$]. Furthermore, the OLSE $\hat{\sigma}^2$ is an unbiased estimator of σ^2 for this model. Therefore, the OLSE $\hat{\sigma}^2$ is often used to estimate σ^2 instead of the MLE $\tilde{\sigma}^2$.

In the following example the MLEs of $\boldsymbol{\beta}$ and σ^2 are derived for a one-way balanced factorial experiment.

Example 6.1.1 Consider the one-way classification described in Examples 1.2.10 and 2.1.4. Rewrite the model $Y_{ij} = \mu_i + R(T)_{(i)j}$ as $\mathbf{Y} = \mathbf{X}\boldsymbol{\beta} + \mathbf{E}$ where $\mathbf{Y} = (Y_{11}, \ldots, Y_{1r}, \ldots, Y_{t1}, \ldots, Y_{tr})'$, $\mathbf{X} = \mathbf{I}_t \otimes \mathbf{1}_r$, $\boldsymbol{\beta} = (\beta_1, \ldots, \beta_t)' = (\mu_1, \ldots, \mu_t)'$, and $\mathrm{cov}(\mathbf{E}) = \sigma^2\mathbf{I}_t \otimes \mathbf{I}_r$. Therefore, the MLE of β is given by

$$\begin{aligned}
\tilde{\beta} &= (\mathbf{X}'\mathbf{X})^{-1}\mathbf{X}'\mathbf{Y} \\
&= [(\mathbf{I}_t \otimes \mathbf{1}_r)'(\mathbf{I}_t \otimes \mathbf{1}_r)]^{-1}(\mathbf{I}_t \otimes \mathbf{1}_r)'\mathbf{Y} \\
&= [\mathbf{I}_t \otimes (1/r)\mathbf{1}_r']\mathbf{Y} \\
&= (\bar{Y}_{1.}, \ldots, \bar{Y}_{t.})'
\end{aligned}$$

where $\bar{Y}_{i.} = \sum_{j=1}^{r} Y_{ij}/r$. That is, the MLEs of $\beta_1, \ldots, \beta_t$ (or $\mu_1, \ldots, \mu_t$) are the observed treatment means. Furthermore, the MLE of σ^2 is

$$\begin{aligned}
\tilde{\sigma}^2 &= \mathbf{Y}'[\mathbf{I}_t \otimes \mathbf{I}_r - \mathbf{X}(\mathbf{X}'\mathbf{X})^{-1}\mathbf{X}']\mathbf{Y}/(rt) \\
&= \mathbf{Y}'\{\mathbf{I}_t \otimes \mathbf{I}_r - (\mathbf{I}_t \otimes \mathbf{1}_r)[(\mathbf{I}_t \otimes \mathbf{1}_r)'(\mathbf{I}_t \otimes \mathbf{1}_r)]^{-1}(\mathbf{I}_t \otimes \mathbf{1}_r)'\}\mathbf{Y}/(rt) \\
&= \mathbf{Y}'\left[\mathbf{I}_t \otimes \mathbf{I}_r - (\mathbf{I}_t \otimes \mathbf{1}_r)\left(\mathbf{I}_t \otimes \frac{1}{r}\right)(\mathbf{I}_t \otimes \mathbf{1}_r')\right]\mathbf{Y}/(rt) \\
&= \mathbf{Y}'\left[\mathbf{I}_t \otimes \left(\mathbf{I}_r - \frac{1}{r}\mathbf{J}_r\right)\right]\mathbf{Y}/(rt).
\end{aligned}$$

Thus, the MLE of σ^2 equals the sum of squares due to the nested replicates divided by tr.

We conclude this section by briefly examining likelihood estimation for the model $\mathbf{Y} = \mathbf{X}\boldsymbol{\beta} + \mathbf{E}$ when $\mathbf{E} \sim \mathrm{N}_n(\mathbf{0}, \sigma^2\mathbf{V})$ and $\mathbf{V}$ is an $n \times n$ positive definite matrix of known constants. Since $\mathbf{V}$ is an $n \times n$ positive definite matrix there exists an $n \times n$ nonsingular matrix $\mathbf{T}$ such that $\mathbf{V} = \mathbf{T}\mathbf{T}'$. Premultiplying the original model by $\mathbf{T}^{-1}$ we obtain

$$\begin{aligned}
\mathbf{T}^{-1}\mathbf{Y} &= \mathbf{T}^{-1}\mathbf{X}\boldsymbol{\beta} + \mathbf{T}^{-1}\mathbf{E} \\
\mathbf{Y}^* &= \mathbf{X}^*\boldsymbol{\beta} + \mathbf{E}^*
\end{aligned}$$

where $\mathbf{Y}^* = \mathbf{T}^{-1}\mathbf{Y}$, $\mathbf{X}^* = \mathbf{T}^{-1}\mathbf{X}$, and $\mathbf{E}^* = \mathbf{T}^{-1}\mathbf{E}$ with $\mathbf{E}^* \sim \mathrm{N}_n(\mathbf{0}, \sigma^2\mathbf{I}_n)$. Therefore, the maximum likelihood estimator of $\boldsymbol{\beta}$ is

$$\begin{aligned}\tilde{\boldsymbol{\beta}} &= (\mathbf{X}^{*\prime}\mathbf{X}^*)^{-1}\mathbf{X}^{*\prime}\mathbf{Y} \\ &= (\mathbf{X}'\mathbf{V}^{-1}\mathbf{X})^{-1}\mathbf{X}'\mathbf{V}^{-1}\mathbf{Y}.\end{aligned}$$

Likewise, the MLE of σ^2 is given by

$$\begin{aligned}\tilde{\sigma}^2 &= (\mathbf{Y}^* - \mathbf{X}^*\tilde{\boldsymbol{\beta}})'(\mathbf{Y}^* - \mathbf{X}^*\tilde{\boldsymbol{\beta}})/n \\ &= \mathbf{Y}'[\mathbf{V}^{-1} - \mathbf{V}^{-1}\mathbf{X}(\mathbf{X}'\mathbf{V}^{-1}\mathbf{X})^{-1}\mathbf{X}'\mathbf{V}^{-1}]\mathbf{Y}/n.\end{aligned}$$

It is left to the reader to find the distributions of $\tilde{\boldsymbol{\beta}}$ and $\tilde{\sigma}^2$ when $\mathbf{E} \sim \mathrm{N}_n(\mathbf{0}, \sigma^2\mathbf{V})$.

6.2 INVARIANCE PROPERTY, SUFFICIENCY, AND COMPLETENESS

In this section the invariance property, sufficiency, completeness, and minimum variance unbiased estimators are discussed.

Definition 6.2.1 *Invariance Property*: Let the $k \times 1$ vector $\tilde{\boldsymbol{\theta}} = (\tilde{\theta}_1, \ldots, \tilde{\theta}_k)'$ be the MLE of the $k \times 1$ vector $\boldsymbol{\theta}$. If $g(\boldsymbol{\theta})$ is a function of $\boldsymbol{\theta}$ then $g(\tilde{\boldsymbol{\theta}})$ is the MLE of $g(\boldsymbol{\theta})$.

Example 6.2.1 Consider the one-way classification in Example 6.1.1. By the invariance property, the MLE of $g(\boldsymbol{\beta}, \sigma^2) = \sum_{i=1}^{t} \beta_i/\sigma$ is

$$\begin{aligned}g(\tilde{\boldsymbol{\beta}}, \tilde{\sigma}^2) &= \left(1/\sqrt{\tilde{\sigma}^2}\right)\mathbf{1}_t'\tilde{\boldsymbol{\beta}} \\ &= \frac{\mathbf{1}_t'(\bar{Y}_{1.}, \ldots, \bar{Y}_{t.})'}{\left\{\mathbf{Y}'\left[\mathbf{I}_t \otimes \left(\mathbf{I}_r - \frac{1}{r}\mathbf{J}_r\right)\right]\mathbf{Y}/(rt)\right\}^{1/2}}.\end{aligned}$$

Sufficiency involves the reduction of data to a concise set of statistics without loss of information about the unknown parameters of the distribution. Thus, if the parameters of the distribution are of interest, attention can be focused on the joint distribution of the "reduced" set of statistics. In this sense, the reduced set of statistics provides sufficient information about the unknown parameters. The topic of sufficiency is addressed in the next theorem.

Theorem 6.2.1 *Factorization Theorem: Let the $n \times 1$ random vector $\mathbf{Y} = (Y_1, \ldots, Y_n)'$ have joint probability distribution function $f_{\mathbf{Y}}(Y_1, \ldots, Y_n, \boldsymbol{\theta})$ where $\boldsymbol{\theta}$ is a $k \times 1$ vector of unknown parameters. Let $\mathbf{S} = (S_1, \ldots, S_r)'$ be a set of r statistics for $r \geq k$. The statistics $S_1, \ldots, S_r$ are jointly sufficient for $\boldsymbol{\theta}$ if and* *only if*

$$f_{\mathbf{Y}}(Y_1, \ldots, Y_n, \boldsymbol{\theta}) = g(\mathbf{S}, \boldsymbol{\theta}) h(Y_1, \ldots, Y_n)$$

where $g(\mathbf{S}, \boldsymbol{\theta})$ *does not depend on* $Y_1, \ldots, Y_n$ *except through* $\mathbf{S}$ *and* $h(Y_1, \ldots, Y_n)$ *does not involve* $\boldsymbol{\theta}$.

Example 6.2.2 Let the $n \times 1$ random vector $\mathbf{Y} = (Y_1, \ldots, Y_n)' \sim \mathrm{N}_n(\alpha \mathbf{1}_n, \sigma^2 \mathbf{I}_n)$. The statistics $S_1 = \mathbf{1}_n' \mathbf{Y}$ and $S_2 = \mathbf{Y}'\mathbf{Y}$ are jointly sufficient for $\theta = (\alpha, \sigma^2)'$ since

$$\begin{aligned} f_{\mathbf{Y}}(Y_1, \ldots, Y_n, \alpha, \sigma^2) &= (2\pi\sigma^2)^{-n/2} e^{-(\mathbf{Y}-\alpha\mathbf{1}_n)'(\mathbf{Y}-\alpha\mathbf{1}_n)/(2\sigma^2)} \\ &= (2\pi\sigma^2)^{-n/2} e^{-[\mathbf{Y}'\mathbf{Y} - 2\alpha(\mathbf{1}_n'\mathbf{Y}) + n\alpha^2]/(2\sigma^2)} \\ &= (2\pi\sigma^2)^{-n/2} e^{-[S_2 - 2\alpha S_1 + n\alpha^2]/(2\sigma^2)}. \end{aligned}$$

The next theorem and example link the ideas of sufficiency and maximum likelihood estimation.

Theorem 6.2.2 *If* $\mathbf{S} = (S_1, \ldots, S_r)'$ *are jointly sufficient for the vector* $\boldsymbol{\theta}$ *and if* $\tilde{\boldsymbol{\theta}}$ *is a unique MLE of* $\boldsymbol{\theta}$, *then* $\tilde{\boldsymbol{\theta}}$ *is a function of* $\mathbf{S}$.

Proof: By the factorization theorem

$$f_{\mathbf{Y}}(Y_1, \ldots, Y_n, \theta) = g(\mathbf{S}, \boldsymbol{\theta})\, h(Y_1, \ldots, Y_n)$$

which means that the value of $\boldsymbol{\theta}$ that maximizes $f_{\mathbf{Y}}(\cdot)$ depends on $\mathbf{S}$. If the MLE is unique, the MLE of $\boldsymbol{\theta}$ must be a function of $\mathbf{S}$. ■

Example 6.2.3 Consider the problem from Example 6.2.2. Rewrite the model as $\mathbf{Y} = \mathbf{X}\alpha + \mathbf{E}$ where the $n \times 1$ matrix $\mathbf{X} = \mathbf{1}_n$ and the $n \times 1$ random vector $\mathbf{E} \sim \mathrm{N}_n(\mathbf{0}, \sigma^2 \mathbf{I}_n)$. Therefore, the MLE of α is given by

$$\begin{aligned} \tilde{\alpha} &= (\mathbf{X}'\mathbf{X})^{-1}\mathbf{X}'\mathbf{Y} \\ &= (\mathbf{1}_n'\mathbf{1}_n)^{-1}\mathbf{1}_n'\mathbf{Y} \\ &= \bar{Y}. \end{aligned}$$

and the MLE of σ^2 is

$$\begin{aligned} \tilde{\sigma}^2 &= \mathbf{Y}'[\mathbf{I}_n - \mathbf{X}(\mathbf{X}'\mathbf{X})^{-1}\mathbf{X}']\mathbf{Y}/n \\ &= \mathbf{Y}'[\mathbf{I}_n - \mathbf{1}_n(\mathbf{1}_n'\mathbf{1}_n)^{-1}\mathbf{1}_n']\mathbf{Y}/n \\ &= \mathbf{Y}'\left[\mathbf{I}_n - \frac{1}{n}\mathbf{J}_n\right]\mathbf{Y}/n. \end{aligned}$$

The MLEs $\tilde{\alpha} = S_1/n$ and $\sigma^2 = [(S_2 - S_1^2/n)/n]$ and jointly sufficient for α and σ^2 where $S_1 = \mathbf{1}_n'\mathbf{Y}$ and $S_2 = \mathbf{Y}'\mathbf{Y}$.

This section concludes with a discussion of completeness and its relation to minimum variance unbiased estimators.

Definition 6.2.2 *Completeness*: A family of probability distribution functions $\{f_{\mathbf{T}}(t, \boldsymbol{\theta}), \boldsymbol{\theta} \in \Theta\}$ is called complete if $\mathrm{E}[u(\mathbf{T})] = 0$ for all $\theta \in \Theta$ implies $u(\mathbf{T}) = 0$ with probability 1 for all $\theta \in \Theta$.

Completeness is a characterization of the joint probability distribution of the statistics $\mathbf{T}$. However, the term *complete* is often linked to the statistics themselves. Therefore, a sufficient statistic whose probability distribution is complete is referred to as a *complete sufficient statistic.*

Completeness implies that two different functions of the statistics $\mathbf{T}$ cannot have the same expectation. To understand this interpretation, let $u_1(\mathbf{T})$ and $u_2(\mathbf{T})$ be two different functions of $\mathbf{T}$ such that $\mathrm{E}[u_1(\mathbf{T})] = \tau(\boldsymbol{\theta})$ and $\mathrm{E}[u_2(\mathbf{T})] = \tau(\boldsymbol{\theta})$. Therefore, $\mathrm{E}[u_1(\mathbf{T}) - u_2(\mathbf{T})] = 0$. If the distribution of $\mathbf{T}$ is complete then $u_1(\mathbf{T}) - u_2(\mathbf{T}) = 0$ or $u_1(\mathbf{T}) = u_2(\mathbf{T})$ with probability 1. That is, an unbiased estimator of any function of θ is unique if the distribution of $\mathbf{T}$ is complete.

Suppose it is of interest to develop an unbiased estimator of $\tau(\theta)$. Let $\mathbf{T}$ be sufficient for $\boldsymbol{\theta}$. Therefore, when searching for an unbiased estimator of $\tau(\boldsymbol{\theta})$, we confine ourselves to functions of $\mathbf{T}$. If $\mathbf{T}$ is a set of complete sufficient statistics, then there is at most one unbiased estimator of $\tau(\boldsymbol{\theta})$ based on $\mathbf{T}$. Since this estimator is the only unbiased estimator of $\tau(\boldsymbol{\theta})$, trivially, it must be the unbiased estimator with the smallest variance.

From the preceding discussion the class of unbiased estimators of $\tau(\boldsymbol{\theta})$ based on complete sufficient statistics has at most one member. Therefore, to call this estimator "best" in its class is in some sense misleading, since there are no competing estimators in the class. With no competition these best estimators could perform very poorly. Surprisingly, minimum variance unbiased estimators based on complete sufficient statistics do in many cases have relatively small variances and, in that sense, do turn out to be good estimators.

The next theorem identifies the complete sufficient statistics of $\boldsymbol{\beta}$ and σ^2 when $\mathbf{Y} = \mathbf{X}\boldsymbol{\beta} + \mathbf{E}$ and $\mathbf{E} \sim \mathrm{N}_n(\mathbf{0}, \sigma^2\mathbf{I}_n)$.

Theorem 6.2.3 *Let* $\mathbf{Y} = \mathbf{X}\boldsymbol{\beta} + \mathbf{E}$ *where* $\mathbf{Y}$ *is an* $n \times 1$ *random vector,* $\mathbf{X}$ *is an* $n \times p$ *matrix of constants,* $\boldsymbol{\beta}$ *is a* $p \times 1$ *vector of unknown parameters, and the* $n \times 1$ *random vector* $\mathbf{E} \sim \mathrm{N}_n(\mathbf{0}, \sigma^2\mathbf{I}_n)$. *The MLEs* $\tilde{\boldsymbol{\beta}} = (\mathbf{X}'\mathbf{X})^{-1}\mathbf{X}'\mathbf{Y}$ *and* $\tilde{\sigma}^2 = \mathbf{Y}'[\mathbf{I}_n - \mathbf{X}(\mathbf{X}'\mathbf{X})^{-1}\mathbf{X}']\mathbf{Y}/n$ *are complete sufficient statistics for* $\boldsymbol{\beta}$ *and* σ^2. *Furthermore, any two linearly independent combinations of* $\tilde{\boldsymbol{\beta}}$ *and* σ^2 *are also complete sufficient statistics for* $\boldsymbol{\beta}$ *and* σ^2.

Example 6.2.4 Consider the problem from Example 6.2.3. By Theorem 6.2.3, $\bar{Y}.$ and $\mathbf{Y}'[\mathbf{I}_n - \frac{1}{n}\mathbf{J}_n]\mathbf{Y}$ are complete sufficient statistics for α and σ^2. Furthermore,

$\bar{Y}.$ and $\mathbf{Y}'[\mathbf{I}_n - \frac{1}{n}\mathbf{J}_n]\mathbf{Y}$ are independent random variables where $\bar{Y}. \sim \mathrm{N}_1(\alpha, \sigma^2/n)$ and $\mathbf{Y}'[\mathbf{I}_n - \frac{1}{n}\mathbf{J}_n]\mathbf{Y} \sim \sigma^2\chi^2_{n-1}(\lambda = 0)$. Therefore, $\mathrm{E}(\bar{Y}.) = \alpha$ and $\mathrm{E}\{(n-3)[\mathbf{Y}'(\mathbf{I}_n - \frac{1}{n}\mathbf{J}_n)\mathbf{Y}]^{-1}\} = \sigma^{-2}$. Thus, the minimum variance unbiased estimator of α/σ^2 is given by $(n-3)(\bar{Y}.)[\mathbf{Y}'(\mathbf{I}_n - \frac{1}{n}\mathbf{J}_n)\mathbf{Y}]^{-1}$.

6.3 ANOVA METHODS FOR FINDING MAXIMUM LIKELIHOOD ESTIMATORS

In some models the ordinary least-squares estimators and certain ANOVA sums of squares provide direct MLE solutions, even when the covariance matrix $\mathrm{cov}(\mathbf{E}) = \Sigma$ is a function of multiple unknown variance parameters. The topic is introduced with an example and then followed by a general theorem.

Example 6.3.1 Consider a two-way balanced factorial experiment with b random blocks and t fixed treatment levels. Let the bt observations be represented by random variables Y_{ij} for $i = 1, \ldots, b$ and $j = 1, \ldots, t$. The model for this experiment is

$$Y_{ij} = \mu_j + B_i + BT_{ij}$$

where μ_j is a constant representing the mean effect of the jth treatment level, B_i is a random variable representing the effect of the ith random block, and BT_{ij} is a random variable representing the interaction of the ith block and the jth treatment. Assume $B_i \sim iid\ \mathrm{N}_1(0, \sigma_B^2)$ and $(\mathbf{I}_b \otimes \mathbf{P}_t')(BT_{11}, \ldots, BT_{bt}) \sim \mathrm{N}_{b(t-1)}(\mathbf{0}, \sigma^2_{BT}\mathbf{I}_b \otimes \mathbf{I}_{t-1})$. Furthermore, assume $(B_1, \ldots, B_b)$ and $[BT_{11}, \ldots, BT_{bt}]$ are mutually independent. Rewrite the model as $\mathbf{Y} = \mathbf{X}\boldsymbol{\beta} + \mathbf{E}$ where the $bt \times 1$ random vector $\mathbf{Y} = (Y_{11}, \ldots, Y_{1t}, \ldots, Y_{b1}, \ldots, Y_{bt})'$, the $bt \times t$ matrix $\mathbf{X} = \mathbf{1}_b \otimes \mathbf{I}_t$, the $t \times 1$ vector $\boldsymbol{\beta} = (\beta_1, \ldots, \beta_t)' = (\mu_1, \ldots, \mu_t)'$, and the $bt \times 1$ error vector $\mathbf{E} = (E_{11}, \ldots, E_{1t}, \ldots, E_{b1}, \ldots, E_{bt})' \sim \mathrm{N}_{bt}(\mathbf{0}, \Sigma)$ with

$$\Sigma = \sigma_B^2[\mathbf{I}_b \otimes \mathbf{J}_t] + \sigma^2_{BT}\left[\mathbf{I}_b \otimes \left(\mathbf{I}_t - \frac{1}{t}\mathbf{J}_t\right)\right].$$

The covariance matrix Σ is derived using the covariance matrix algorithm in Chapter 4. The problem is to derive the maximum likelihood estimators of $\beta_1, \ldots, \beta_t, \sigma_B^2$ and σ^2_{BT}. First, let the $bt \times bt$ matrix

$$\mathbf{P}' = \begin{bmatrix} \mathbf{I}_b \otimes \mathbf{1}_t' \\ \mathbf{I}_b \otimes \mathbf{P}_t' \end{bmatrix}$$

where $\mathbf{P}_t'$ is the $(t-1) \times t$ lower portion of the t-dimensional Helmert matrix.

Recall $\mathbf{P}_t\mathbf{P}_t' = \mathbf{I}_t - \frac{1}{t}\mathbf{J}_t$, $\mathbf{P}_t'\mathbf{P}_t = \mathbf{I}_{t-1}$, and $\mathbf{1}_t'\mathbf{P}_t = \mathbf{0}_{1\times(t-1)}$. By Theorem 2.1.2,

$$\mathbf{P}'\mathbf{Y} = \begin{bmatrix} \mathbf{I}_b \otimes \mathbf{1}_t' \\ \mathbf{I}_b \otimes \mathbf{P}_t' \end{bmatrix} \mathbf{Y} \sim \mathrm{N}_{bt}(\mathbf{P}'\mathbf{X}\beta, \mathbf{P}'\Sigma\mathbf{P})$$

where

$$\mathbf{P}'\mathbf{X}\beta = \begin{bmatrix} \mathbf{1}_b \otimes \mathbf{1}_t'\boldsymbol{\beta} \\ \mathbf{1}_b \otimes \mathbf{P}_t'\boldsymbol{\beta} \end{bmatrix} \quad \text{and} \quad \mathbf{P}'\Sigma\mathbf{P} = \begin{bmatrix} t^2\sigma_B^2\mathbf{I}_b & \mathbf{0} \\ \mathbf{0} & \sigma_{BT}^2\mathbf{I}_b \otimes \mathbf{I}_{t-1} \end{bmatrix}.$$

The distribution of the $b \times 1$ vector $(\mathbf{I}_b \otimes \mathbf{1}_t')\mathbf{Y}$ is a function of the two unknown parameters $\mathbf{1}_t'\boldsymbol{\beta}$ and σ_B^2. The distribution of the $b(t-1) \times 1$ vector $(\mathbf{I}_b \otimes P_t')\mathbf{Y}$ is a function of the t unknown parameters $\mathbf{P}_t'\boldsymbol{\beta}$ and σ_{BT}^2. Furthermore, by Theorem 2.1.4, the two vectors are independent. Therefore, the MLEs of $\mathbf{1}_t'\boldsymbol{\beta}$ and σ_B^2 can be calculated separately from the MLEs of $\mathbf{P}_t'\boldsymbol{\beta}$ and σ_{BT}^2. First, model the $b \times 1$ vector $(\mathbf{I}_b \otimes \mathbf{1}_t')\mathbf{Y}$ as

$$\begin{aligned} (\mathbf{I}_b \otimes \mathbf{1}_t')\mathbf{Y} &= \mathbf{1}_b \otimes \mathbf{1}_t'\boldsymbol{\beta} + \mathbf{E}_1 \\ \mathbf{Y}_1 &= \mathbf{K}_1\theta_1 + \mathbf{E}_1 \end{aligned}$$

where the $b \times 1$ vector $\mathbf{Y}_1 = (\mathbf{I}_b \otimes \mathbf{1}_t')\mathbf{Y}$, the $b \times 1$ matrix $\mathbf{K}_1 = \mathbf{1}_b$, the unknown scalar $\theta_1 = \mathbf{1}_t'\boldsymbol{\beta} = \sum_{j=1}^t \beta_j$ and the $b \times 1$ random vector $\mathbf{E}_1 \sim \mathrm{N}_b(\mathbf{0}, t^2\sigma_B^2\mathbf{I}_b)$. Therefore, the MLE of θ_1 is

$$\begin{aligned} \tilde{\theta}_1 &= (\mathbf{K}_1'\mathbf{K}_1)^{-1}\mathbf{K}_1'\mathbf{Y}_1 \\ &= (\mathbf{1}_b'\mathbf{1}_b)^{-1}(\mathbf{1}_b \otimes 1)'(\mathbf{I}_b \otimes \mathbf{1}_t')\mathbf{Y} \\ &= \sum_{i=1}^{b}\sum_{j=1}^{t} Y_{ij}/b. \end{aligned}$$

The MLE of $t^2\sigma_B^2$ is given by

$$t^2\tilde{\sigma}_B^2 = \mathbf{Y}_1'[\mathbf{I}_b - \mathbf{K}_1(\mathbf{K}_1'\mathbf{K}_1)^{-1}\mathbf{K}_1']\mathbf{Y}_1/b.$$

Therefore, the MLE of σ_B^2 is

$$\begin{aligned} \tilde{\sigma}_B^2 &= \mathbf{Y}'(\mathbf{I}_b \otimes \mathbf{1}_t')'[\mathbf{I}_b - \mathbf{1}_b(\mathbf{1}_b'\mathbf{1}_b)^{-1}\mathbf{1}_b'](\mathbf{I}_b \otimes \mathbf{1}_t')\mathbf{Y}/(t^2b) \\ &= \mathbf{Y}'\left[\left(\mathbf{I}_b - \frac{1}{b}\mathbf{J}_b\right) \otimes \frac{1}{t}\mathbf{J}_t\right]\mathbf{Y}/(bt). \end{aligned}$$

Note that the MLE of σ_B^2 equals the sum of squares for blocks divided by bt. Now model the $b(t-1) \times 1$ vector $(\mathbf{I}_b \otimes \mathbf{P}_t')\mathbf{Y}$ as

$$\begin{aligned} (\mathbf{I}_b \otimes \mathbf{P}_t')\mathbf{Y} &= \mathbf{I}_b \otimes \mathbf{P}_t'\boldsymbol{\beta} + \mathbf{E}_2 \\ \mathbf{Y}_2 &= \mathbf{K}_2\theta_2 + \mathbf{E}_2 \end{aligned}$$

where the $b(t-1) \times 1$ vector $\mathbf{Y}_2 = (\mathbf{I}_b \otimes \mathbf{P}_t')\mathbf{Y}$, the $b(t-1) \times (t-1)$ matrix $\mathbf{K}_2 = \mathbf{1}_b \otimes \mathbf{I}_{t-1}$, the $(t-1) \times 1$ vector of unknown parameters $\boldsymbol{\theta}_2 = \mathbf{P}_t'\beta$, and the $b(t-1) \times 1$ random vector $\mathbf{E}_2 \sim \mathrm{N}_{b(t-1)}(\mathbf{0}, \sigma_{BT}^2 \mathbf{I}_b \otimes \mathbf{I}_{t-1})$. Therefore, the MLE of the $(t-1) \times 1$ vector $\boldsymbol{\theta}_2$ is

$$\begin{aligned}\tilde{\boldsymbol{\theta}}_2 &= (\mathbf{K}_2'\mathbf{K}_2)^{-1}\mathbf{K}_2'\mathbf{Y}_2 \\ &= [(\mathbf{1}_b \otimes \mathbf{I}_{t-1})'(\mathbf{1}_b \otimes \mathbf{I}_{t-1})]^{-1}(\mathbf{1}_b \otimes \mathbf{I}_{t-1})'(\mathbf{I}_b \otimes \mathbf{P}_t')\mathbf{Y} \\ &= [(1/b)\mathbf{1}_b' \otimes \mathbf{P}_t']\mathbf{Y}.\end{aligned}$$

The MLE of σ_{BT}^2 is given by

$$\begin{aligned}\tilde{\sigma}_{BT}^2 &= \mathbf{Y}_2'[\mathbf{I}_b \otimes \mathbf{I}_{t-1} - \mathbf{K}_2(\mathbf{K}_2'\mathbf{K}_2)^{-1}\mathbf{K}_2']\mathbf{Y}_2/[b(t-1)] \\ &= \mathbf{Y}'(\mathbf{I}_b \otimes \mathbf{P}_t')'\{\mathbf{I}_b \otimes \mathbf{I}_{t-1} - (\mathbf{1}_b \otimes \mathbf{I}_{t-1})[(\mathbf{1}_b \otimes \mathbf{I}_{t-1})'(\mathbf{1}_b \otimes \mathbf{I}_{t-1})]^{-1} \\ &\quad (\mathbf{1}_b \otimes \mathbf{I}_{t-1})'\}(\mathbf{I}_b \otimes \mathbf{P}_t')\mathbf{Y}/[b(t-1)] \\ &= \mathbf{Y}'(\mathbf{I}_b \otimes \mathbf{P}_t)\left[(\mathbf{I}_b \otimes \mathbf{I}_{t-1}) - \left(\frac{1}{b}\mathbf{J}_b \otimes \mathbf{I}_{t-1}\right)\right](\mathbf{I}_b \otimes \mathbf{P}_t')\mathbf{Y}/[b(t-1)] \\ &= \mathbf{Y}'(\mathbf{I}_b \otimes \mathbf{P}_t)\left[\left(\mathbf{I}_b - \frac{1}{b}\mathbf{J}_b\right) \otimes \mathbf{I}_{t-1}\right](\mathbf{I}_b \otimes \mathbf{P}_t')\mathbf{Y}/[b(t-1)] \\ &= \mathbf{Y}'\left[\left(\mathbf{I}_b - \frac{1}{b}\mathbf{J}_b\right) \otimes \mathbf{P}_t\mathbf{P}_t'\right]\mathbf{Y}/[b(t-1)] \\ &= \mathbf{Y}'\left[\left(\mathbf{I}_b - \frac{1}{b}\mathbf{J}_b\right) \otimes \left(\mathbf{I}_t - \frac{1}{t}\mathbf{J}_t\right)\right]\mathbf{Y}/[b(t-1)].\end{aligned}$$

Note that the MLE of σ_{BT}^2 is the sum of squares for the block by treatment interaction divided by $b(t-1)$. Now the maximum likelihood estimators of $(\theta_1, \boldsymbol{\theta}_2')$ and the invariance property are used to derive the MLEs of the original parameters $\beta = (\beta_1, \ldots, \beta_t)'$. Note that

$$\begin{bmatrix}\mathbf{1}_t' \\ \mathbf{P}_t'\end{bmatrix}\beta = \begin{bmatrix}\theta_1 \\ \boldsymbol{\theta}_2\end{bmatrix}.$$

Premultiplying by the $t \times t$ matrix $(\frac{1}{t}\mathbf{1}_t|\mathbf{P}_t)$ we obtain

$$\begin{aligned}\left(\frac{1}{t}\mathbf{1}_t|\mathbf{P}_t\right)\begin{bmatrix}\mathbf{1}_t' \\ \mathbf{P}_t'\end{bmatrix}\beta &= \left(\frac{1}{t}\mathbf{1}_t|\mathbf{P}_t\right)\begin{bmatrix}\theta_1 \\ \boldsymbol{\theta}_2\end{bmatrix} \\ \left(\frac{1}{t}\mathbf{J}_t + \mathbf{P}_t\mathbf{P}_t'\right)\beta &= \frac{1}{t}\mathbf{1}_t\theta_1 + \mathbf{P}_t\boldsymbol{\theta}_2 \\ \left[\frac{1}{t}\mathbf{J}_t + \left(\mathbf{I}_t - \frac{1}{t}\mathbf{J}_t\right)\right]\beta &= \frac{1}{t}\mathbf{1}_t\theta_1 + \mathbf{P}_t\boldsymbol{\theta}_2\end{aligned}$$

or

$$\boldsymbol{\beta} = \frac{1}{t}\mathbf{1}_t\theta_1 + \mathbf{P}_t\boldsymbol{\theta}_2.$$

Therefore, by the invariance property, the MLE of β is given by

$$\begin{aligned}
\tilde{\boldsymbol{\beta}} &= \frac{1}{t}\mathbf{1}_t\tilde{\theta}_1 + \mathbf{P}_t\tilde{\boldsymbol{\theta}}_2 \\
&= \frac{1}{t}\mathbf{1}_t[(1/b)\mathbf{1}_b' \otimes \mathbf{1}_t']\mathbf{Y} + \mathbf{P}_t[(1/b)\mathbf{1}_b' \otimes \mathbf{P}_t']\mathbf{Y} \\
&= \left[(1/b)\mathbf{1}_b' \otimes \frac{1}{t}\mathbf{J}_t + (1/b)\mathbf{1}_b' \otimes \mathbf{P}_t\mathbf{P}_t'\right]\mathbf{Y} \\
&= \left[(1/b)\mathbf{1}_b' \otimes \left(\frac{1}{t}\mathbf{J}_t + \mathbf{I}_t - \frac{1}{t}\mathbf{J}_t\right)\right]\mathbf{Y} \\
&= [(1/b)\mathbf{1}_b' \otimes \mathbf{I}_t]\mathbf{Y} \\
&= [(\mathbf{1}_b \otimes \mathbf{I}_t)'(\mathbf{1}_b \otimes \mathbf{I}_t)]^{-1}(\mathbf{1}_b \otimes \mathbf{I}_t)'\mathbf{Y} \\
&= (\mathbf{X}'\mathbf{X})^{-1}\mathbf{X}'\mathbf{Y}.
\end{aligned}$$

The model from Example 6.3.1 belongs to a class of linear models where the MLE of β equals the ordinary least-squares estimator of β and the MLEs of the variance parameters in Σ are linear combinations of the ANOVA mean squares for the random effects. The next theorem provides formulas for the MLEs of a broad class of models, including the model from Example 6.2.1.

Theorem 6.3.1 *Let $\mathbf{Y} = \mathbf{X}\boldsymbol{\beta} + \mathbf{E}$ where $\mathbf{Y}$ is an $n \times 1$ random vector, $\mathbf{X}$ is an $n \times p$ matrix of rank p, $\boldsymbol{\beta}$ is a $p \times 1$ vector of unknown paramters, and the $n \times 1$ random vector $\mathbf{E} \sim \mathrm{N}_n(\mathbf{0}, \Sigma)$. For $i = 1, \ldots, m$, let $\mathbf{Y}'\mathbf{B}_i\mathbf{Y}$ and $\mathbf{Y}'\mathbf{C}_i\mathbf{Y}$ be sums of squares corresponding to the various fixed and random effects, respectively, such that $\mathbf{I}_n = \sum_{i=1}^m (B_i + C_i)$ where rank$(\mathbf{B}_i) = p_i \geq 0$, rank $(\mathbf{C}_i) = r_i > 0$, $p = \sum_{i=1}^m p_i$ and $n = \sum_{i=1}^m (p_i + r_i)$. If there exist unique constants $a_i > 0$ such that $\Sigma = \sum_{i=1}^m a_i(\mathbf{B}_i + \mathbf{C}_i)$ then*

i) the maximum likelihood estimator of $\boldsymbol{\beta}$ is given by $\tilde{\boldsymbol{\beta}} = (\mathbf{X}'\mathbf{X})^{-1}\mathbf{X}'\mathbf{Y}$ if and only if $\sum_{i=1}^m \mathbf{B}_i\mathbf{X} = \mathbf{X}$, and

ii) under the conditions that induce i), the maximum likelihood estimator of a_i is given by $\tilde{a}_i = \mathbf{Y}'\mathbf{C}_i\mathbf{Y}/(p_i + r_i)$.

Proof: By Theorem 1.1.7, $\mathbf{B}_i$ and $\mathbf{C}_i$ are idempotent matrices for $i = 1, \ldots, m$; $\mathbf{B}_i\mathbf{C}_j = \mathbf{0}_{n\times n}$ for any $i, j = 1, \ldots, m$; $\mathbf{B}_i\mathbf{B}_j = \mathbf{0}_{n\times n}$ for any $i \neq j$; and therefore, $\mathbf{B}_i + \mathbf{C}_i$ is an idempotent matrix of rank $p_i + r_i$. These conditions imply $\boldsymbol{\Sigma}^{-1} = \sum_{i=1}^m a_i^{-1}(\mathbf{B}_i + \mathbf{C}_i)$.

i) Assume $\sum_{i=1}^{m} \mathbf{B}_i\mathbf{X} = \mathbf{X}$. Then $\mathbf{B}_i\mathbf{C}_j = \mathbf{0}_{n\times n}$ for all i, j implies $\mathbf{C}_j\mathbf{X} = \sum_{i=1}^{m} \mathbf{C}_j\mathbf{B}_i\mathbf{X} = \mathbf{0}_{n\times p}$ for all j. Now let the $n \times p$ matrix $\mathbf{Q} = [\mathbf{Q}_1|\mathbf{Q}_2|\cdots|\mathbf{Q}_m]$ where $\mathbf{Q}_i$ is an $n \times p_i$ matrix of rank p_i such that $\mathbf{B}_i = \mathbf{Q}_i\mathbf{Q}_i'$ for each $i = 1, \ldots, m$. Thus, $\mathbf{QQ}' = \sum_{i=1}^{m} \mathbf{B}_i$ and $\sum_{i=1}^{m} a_i^{-1}\mathbf{B}_i = \mathbf{QA}^{-1}\mathbf{Q}'$ where $\mathbf{A}$ is a $p\times p$ nonsingular block diagonal matrix with $a_i\mathbf{I}_{p_i}$ on the diagonal for $i = 1, \ldots, m$. Therefore, $\mathbf{X} = \sum_{i=1}^{m} \mathbf{B}_i\mathbf{X} = \mathbf{QQ}'\mathbf{X}$, which implies $\mathbf{Q} = \mathbf{X}(\mathbf{Q}'\mathbf{X})^{-1}$. The maximum likelihood estimator of β is given by

$$
\begin{aligned}
\tilde{\beta} &= (\mathbf{X}'\tilde{\Sigma}^{-1}\mathbf{X})^{-1}\mathbf{X}'\tilde{\Sigma}^{-1}\mathbf{Y} \\
&= \left\{\mathbf{X}'\left[\sum_{i=1}^{m} a_i^{-1}(\mathbf{B}_i + \mathbf{C}_i)\right]\mathbf{X}\right\}^{-1}\mathbf{X}'\left[\sum_{i=1}^{m} a_i^{-1}(\mathbf{B}_i + \mathbf{C}_i)\right]\mathbf{Y} \\
&= \left\{\mathbf{X}'\left[\sum_{i=1}^{m} a_i^{-1}\mathbf{B}_i\right]\mathbf{X}\right\}^{-1}\mathbf{X}'\left[\sum_{i=1}^{m} a_i^{-1}\mathbf{B}_i\right]\mathbf{Y} \\
&= (\mathbf{X}'\mathbf{QA}^{-1}\mathbf{Q}'\mathbf{X})^{-1}\mathbf{X}'\mathbf{QA}^{-1}\mathbf{Q}'\mathbf{Y} \\
&= (\mathbf{Q}'\mathbf{X})^{-1}\mathbf{A}(\mathbf{X}'\mathbf{Q})^{-1}\mathbf{X}'\mathbf{QA}^{-1}\mathbf{Q}'\mathbf{Y} = (\mathbf{Q}'\mathbf{X})^{-1}\mathbf{Q}'\mathbf{Y} \\
&= \{[(\mathbf{Q}'\mathbf{X})^{-1}]'\mathbf{X}'\mathbf{X}\}^{-1}[(\mathbf{Q}'\mathbf{X})^{-1}]'\mathbf{X}'\mathbf{Y} = (\mathbf{X}'\mathbf{X})^{-1}\mathbf{X}'\mathbf{Y}.
\end{aligned}
$$

The "only if" portion of the proof of i) is omitted, but can be found in Moser and McCann (1995).

ii) Before deriving the MLE of a_i, we need to derive a particular relationship between the matrices $\mathbf{X}$ and $\mathbf{B}_i$. From the proof of i), $\mathbf{Q} = \mathbf{X}(\mathbf{Q}'\mathbf{X})^{-1}$. Therefore, $\mathbf{X}(\mathbf{X}'\mathbf{X})^{-1}\mathbf{X}\mathbf{Q}_i = \mathbf{Q}_i$ for $i = 1, \ldots, m$. Premultiplying by $\mathbf{Q}_i'$ produces $\mathbf{X}(\mathbf{X}'\mathbf{X})^{-1}\mathbf{X}\mathbf{B}_i = \mathbf{B}_i$. Now, to find the MLE of a_i, write the likelihood function with β replaced by $\tilde{\beta}$ and $\Sigma^{-1} = \sum_{i=1}^{m} a_i^{-1}(\mathbf{B}_i + \mathbf{C}_i)$. Note that a_i are the eigenvalues of Σ with multiplicity $p_i + r_i$ for $i = 1, \ldots, m$. Therefore,

$$
|\Sigma| = \prod_{i=1}^{m} a_i^{(p_i+r_i)}
$$

and

$$
\begin{aligned}
f(a_1, \ldots, a_m, \tilde{\beta}, \mathbf{Y}) &= (2\pi)^{-n/2}|\Sigma|^{-1/2}e^{-[(\mathbf{Y}-\mathbf{X}\tilde{\beta})'\Sigma^{-1}(\mathbf{Y}-\mathbf{X}\tilde{\beta})]/2} \\
&= (2\pi)^{-n/2}\left[\prod_{i=1}^{m} a_i^{-(p_i+r_i)/2}\right] \\
&\quad \times e^{-(\mathbf{Y}-\mathbf{X}\tilde{\beta})'\left[\sum_{i=1}^{m}(\mathbf{B}_i+\mathbf{C}_i)/a_i\right](\mathbf{Y}-\mathbf{X}\tilde{\beta})/2}
\end{aligned}
$$

Take derivatives of $\log f(a_1, \ldots, a_m, \tilde{\boldsymbol{\beta}}, \mathbf{Y})$ with respect to a_j for $j = 1, \ldots, m$, set the derivatives equal to zero, and solve for a_j:

$$\begin{aligned}\log f(a_1, \ldots, a_m, \tilde{\boldsymbol{\beta}}, \mathbf{Y}) = -(n/2)\log(2\pi) & \\ & + \sum_{i=1}^{m}[-(p_i + r_i)/2]\log(a_i) - (\mathbf{Y} - \mathbf{X}\tilde{\boldsymbol{\beta}})' \\ & \left[\sum_{i=1}^{m}(1/a_i)(\mathbf{B}_i + \mathbf{C}_i)\right](\mathbf{Y} - \mathbf{X}\tilde{\boldsymbol{\beta}})/2\end{aligned}$$

and

$$\begin{aligned}\partial \log f(a_1, \ldots, a_m, \tilde{\boldsymbol{\beta}}, \mathbf{Y})/\partial a_j = & -(p_j + r_j)/(2a_j) \\ & + (\mathbf{Y} - \mathbf{X}\tilde{\boldsymbol{\beta}})'(\mathbf{B}_j + \mathbf{C}_j) \\ & (\mathbf{Y} - \mathbf{X}\tilde{\boldsymbol{\beta}})/(2a_j^2) = 0.\end{aligned}$$

Therefore,

$$\begin{aligned}\tilde{a}_j &= (\mathbf{Y} - \mathbf{X}\tilde{\boldsymbol{\beta}})'(\mathbf{B}_j + \mathbf{C}_j)(\mathbf{Y} - \mathbf{X}\tilde{\boldsymbol{\beta}})/(p_j + r_j) \\ &= \mathbf{Y}'[\mathbf{I}_n - \mathbf{X}(\mathbf{X}'\mathbf{X})^{-1}\mathbf{X}'](\mathbf{B}_j + \mathbf{C}_j) \\ &\quad [\mathbf{I}_n - \mathbf{X}(\mathbf{X}'\mathbf{X})^{-1}\mathbf{X}']\mathbf{Y}/(p_j + r_j) \\ &= \mathbf{Y}'[(\mathbf{B}_j + \mathbf{C}_j) - \mathbf{X}(\mathbf{X}'\mathbf{X})^{-1}\mathbf{X}'\mathbf{B}_j - \mathbf{B}_j\mathbf{X}(\mathbf{X}'\mathbf{X})^{-1}\mathbf{X}' \\ &\quad + \mathbf{X}(\mathbf{X}'\mathbf{X})^{-1}\mathbf{X}'\mathbf{B}_j\mathbf{X}(\mathbf{X}'\mathbf{X})^{-1}\mathbf{X}']\mathbf{Y}/(p_j + r_j) \\ &= \mathbf{Y}'\mathbf{C}_j\mathbf{Y}/(p_j + r_j). \quad \blacksquare\end{aligned}$$

In Theorem 6.3.1 the lower limit on p_i is zero. Therefore, the theorem allows $\mathbf{B}_i$ to equal $\mathbf{0}_{n\times n}$ and thus admits situations where the number of sums of squares for fixed effects differs from the number of random effects sums of squares. Furthermore, note that the lower limit on r_i is strictly positive, implying that the number of matrices $\mathbf{C}_i$ and the number of unknown variance parameters both equal m. This restriction is imposed so that the MLE estimators are defined, rather than being nonestimable.

Example 6.3.1 is now reworked using Theorem 6.3.1.

Example 6.3.2 Consider the two-way balanced factorial experiment given in Example 6.3.1. The model can be written as $\mathbf{Y} = \mathbf{X}\boldsymbol{\beta} + \mathbf{E}$ where $\mathbf{X} = \mathbf{1}_b \otimes \mathbf{I}_t$, $\boldsymbol{\beta} = (\beta_1 \ldots, \beta_t)'$ and

$$\Sigma = \sigma_B^2[\mathbf{I}_b \otimes \mathbf{J}_t] + \sigma_{BT}^2\left[\mathbf{I}_b \otimes \left(\mathbf{I}_t - \frac{1}{t}\mathbf{J}_t\right)\right].$$

Let the sums of squares due to the overall mean, the random blocks, the fixed treatments, and the random interaction of blocks and treatments be represented by $\mathbf{Y}'\mathbf{B}_1\mathbf{Y}$, $\mathbf{Y}'\mathbf{C}_1\mathbf{Y}$, $\mathbf{Y}'\mathbf{B}_2\mathbf{Y}$, and $\mathbf{Y}'\mathbf{C}_2\mathbf{Y}$, respectively, where $p_1 = 1, r_1 = b-1, p_2 = t-1, r_2 = (b-1)(t-1)$ and

$$\begin{aligned}
\mathbf{B}_1 &= \frac{1}{b}\mathbf{J}_b \otimes \frac{1}{t}\mathbf{J}_t \\
\mathbf{C}_1 &= \left(\mathbf{I}_b - \frac{1}{b}\mathbf{J}_b\right) \otimes \frac{1}{t}\mathbf{J}_t \\
\mathbf{B}_2 &= \frac{1}{b}\mathbf{J}_b \otimes \left(\mathbf{I}_t - \frac{1}{t}\mathbf{J}_t\right) \\
\mathbf{C}_2 &= \left(\mathbf{I}_b - \frac{1}{b}\mathbf{J}_b\right) \otimes \left(\mathbf{I}_t - \frac{1}{t}\mathbf{J}_t\right).
\end{aligned}$$

Note that $\mathbf{I}_b \otimes \mathbf{I}_t = \mathbf{B}_1 + \mathbf{C}_1 + \mathbf{B}_2 + \mathbf{C}_2$. Furthermore, $\mathbf{B}_1\Sigma = t\sigma_B^2\mathbf{B}_1$, $\mathbf{C}_1\Sigma = t\sigma_B^2\mathbf{C}_1$, $\mathbf{B}_2\Sigma = \sigma_{BT}^2\mathbf{B}_2$, and $\mathbf{C}_2\Sigma = \sigma_{BT}^2\mathbf{C}_2$. Therefore, $\sum_{i=1}^{2} a_i(\mathbf{B}_i + \mathbf{C}_i) = t\sigma_B^2[\mathbf{I}_b \otimes \frac{1}{t}\mathbf{J}_t] + \sigma_{BT}^2[\mathbf{I}_b \otimes (\mathbf{I}_t - \frac{1}{t}\mathbf{J}_t)] = \Sigma$ where $a_1 = t\sigma_B^2$ and $a_2 = \sigma_{BT}^2$. Furthermore,

$$\begin{aligned}
\sum_{i=1}^{2}\mathbf{B}_i\mathbf{X} &= \left[\left(\frac{1}{b}\mathbf{J}_b \otimes \frac{1}{t}\mathbf{J}_t\right) + \left(\frac{1}{b}\mathbf{J}_b \otimes \left(\mathbf{I}_t - \frac{1}{t}\mathbf{J}_t\right)\right)\right][\mathbf{1}_b \otimes \mathbf{I}_t] \\
&= \mathbf{1}_b \otimes \mathbf{I}_t \\
&= \mathbf{X}.
\end{aligned}$$

Therefore, by Theorem 6.3.1, the MLE of $\boldsymbol{\beta}$ is given by

$$\begin{aligned}
\tilde{\boldsymbol{\beta}} &= (\mathbf{X}'\mathbf{X})^{-1}\mathbf{X}'\mathbf{Y} \\
&= [(\mathbf{1}_b \otimes \mathbf{I}_t)'(\mathbf{1}_b \otimes \mathbf{I}_t)]^{-1}(\mathbf{1}_b \otimes \mathbf{I}_t)'\mathbf{Y} \\
&= (\bar{Y}_{.1}, \ldots, \bar{Y}_{.t})'
\end{aligned}$$

and the MLEs of a_1 and a_2 are

$$\tilde{a}_1 = t\tilde{\sigma}_B^2 = \mathbf{Y}'\mathbf{C}_1\mathbf{Y}/b$$

and

$$\tilde{a}_2 = \tilde{\sigma}_{BT}^2 = \mathbf{Y}'\mathbf{C}_2\mathbf{Y}/[b(t-1)].$$

Therefore, the MLEs of σ_B^2 and σ_{BT}^2 are

$$\tilde{\sigma}_B^2 = \mathbf{Y}'\left[\left(\mathbf{I}_b - \frac{1}{b}\mathbf{J}_b\right) \otimes \frac{1}{t}\mathbf{J}_t\right]\mathbf{Y}/(bt)$$

and

$$\tilde{\sigma}^2_{BT} = \mathbf{Y}'\left[\left(\mathbf{I}_b - \frac{1}{b}\mathbf{J}_b\right) \otimes \left(\mathbf{I}_t - \frac{1}{t}\mathbf{J}_t\right)\right]\mathbf{Y}/[b(t-1)].$$

These are the same MLEs derived in Example 6.3.1.

Although this chapter deals mainly with maximum likelihood estimators of multivariate normal models, Theorem 6.3.1 also motivates a further generalization of the Gauss–Markov theorem. The Gauss–Markov theorem was introduced in Section 5.2 for the model $\mathbf{Y} = \mathbf{X}\boldsymbol{\beta} + \mathbf{E}$ when the $n \times 1$ error vector $\mathbf{E}$ had a distribution with $\mathrm{E}(\mathbf{E}) = \mathbf{0}$ and $\mathrm{cov}(\mathbf{E}) = \sigma^2\mathbf{I}_n$. In Section 5.4 the Gauss–Markov theorem was extended to include the model $\mathbf{Y} = \mathbf{X}\boldsymbol{\beta} + \mathbf{E}$ with $\mathrm{E}(\mathbf{E}) = \mathbf{0}$ and $\mathrm{cov}(\mathbf{E}) = \sigma^2\mathbf{V}$ where $\mathbf{V}$ is an $n \times n$ positive definite matrix of known constants. In the next theorem, the Gauss–Markov theorem is again extended to include an even broader class of covariance matrices.

Theorem 6.3.2 *Let $\mathbf{Y} = \mathbf{X}\boldsymbol{\beta} + \mathbf{E}$ where $\mathbf{Y}$ is an $n \times 1$ random vector, $\mathbf{X}$ is an $n \times p$ matrix of rank p, $\boldsymbol{\beta}$ is a $p \times 1$ vector of unknown parameters, and $\mathbf{E}$ is an $n \times 1$ random vector with $\mathrm{E}(\mathbf{E}) = \mathbf{0}$ and $\mathrm{cov}(\mathbf{E}) = \Sigma$. For $i = 1, \ldots, m$, let $\mathbf{Y}'\mathbf{B}_i\mathbf{Y}$ and $\mathbf{Y}'\mathbf{C}_i\mathbf{Y}$ be the sums of squares corresponding to the various fixed and random effects, respectively, such that $\mathbf{I}_n = \sum_{i=1}^m(\mathbf{B}_i + \mathbf{C}_i)$ where $rank(\mathbf{B}_i) = p_i \geq 0$, rank $(\mathbf{C}_i) = r_i > 0$, $p = \sum_{i=1}^m p_i$ and $n = \sum_{i=1}^m(p_i + r_i)$. If there exist unique constants $a_i > 0$ such that $\Sigma = \sum_{i=1}^m a_i(\mathbf{B}_i + \mathbf{C}_i)$ then the BLUE of $\mathbf{t}'\boldsymbol{\beta}$ is given by $\mathbf{t}'(\mathbf{X}'\mathbf{X})^{-1}\mathbf{X}'\mathbf{Y}$ if and only if $\sum_{i=1}^m \mathbf{B}_i\mathbf{X} = \mathbf{X}$.*

Proof: (Sufficiency) Assume $\sum_{i=1}^m \mathbf{B}_i\mathbf{X} = \mathbf{X}$. The BLUE of $\mathbf{t}'\boldsymbol{\beta}$ is given by

$$\begin{aligned}\mathbf{t}'\tilde{\boldsymbol{\beta}} &= \mathbf{t}'[\mathbf{X}'\Sigma^{-1}\mathbf{X}]^{-1}\mathbf{X}'\Sigma^{-1}\mathbf{Y} \\ &= \mathbf{t}'\left\{\mathbf{X}'\left[\sum_{i=1}^m a_i^{-1}(\mathbf{B}_i + \mathbf{C}_i)\right]\mathbf{X}\right\}^{-1}\mathbf{X}'\left[\sum_{i=1}^m a_i^{-1}(\mathbf{B}_i + \mathbf{C}_i)\right]\mathbf{Y} \\ &= \mathbf{t}'(\mathbf{X}'\mathbf{X})^{-1}\mathbf{X}'\mathbf{Y}.\end{aligned}$$

(Necessity) The necessity proof is given in Moser (1995). ■

Theorem 6.3.2 is applied in the next example.

Example 6.3.3 Consider the model in Example 6.3.2. Since $\Sigma = \sum_{i=1}^m a_i(\mathbf{B}_i + \mathbf{C}_i)$ and $\sum_{i=1}^m \mathbf{B}_i\mathbf{X} = \mathbf{X}$, by Theorem 6.3.2, the BLUE of $\mathbf{s}'\boldsymbol{\beta}$ is

$$\mathbf{s}'(\mathbf{X}'\mathbf{X})^{-1}\mathbf{X}'\mathbf{Y} = \mathbf{s}'(\bar{Y}_{.1}, \ldots, \bar{Y}_{.t})' = \sum_{i=1}^t s_i\bar{Y}_{.i}$$

where the $t \times 1$ vector $\mathbf{s} = (s_1, \ldots, s_t)'$.

In the next section the likelihood ratio test in derived for hypotheses of the form $\mathbf{H}\boldsymbol{\beta} = \mathbf{h}$ where $\mathbf{H}$ is a $q \times p$ matrix of constants and $\mathbf{h}$ is a $q \times 1$ vector of constants.

6.4 THE LIKELIHOOD RATIO TEST FOR $\mathbf{H}\boldsymbol{\beta} = \mathbf{h}$

Tests of the hypothesis $\mathbf{H}\boldsymbol{\beta} = \mathbf{h}$ are developed using the likelihood ratio statistic. For the model $\mathbf{Y} = \mathbf{X}\boldsymbol{\beta} + \mathbf{E}$ where $\mathbf{E} \sim \mathrm{N}_n(\mathbf{0}, \sigma^2\mathbf{I}_n)$ the likelihood function is given by

$$\ell(\boldsymbol{\beta}, \sigma^2, \mathbf{Y}) = (2\pi\sigma^2)^{-n/2} e^{-(\mathbf{Y}-\mathbf{X}\boldsymbol{\beta})'(\mathbf{Y}-\mathbf{X}\boldsymbol{\beta})/(2\sigma^2)}.$$

The likelihood ratio statistic is a function of two values:

1. the maximum value of $\ell(\boldsymbol{\beta}, \sigma^2, \mathbf{Y})$ maximized over all possible values of $\boldsymbol{\beta}$ and σ^2, that is, over all $0 < \sigma^2 < \infty$ and $-\infty < \beta_i < \infty$ for $\boldsymbol{\beta} = (\beta_1, \ldots, \beta_p)'$ where β_i is a scalar and
2. the maximum value of $\ell(\boldsymbol{\beta}, \sigma^2, \mathbf{Y})$ maximized over the parameter space defined by $\mathbf{H}\boldsymbol{\beta} = \mathbf{h}$.

The likelihood ratio statistic, L is the ratio of these two values.

$$L = \frac{\max\limits_{\boldsymbol{\beta}, \sigma^2, \mathbf{H}\boldsymbol{\beta}=\mathbf{h}} \ell(\boldsymbol{\beta}, \sigma^2, \mathbf{Y})}{\max\limits_{\boldsymbol{\beta}, \sigma^2} \ell(\boldsymbol{\beta}, \sigma^2, \mathbf{Y})}.$$

The denominator of L is maximized when the maximum likelihood estimators of $\boldsymbol{\beta}$ and σ^2 are used in $\ell(\boldsymbol{\beta}, \sigma^2, \mathbf{Y})$. That is,

$$\begin{aligned}\max_{\boldsymbol{\beta}, \sigma^2} \ell(\boldsymbol{\beta}, \sigma^2, \mathbf{Y}) &= \left(2\pi\tilde{\sigma}_{\mathrm{D}}^2\right)^{-n/2} e^{-(\mathbf{Y}-\mathbf{X}\tilde{\boldsymbol{\beta}}_{\mathrm{D}})'(\mathbf{Y}-\mathbf{X}\tilde{\boldsymbol{\beta}}_{\mathrm{D}})/(2\tilde{\sigma}_{\mathrm{D}}^2)} \\ &= (2\pi)^{-n/2}\left(\tilde{\sigma}_{\mathrm{D}}^2\right)^{-n/2} e^{-n/2}\end{aligned}$$

where $\tilde{\boldsymbol{\beta}}_D = (\mathbf{X}'\mathbf{X})^{-1}\mathbf{X}'\mathbf{Y}$ and $\tilde{\sigma}_{\mathrm{D}}^2 = \mathbf{Y}'(\mathbf{I}_n - \mathbf{X}(\mathbf{X}'\mathbf{X})^{-1}\mathbf{X}')\mathbf{Y}/n$.

For the numerator of L, the likelihood function is maximized with respect to $\boldsymbol{\beta}$ and σ^2 under the restriction $\mathbf{H}\boldsymbol{\beta} = \mathbf{h}$. To this end, let $\mathbf{G}$ be a $(p-q) \times p$ matrix of rank $p - q$ chosen such that the $p \times p$ matrix $\left[\begin{smallmatrix}\mathbf{H}\\ \mathbf{G}\end{smallmatrix}\right]$ has full rank with $\mathbf{H}\mathbf{G}' = \mathbf{0}_{q\times(p-q)}$. Let $\left[\begin{smallmatrix}\mathbf{H}\\ \mathbf{G}\end{smallmatrix}\right]^{-1} = [\mathbf{R}|\mathbf{S}]$ where $\mathbf{R}$ is a $p \times q$ matrix and $\mathbf{S}$ is a $p \times (p-q)$ matrix. Therefore, $\mathbf{HR} = \mathbf{I}_q$, $\mathbf{GS} = \mathbf{I}_{p-q}$, $\mathbf{HS} = \mathbf{0}_{q\times(p-q)}$, and $\mathbf{GR} = \mathbf{0}_{(p-q)\times q}$. Note that

$$\begin{bmatrix}\mathbf{H}\\ \mathbf{G}\end{bmatrix}[\mathbf{H}'(\mathbf{HH}')^{-1}|\mathbf{G}'(\mathbf{GG}')^{-1}] = \mathbf{I}_p$$

so $\mathbf{R} = \mathbf{H}'(\mathbf{H}\mathbf{H}')^{-1}$ and $\mathbf{S} = \mathbf{G}'(\mathbf{G}\mathbf{G}')^{-1}$. Also,

$$[\mathbf{H}'(\mathbf{H}\mathbf{H}')^{-1}|\mathbf{G}'(\mathbf{G}\mathbf{G}')^{-1}]\begin{bmatrix}\mathbf{H}\\ \mathbf{G}\end{bmatrix} = \mathbf{I}_p$$

or

$$\mathbf{H}'(\mathbf{H}\mathbf{H}')^{-1}\mathbf{H} + \mathbf{G}'(\mathbf{G}\mathbf{G}')^{-1}\mathbf{G} = \mathbf{I}_p.$$

Rewrite $\mathbf{Y} = \mathbf{X}\boldsymbol{\beta} + \mathbf{E}$ under $\mathbf{H}\boldsymbol{\beta} = \mathbf{h}$ as

$$\mathbf{Y} = \mathbf{X}(\mathbf{R}\ \mathbf{S})\begin{bmatrix}\mathbf{H}\\ \mathbf{G}\end{bmatrix}\boldsymbol{\beta} + \mathbf{E}$$

$$\mathbf{Y} = \mathbf{X}\mathbf{R}\mathbf{H}\boldsymbol{\beta} + \mathbf{X}\mathbf{S}\mathbf{G}\boldsymbol{\beta} + \mathbf{E}$$

$$\mathbf{Y} = \mathbf{X}\mathbf{R}\mathbf{h} + \mathbf{X}\mathbf{S}\mathbf{G}\boldsymbol{\beta} + \mathbf{E}.$$

Since $\mathbf{XRh}$ is known, rewrite the above model as

$$\mathbf{Z} = \mathbf{K}\boldsymbol{\theta} + \mathbf{E}$$

where the $n \times 1$ random vector $\mathbf{Z} = \mathbf{Y} - \mathbf{XRh}$, the $n \times (p-q)$ matrix $\mathbf{K} = \mathbf{XS}$, the $(p-q) \times 1$ vector $\boldsymbol{\theta} = \mathbf{G}\boldsymbol{\beta}$, and the $n \times 1$ random vector $\mathbf{E} \sim \mathrm{N}_n(\mathbf{0}, \sigma^2\mathbf{I}_n)$. Therefore, the likelihood in the numerator is maximized under $\mathbf{H}\boldsymbol{\beta} = \mathbf{h}$ when the maximum likelihood estimators of $\boldsymbol{\theta}$ and σ^2 for the model $\mathbf{Z} = \mathbf{K}\boldsymbol{\theta} + \mathbf{E}$ are used in the likelihood function. That is,

$$\begin{aligned}\max_{\boldsymbol{\beta},\sigma^2,\mathbf{H}\boldsymbol{\beta}=\mathbf{h}} \ell(\boldsymbol{\beta},\sigma^2,\mathbf{Y}) &= \max_{\boldsymbol{\theta},\sigma^2} \ell(\boldsymbol{\theta},\sigma^2,\mathbf{Z})\\ &= \left(2\pi\tilde{\sigma}_{\mathrm{N}}^2\right)^{-n/2} e^{-(\mathbf{Z}-\mathbf{K}\tilde{\boldsymbol{\theta}}_{\mathrm{N}})'(\mathbf{Z}-\mathbf{K}\tilde{\boldsymbol{\theta}}_{\mathrm{N}})/(2\tilde{\sigma}_{\mathrm{N}}^2)}\\ &= (2\pi)^{-n/2}\left(\tilde{\sigma}_{\mathrm{N}}^2\right)^{-n/2} e^{-n/2}\end{aligned}$$

where $\hat{\boldsymbol{\theta}}_{\mathrm{N}} = (\mathbf{K}'\mathbf{K})^{-1}\mathbf{K}'\mathbf{Z}$ and $\tilde{\sigma}_{\mathrm{N}}^2 = \mathbf{Z}'(\mathbf{I}_n - \mathbf{K}(\mathbf{K}'\mathbf{K})^{-1}\mathbf{K}')\mathbf{Z}/n$. Therefore, the likelihood ratio statistic is given by

$$L = \frac{(2\pi)^{-n/2}\left(\tilde{\sigma}_{\mathrm{N}}^2\right)^{-n/2} e^{-n/2}}{(2\pi)^{-n/2}\left(\tilde{\sigma}_{\mathrm{D}}^2\right)^{-n/2} e^{-n/2}} = \left(\tilde{\sigma}_{\mathrm{D}}^2/\tilde{\sigma}_{\mathrm{N}}^2\right)^{n/2}.$$

Instead of using L as the test statistic, use the following monotonic function of L:

$$\begin{aligned}V &= (L^{-2/n} - 1)(n-p)/q = \left(\tilde{\sigma}_{\mathrm{N}}^2/\tilde{\sigma}_{\mathrm{D}}^2 - 1\right)(n-p)/q\\ &= \frac{\left(\tilde{\sigma}_{\mathrm{N}}^2 - \tilde{\sigma}_{\mathrm{D}}^2\right)/q}{\tilde{\sigma}_{\mathrm{D}}^2/(n-p)}\\ &= \frac{\{\mathbf{Z}'(\mathbf{I}_n - \mathbf{K}(\mathbf{K}'\mathbf{K})^{-1}\mathbf{K}')\mathbf{Z} - \mathbf{Y}'(\mathbf{I}_n - \mathbf{X}(\mathbf{X}'\mathbf{X})^{-1}\mathbf{X}')\mathbf{Y}\}/q}{\mathbf{Y}'(\mathbf{I}_n - \mathbf{X}(\mathbf{X}'\mathbf{X})^{-1}\mathbf{X}')\mathbf{Y}/(n-p)}.\end{aligned}$$

A second form of V can be generated by noting that

$$\begin{aligned}[\mathbf{I}_n - \mathbf{X}(\mathbf{X}'\mathbf{X})^{-1}\mathbf{X}']\mathbf{Z} &= [\mathbf{I}_n - \mathbf{X}(\mathbf{X}'\mathbf{X})^{-1}\mathbf{X}'][\mathbf{Y} - \mathbf{X}\mathbf{H}'(\mathbf{H}\mathbf{H}')^{-1}\mathbf{h}] \\ &= \mathbf{Y} - \mathbf{X}(\mathbf{X}'\mathbf{X})^{-1}\mathbf{X}'\mathbf{Y} - \mathbf{X}\mathbf{H}'(\mathbf{H}\mathbf{H}')^{-1}\mathbf{h} \\ &\quad + \mathbf{X}(\mathbf{X}'\mathbf{X})^{-1}\mathbf{X}'\mathbf{X}\mathbf{H}'(\mathbf{H}\mathbf{H}')^{-1}\mathbf{h} \\ &= [\mathbf{I}_n - \mathbf{X}(\mathbf{X}'\mathbf{X})^{-1}\mathbf{X}']\mathbf{Y}\end{aligned}$$

or

$$\mathbf{Z}'[\mathbf{I}_n - \mathbf{X}(\mathbf{X}'\mathbf{X})^{-1}\mathbf{X}']\mathbf{Z} = \mathbf{Y}'[\mathbf{I}_n - \mathbf{X}(\mathbf{X}'\mathbf{X})^{-1}\mathbf{X}']\mathbf{Y}.$$

Therefore, V can be rewritten in a second form as

$$V = \frac{\mathbf{Z}'[\mathbf{X}(\mathbf{X}'\mathbf{X})^{-1}\mathbf{X}' - \mathbf{K}(\mathbf{K}'\mathbf{K})^{-1}\mathbf{K}']\mathbf{Z}/q}{\mathbf{Y}'(\mathbf{I}_n - \mathbf{X}(\mathbf{X}'\mathbf{X})^{-1}\mathbf{X}')\mathbf{Y}/(n-p)}.$$

Another form of the likelihood ratio statistic, V, can be generated using Lagrange multipliers. This third form of V equals

$$V = \frac{(\mathbf{H}\hat{\boldsymbol{\beta}} - \mathbf{h})'[\mathbf{H}(\mathbf{X}'\mathbf{X})^{-1}\mathbf{H}']^{-1}(\mathbf{H}\hat{\boldsymbol{\beta}} - \mathbf{h})/q}{\mathbf{Y}'(\mathbf{I}_n - \mathbf{X}(\mathbf{X}'\mathbf{X})^{-1}\mathbf{X}')\mathbf{Y}/(n-p)}$$

where $\hat{\boldsymbol{\beta}} = (\mathbf{X}'\mathbf{X})^{-1}\mathbf{X}'\mathbf{Y}$. The denominator of the third form of L is the same as the other two forms. That is, the denominator equals

$$\max_{\boldsymbol{\beta},\sigma^2} \ell(\boldsymbol{\beta}, \sigma^2, \mathbf{Y}) = (2\pi)^{-n/2}\left(\tilde{\sigma}_{\mathrm{D}}^2\right)^{-n/2} e^{-n/2}$$

where $\tilde{\sigma}_{\mathrm{D}}^2$ was defined earlier. Lagrange multipliers are used to derive the numerator of the third form of L. By the Lagrange multiplier technique, the numerator of L is given by

$$\max_{\boldsymbol{\beta},\sigma^2,\mathbf{H}\boldsymbol{\beta}=\mathbf{h}} \ell(\boldsymbol{\beta}, \sigma^2, \mathbf{Y}) = \max_{\boldsymbol{\beta},\sigma^2,\lambda} [\ell(\boldsymbol{\beta}, \sigma^2, \mathbf{Y}) - \lambda'(\mathbf{H}\boldsymbol{\beta} - \mathbf{h})]$$

where λ is a $q \times 1$ vector. To maximize the right side of this expression, take derivatives with respect to $\boldsymbol{\beta}, \sigma^2$, and λ, set the resulting expression equal to zero, solve for $\boldsymbol{\beta}, \sigma^2$, and λ, and substitute these solutions back into the original expression. Let $\ell^*(\boldsymbol{\beta}, \sigma^2, \lambda, \mathbf{Y}) = \ell(\boldsymbol{\beta}, \sigma^2, \mathbf{Y}) - \lambda'(\mathbf{H}\boldsymbol{\beta} - \mathbf{h})$. Then the derivatives are

$$\partial\ell^*(\boldsymbol{\beta}, \sigma^2, \lambda, \mathbf{Y})/\partial\boldsymbol{\beta} = (-2\sigma^2)^{-1}(2\mathbf{X}'\mathbf{X}\boldsymbol{\beta} - 2\mathbf{X}'\mathbf{Y})\,\ell(\boldsymbol{\beta}, \sigma^2, \mathbf{Y}) - \mathbf{H}'\lambda \quad (1)$$

$$\partial\ell^*(\boldsymbol{\beta}, \sigma^2, \lambda, \mathbf{Y})/\partial\sigma^2 = \left[-\frac{n}{2\sigma^2} + \frac{1}{2\sigma^4}(\mathbf{Y} - \mathbf{X}\boldsymbol{\beta})'(\mathbf{Y} - \mathbf{X}\boldsymbol{\beta})\right]\ell(\boldsymbol{\beta}, \sigma^2, \mathbf{Y}) \quad (2)$$

$$\partial\ell^*(\boldsymbol{\beta}, \sigma^2, \lambda, \mathbf{Y})/\partial\lambda = -(\mathbf{H}\boldsymbol{\beta} - \mathbf{h}). \quad (3)$$

Set equations (1), (2), and (3) equal to zero and solve for $\boldsymbol{\beta}$, σ^2, and λ. Let the solutions of $\boldsymbol{\beta}$, σ^2, and λ, be designated as $\tilde{\boldsymbol{\beta}}_N$, $\tilde{\sigma}_N^2$, and $\tilde{\lambda}_N$. From equations (1), (2), and (3):

$$-\mathbf{X}'\mathbf{X}\tilde{\boldsymbol{\beta}}_N + \mathbf{X}'\mathbf{Y} = \mathbf{H}'\tilde{\lambda}_N^* \text{ where } \tilde{\lambda}_N^* = \tilde{\sigma}_N^2\lambda/\ell(\tilde{\boldsymbol{\beta}}_N, \tilde{\sigma}_N^2, \mathbf{Y}) \tag{4}$$

$$\tilde{\sigma}_N^2 = (\mathbf{Y} - \mathbf{X}\tilde{\boldsymbol{\beta}}_N)'(\mathbf{Y} - \mathbf{X}\tilde{\boldsymbol{\beta}}_N)/n \tag{5}$$

$$\mathbf{H}\tilde{\boldsymbol{\beta}}_N = \mathbf{h}. \tag{6}$$

From (4):

$$\tilde{\boldsymbol{\beta}}_N = (\mathbf{X}'\mathbf{X})^{-1}\mathbf{X}'\mathbf{Y} - (\mathbf{X}'\mathbf{X})^{-1}\mathbf{H}'\tilde{\lambda}_N^*. \tag{7}$$

Premultiplying each side of (7) by $\mathbf{H}$ and noting from (6) that $\mathbf{H}\tilde{\boldsymbol{\beta}}_N = \mathbf{h}$, we obtain

$$\mathbf{H}\tilde{\boldsymbol{\beta}}_N = \mathbf{H}(\mathbf{X}'\mathbf{X})^{-1}\mathbf{X}'\mathbf{Y} - \mathbf{H}(\mathbf{X}'\mathbf{X})^{-1}\mathbf{H}'\tilde{\lambda}_N^* = \mathbf{h}. \tag{8}$$

Solving the right-hand equality in (8) for $\tilde{\lambda}_N^*$, we have

$$\tilde{\lambda}_N^* = [\mathbf{H}(\mathbf{X}'\mathbf{X})^{-1}\mathbf{H}']^{-1}[\mathbf{H}(\mathbf{X}'\mathbf{X})^{-1}\mathbf{X}'\mathbf{Y} - \mathbf{h}]. \tag{9}$$

Substituting $\tilde{\lambda}_N^*$ into (4) and solving for $\tilde{\boldsymbol{\beta}}_N$, we obtain

$$\begin{aligned} -(\mathbf{X}'\mathbf{X})\tilde{\boldsymbol{\beta}}_N + \mathbf{X}'\mathbf{Y} &= \mathbf{H}'[\mathbf{H}(\mathbf{X}'\mathbf{X})^{-1}\mathbf{H}']^{-1}[\mathbf{H}(\mathbf{X}'\mathbf{X})^{-1}\mathbf{X}'\mathbf{Y} - \mathbf{h}] \\ \tilde{\boldsymbol{\beta}}_N &= (\mathbf{X}'\mathbf{X})^{-1}\mathbf{X}'\mathbf{Y} - (\mathbf{X}'\mathbf{X})^{-1}\mathbf{H}'[\mathbf{H}(\mathbf{X}'\mathbf{X})^{-1}\mathbf{H}']^{-1} \\ &\quad \times [\mathbf{H}(\mathbf{X}'\mathbf{X})^{-1}\mathbf{X}'\mathbf{Y} - \mathbf{h}] \end{aligned}$$

or

$$\tilde{\boldsymbol{\beta}}_N = \tilde{\boldsymbol{\beta}}_D - (\mathbf{X}'\mathbf{X})^{-1}\mathbf{H}'\,[\mathbf{H}(\mathbf{X}'\mathbf{X})^{-1}\mathbf{H}']^{-1}[\mathbf{H}(\mathbf{X}'\mathbf{X})^{-1}\mathbf{X}'\mathbf{Y} - \mathbf{h}]. \tag{10}$$

Substituting (10) into (5) and rearranging terms, we have

$$\begin{aligned} \tilde{\sigma}_N^2 = \{&(\mathbf{Y} - \mathbf{X}\tilde{\boldsymbol{\beta}}_D) - \mathbf{X}(\mathbf{X}'\mathbf{X})^{-1}\mathbf{H}'[\mathbf{H}(\mathbf{X}'\mathbf{X})^{-1}\mathbf{H}']^{-1}[\mathbf{H}\tilde{\boldsymbol{\beta}}_D - \mathbf{h}]\}' \\ &\{(\mathbf{Y} - \mathbf{X}\tilde{\boldsymbol{\beta}}_D) - \mathbf{X}(\mathbf{X}'\mathbf{X})^{-1}\mathbf{H}'[\mathbf{H}(\mathbf{X}'\mathbf{X})^{-1}\mathbf{H}']^{-1}[\mathbf{H}\tilde{\boldsymbol{\beta}}_D - \mathbf{h}]\}/n. \end{aligned} \tag{11}$$

Since

$$(\mathbf{Y} - \mathbf{X}\tilde{\boldsymbol{\beta}}_D)'\mathbf{X}(\mathbf{X}'\mathbf{X})^{-1}\mathbf{H}'[\mathbf{H}(\mathbf{X}'\mathbf{X})^{-1}\mathbf{H}']^{-1}[\mathbf{H}\tilde{\boldsymbol{\beta}}_D - \mathbf{h}] = 0$$

and

$$\begin{aligned} &\{\mathbf{X}(\mathbf{X}'\mathbf{X})^{-1}\mathbf{H}'[\mathbf{H}(\mathbf{X}'\mathbf{X})^{-1}\mathbf{H}']^{-1}[\mathbf{H}\tilde{\boldsymbol{\beta}}_D - \mathbf{h}]\}'\{\mathbf{X}(\mathbf{X}'\mathbf{X})^{-1}\mathbf{H}'[\mathbf{H}(\mathbf{X}'\mathbf{X})^{-1}\mathbf{H}']^{-1} \\ &\quad \times [\mathbf{H}\tilde{\boldsymbol{\beta}}_D - \mathbf{h}]\} = [\mathbf{H}\tilde{\boldsymbol{\beta}}_D - \mathbf{h}]'[\mathbf{H}(\mathbf{X}'\mathbf{X})^{-1}\mathbf{H}']^{-1}[\mathbf{H}\tilde{\boldsymbol{\beta}}_D - \mathbf{h}], \end{aligned}$$

equation (11) reduces to

$$\begin{aligned}\tilde{\sigma}_{\mathrm{N}}^2 &= \{(\mathbf{Y}-\mathbf{X}\tilde{\boldsymbol{\beta}}_{\mathrm{D}})'(\mathbf{Y}-\mathbf{X}\tilde{\boldsymbol{\beta}}_{\mathrm{D}}) \\ &\quad + (\mathbf{H}\tilde{\boldsymbol{\beta}}_{\mathrm{D}}-\mathbf{h})'[\mathbf{H}(\mathbf{X}'\mathbf{X})^{-1}\mathbf{H}']^{-1}(\mathbf{H}\tilde{\boldsymbol{\beta}}_{\mathrm{D}}-\mathbf{h})\}/n \\ &= \tilde{\sigma}_{\mathrm{D}}^2 + \frac{1}{n}(\mathbf{H}\tilde{\boldsymbol{\beta}}_{\mathrm{D}}-\mathbf{h})'[\mathbf{H}(\mathbf{X}'\mathbf{X})^{-1}\mathbf{H}']^{-1}(\mathbf{H}\tilde{\boldsymbol{\beta}}_{\mathrm{D}}-\mathbf{h}).\end{aligned} \tag{12}$$

Therefore,

$$\begin{aligned}\max_{\boldsymbol{\beta},\sigma^2,\lambda}[\ell(\boldsymbol{\beta},\sigma^2,\mathbf{Y})-\lambda'(\mathbf{H}\boldsymbol{\beta}-\mathbf{h})] &= \ell(\tilde{\boldsymbol{\beta}}_{\mathrm{N}},\tilde{\sigma}_{\mathrm{N}}^2\mathbf{Y}) \\ &= (2\pi)^{-n/2}\left(\tilde{\sigma}_{\mathrm{N}}^2\right)^{-n/2}e^{-n/2}.\end{aligned}$$

The likelihood ratio statistic L therefore equals

$$L = \left(\tilde{\sigma}_{\mathrm{D}}^2/\tilde{\sigma}_{\mathrm{N}}^2\right)^{n/2}.$$

Instead of using L as the test statistic, use a monotonic function of L:

$$\begin{aligned}V &= (L^{-2/n}-1)(n-p)/q \\ &= \frac{\left(\tilde{\sigma}_{\mathrm{N}}^2-\tilde{\sigma}_{\mathrm{D}}^2\right)/q}{\tilde{\sigma}_{\mathrm{D}}^2/(n-p)} \\ &= \frac{(\mathbf{H}\tilde{\boldsymbol{\beta}}_{\mathrm{D}}-\mathbf{h})'[\mathbf{H}(\mathbf{X}'\mathbf{X})^{-1}\mathbf{H}']^{-1}(\mathbf{H}\tilde{\boldsymbol{\beta}}_{\mathrm{D}}-\mathbf{h})/q}{\mathbf{Y}'(\mathbf{I}_n-\mathbf{X}(\mathbf{X}'\mathbf{X})^{-1}\mathbf{X}')\mathbf{Y}/(n-p)}.\end{aligned}$$

The MLE $\tilde{\boldsymbol{\beta}}_{\mathrm{D}}$ will subsequently be referred to as the least-squares estimator $\hat{\boldsymbol{\beta}}$ since $\tilde{\boldsymbol{\beta}}_{\mathrm{D}} = \hat{\boldsymbol{\beta}}$. The derivations of all three forms of V essentially follow the derivations presented by Graybill (1976).

The distribution of V is needed before the statistic V can be used to test the hypothesis $\mathbf{H}\boldsymbol{\beta} = \mathbf{h}$. Note that $\mathbf{H}\hat{\boldsymbol{\beta}} = \mathbf{H}(\mathbf{X}'\mathbf{X})^{-1}\mathbf{X}'\mathbf{Y}$; therefore, by Theorem 2.1.2, $\mathbf{H}\hat{\boldsymbol{\beta}} \sim \mathrm{N}_q(\mathbf{H}\boldsymbol{\beta}, \sigma^2\mathbf{H}(\mathbf{X}'\mathbf{X})^{-1}\mathbf{H}')$. Under $\mathrm{H}_0 : \mathbf{H}\boldsymbol{\beta} = \mathbf{h}, \mathbf{H}\hat{\boldsymbol{\beta}} - \mathbf{h} \sim \mathrm{N}_q(\mathbf{0}, \sigma^2\mathbf{H}(\mathbf{X}'\mathbf{X})^{-1}\mathbf{H}')$. By Corollary 3.1.2(b), $(\mathbf{H}\hat{\boldsymbol{\beta}}-\mathbf{h})'[\mathbf{H}(\mathbf{X}'\mathbf{X})^{-1}\mathbf{H}']^{-1}(\mathbf{H}\hat{\boldsymbol{\beta}}-\mathbf{h}) \sim \sigma^2\chi_q^2(\lambda)$ where $\lambda = 0$ if $\mathbf{H}\boldsymbol{\beta} = \mathbf{h}$ and $\lambda > 0$ if $\mathbf{H}\boldsymbol{\beta} \neq \mathbf{h}$. Furthermore, by Theorem 3.2.1, $(\mathbf{H}\hat{\boldsymbol{\beta}})'[\mathbf{H}(\mathbf{X}'\mathbf{X})^{-1}\mathbf{H}']^{-1}\mathbf{H}\hat{\boldsymbol{\beta}}$ is independent of $\mathbf{Y}'[\mathbf{I}_n - \mathbf{X}(\mathbf{X}'\mathbf{X})^{-1}\mathbf{X}']\mathbf{Y}$ since

$$(\mathbf{H}\hat{\boldsymbol{\beta}})'[\mathbf{H}(\mathbf{X}'\mathbf{X})^{-1}\mathbf{H}']^{-1}\mathbf{H}\hat{\boldsymbol{\beta}} = \mathbf{Y}'\mathbf{X}(\mathbf{X}'\mathbf{X})^{-1}\mathbf{H}'[\mathbf{H}(\mathbf{X}'\mathbf{X})^{-1}\mathbf{H}']^{-1}\mathbf{H}(\mathbf{X}'\mathbf{X})^{-1}\mathbf{X}'\mathbf{Y}$$

and

$$[\mathbf{I}_n - \mathbf{X}(\mathbf{X}'\mathbf{X})^{-1}\mathbf{X}'](\sigma^2\mathbf{I}_n)[\mathbf{X}(\mathbf{X}'\mathbf{X})^{-1}\mathbf{H}'[\mathbf{H}(\mathbf{X}'\mathbf{X})^{-1}\mathbf{H}']^{-1}\mathbf{H}(\mathbf{X}'\mathbf{X})^{-1}\mathbf{X}'] = \mathbf{0}_{n\times n}.$$

Since $\mathbf{H}\hat{\boldsymbol{\beta}}-\mathbf{h}$ has the same covariance matrix as $\mathbf{H}\hat{\boldsymbol{\beta}}$ and both vectors are normally distributed, $(\mathbf{H}\hat{\boldsymbol{\beta}}-\mathbf{h})'[\mathbf{H}(\mathbf{X}'\mathbf{X})^{-1}\mathbf{H}']^{-1}(\mathbf{H}\hat{\boldsymbol{\beta}}-\mathbf{h})$ is also independent of $\mathbf{Y}'[\mathbf{I}_n - \mathbf{X}(\mathbf{X}'\mathbf{X})^{-1}\mathbf{X}']\mathbf{Y}$. Therefore, by Definition 3.3.3, $V \sim F_{q,n-p}(\lambda)$. Note that all three forms of the statistic V are equal and therefore they all have the same F distribution. The various forms of the likelihood ratio statistic are now applied to a number of example problems.

Example 6.4.1 Consider the model $\mathbf{Y} = \mathbf{X}\boldsymbol{\beta} + \mathbf{E}$ where $\mathbf{Y}$ is an $n \times 1$ random vector, $\mathbf{X} = (\mathbf{X}_1|\mathbf{X}_2)$ is an $n \times p$ matrix where $\mathbf{X}_1$ is an $n \times (p-q)$ matrix and $\mathbf{X}_2$ is an $n \times q$ matrix, $\boldsymbol{\beta} = (\beta_0, \ldots, \beta_{p-q-1}|\beta_{p-q}, \ldots, \beta_{p-1})'$ is a $p \times 1$ vector and the $n \times 1$ error vector $\mathbf{E} \sim \mathrm{N}_n(\mathbf{0}, \sigma^2\mathbf{I}_n)$. The objective is to construct a likelihood ratio test for the hypothesis $\mathrm{H}_0 : \beta_{p-q} = \cdots = \beta_{p-1} = 0$ versus H_1 : not all $\beta_i = 0$ for $i = p-q, \ldots, p-1$. The hypothesis H_0 is equivalent to the hypothesis $\mathbf{H}\boldsymbol{\beta} = \mathbf{h}$ where the $q \times p$ matrix $\mathbf{H} = [\mathbf{0}_{q\times(p-q)}|\mathbf{I}_q]$ and the $q \times 1$ vector $\mathbf{h} = \mathbf{0}$. Since $\mathbf{h} = \mathbf{0}, \mathbf{Z} = \mathbf{Y} - \mathbf{XH}(\mathbf{HH}')^{-1}\mathbf{h} = \mathbf{Y}$. Furthermore, let the $(p-q) \times p$ matrix $\mathbf{G} = [\mathbf{I}_{p-q}|\mathbf{0}_{(p-q)\times q}]$ so that $\left[\begin{smallmatrix}\mathbf{H}\\ \mathbf{G}\end{smallmatrix}\right]$ has rank p and $\mathbf{HG}' = \mathbf{0}_{q\times p}$. Therefore,

$$\begin{aligned}
\mathbf{K} &= \mathbf{XG}'(\mathbf{GG}')^{-1} \\
&= (\mathbf{X}_1|\mathbf{X}_2)\begin{bmatrix}\mathbf{I}_{p-q}\\ \mathbf{0}\end{bmatrix}\left\{[\mathbf{I}_{p-q}|\mathbf{0}]\begin{bmatrix}\mathbf{I}_{p-q}\\ \mathbf{0}\end{bmatrix}\right\}^{-1} \\
&= \mathbf{X}_1
\end{aligned}$$

and the first form of the likelihood ratio statistic V equals

$$V = \frac{\{\mathbf{Y}'(\mathbf{I}_n - \mathbf{X}_1(\mathbf{X}_1'\mathbf{X}_1)^{-1}\mathbf{X}_1')\mathbf{Y} - \mathbf{Y}'(\mathbf{I}_n - \mathbf{X}(\mathbf{X}'\mathbf{X})^{-1}\mathbf{X}')\mathbf{Y}\}/q}{\mathbf{Y}'(\mathbf{I}_n - \mathbf{X}(\mathbf{X}'\mathbf{X})^{-1}\mathbf{X}')\mathbf{Y}/(n-p)}.$$

Label $\mathbf{Y} = \mathbf{X}\boldsymbol{\beta} + \mathbf{E}$ as the full model with p unknown parameters in $\boldsymbol{\beta}$. Label $\mathbf{Y} = \mathbf{X}_1\boldsymbol{\beta}^{(1)} + \mathbf{E}_1$ as the reduced model with $p-q$ unknown parameters in $\boldsymbol{\beta}^{(1)} = (\beta_0, \ldots, \beta_{p-q-1})'$. The first form of the likelihood ratio statistic equals

$$V = \frac{(\mathrm{SSE_R} - \mathrm{SSE_F})/q}{\mathrm{SSE_F}/(n-p)} \sim F_{q,n-p}(\lambda)$$

where $\mathrm{SSE_R}$ is the sum of squares residual for the reduced model and $\mathrm{SSE_F}$ is the sum of squares residual for the full model. A γ level test of H_0 versus H_1 is to reject H_0 if $V > F^{\gamma}_{q,n-p}$. Note that $\mathbf{Y}'(\mathbf{I}_n - \frac{1}{n}\mathbf{J}_n)\mathbf{Y} = \mathrm{SSReg_R} + \mathrm{SSE_R} = \mathrm{SSReg_F} + \mathrm{SSE_F}$, which implies $\mathrm{SSE_R} - \mathrm{SSE_F} = \mathrm{SSReg_F} - \mathrm{SSReg_R}$ where $\mathrm{SSReg_F}$ and $\mathrm{SSReg_R}$ are the sum of squares regression for the full and reduced models, respectively. Therefore, the likelihood ratio statistic for this problem also takes the form

$$V = \frac{(\mathrm{SSReg_F} - \mathrm{SSReg_R})/q}{\mathrm{SSE_F}/(n-p)} \sim F_{q,n-p}(\lambda).$$

Example 6.4.2 Consider a special case of the problem posed in Example 6.4.1. From Example 6.4.1 let the $n \times p$ matrix $\mathbf{X} = (\mathbf{X}_1|\mathbf{X}_2)$ where $\mathbf{X}_1$ equals the $n \times 1$ vector $\mathbf{1}_n$ and $\mathbf{X}_2$ equals the $n \times (p-1)$ matrix $\mathbf{X}_c$ such that $\mathbf{1}_n'\mathbf{X}_c = \mathbf{0}_{1\times(p-1)}$, and partition the $p \times 1$ vector $\boldsymbol{\beta}$ as $(\beta_0|\beta_1, \ldots, \beta_{p-1})'$. The objective is to test the hypothesis $H_0 : \beta_1 = \cdots = \beta_{p-1} = 0$ versus H_1 : not all $\beta_i = 0$ for $i = 1, \ldots, p-1$. From Example 6.4.1, the likelihood ratio statistic equals

$$\begin{aligned} V &= \frac{\{\mathbf{Y}'(\mathbf{I}_n - \mathbf{1}_n(\mathbf{1}_n'\mathbf{1}_n)^{-1}\mathbf{1}_n')\mathbf{Y} - \mathbf{Y}'(\mathbf{I}_n - \mathbf{X}(\mathbf{X}'\mathbf{X})^{-1}\mathbf{X}')\mathbf{Y}\}/(p-1)}{\mathbf{Y}'(\mathbf{I}_n - \mathbf{X}(\mathbf{X}'\mathbf{X})^{-1}\mathbf{X}')\mathbf{Y}/(n-p)} \\ &= \frac{\mathbf{Y}'[\mathbf{X}(\mathbf{X}'\mathbf{X})^{-1}\mathbf{X}' - \frac{1}{n}\mathbf{J}_n]\mathbf{Y}/(p-1)}{\mathbf{Y}'(\mathbf{I}_n - \mathbf{X}(\mathbf{X}'\mathbf{X})^{-1}\mathbf{X}')\mathbf{Y}/(n-p)} \\ &= \frac{\mathbf{Y}'[\mathbf{X}_c(\mathbf{X}_c'\mathbf{X}_c)^{-1}\mathbf{X}_c']\mathbf{Y}/(p-1)}{\mathbf{Y}'(\mathbf{I}_n - \mathbf{X}(\mathbf{X}'\mathbf{X})^{-1}\mathbf{X}')\mathbf{Y}/(n-p)} \sim F_{p-1,n-p}(\lambda). \end{aligned}$$

Therefore, V equals the mean square regression due to $\beta_1, \ldots, \beta_{p-1}$ divided by the mean square residual, as depicted in Table 5.3.1.

Example 6.4.3 Consider the one-way classification from Example 2.1.4. The experiment can be modeled as $\mathbf{Y} = \mathbf{X}\beta + \mathbf{E}$ where the $tr \times 1$ vector $\mathbf{Y} = (Y_{11}, \ldots, Y_{1r}, \ldots, Y_{t1}, \ldots, Y_{tr})'$, the $tr \times t$ matrix $\mathbf{X} = [\mathbf{I}_t \otimes \mathbf{1}_r]$, the $t \times 1$ vector $\beta = (\beta_1, \ldots, \beta_t)' = (\mu_1, \ldots, \mu_t)'$ where $\mu_1, \ldots, \mu_t$ are defined in Example 2.1.4, and the $n \times 1$ random vector $\mathbf{E} \sim \mathrm{N}_{tr}(\mathbf{0}, \sigma^2_{R(T)}\mathbf{I}_t \otimes \mathbf{I_r})$. The objective is to construct the likelihood ratio test for the hypothesis $H_0 : \beta_1 = \cdots = \beta_t$ versus H_1 : not all β_i equal. The hypothesis H_0 is equivalent to the hypothesis $\mathbf{H}\beta = \mathbf{h}$ with the $(t-1) \times t$ matrix $\mathbf{H} = \mathbf{P}_t'$ and the $(p-1) \times 1$ vector $\mathbf{h} = \mathbf{0}$ where $\mathbf{P}_t'$ is the $(t-1) \times t$ lower portion of a t-dimensional Helmert matrix with $\mathbf{P}_t'\mathbf{P}_t = \mathbf{I}_{t-1}$, $\mathbf{P}_t\mathbf{P}_t' = \mathbf{I}_t - \frac{1}{t}\mathbf{J}_t$, and $\mathbf{1}_t'\mathbf{P}_t = \mathbf{0}_{1\times(t-1)}$. Note

$$\begin{aligned} \hat{\beta} &= (\mathbf{X}'\mathbf{X})^{-1}\mathbf{X}'\mathbf{Y} = [(\mathbf{I}_t \otimes \mathbf{1}_r)'(\mathbf{I}_t \otimes \mathbf{1}_r)]^{-1}(\mathbf{I}_t \otimes \mathbf{1}_r)'\mathbf{Y} \\ &= \left(\mathbf{I}_t \otimes \frac{1}{r}\mathbf{1}_r'\right)\mathbf{Y} \end{aligned}$$

$$\mathbf{H}\hat{\beta} - \mathbf{h} = \mathbf{P}_t'\left(\mathbf{I}_t \otimes \frac{1}{r}\mathbf{1}_r'\right)\mathbf{Y} = \left(\mathbf{P}_t' \otimes \frac{1}{r}\mathbf{1}_r'\right)\mathbf{Y}$$

$$\begin{aligned} [\mathbf{H}(\mathbf{X}'\mathbf{X})^{-1}\mathbf{H}']^{-1} &= \{\mathbf{P}_t'[(\mathbf{I}_t \otimes \mathbf{1}_r)'(\mathbf{I}_t \otimes \mathbf{1}_r)]^{-1}\mathbf{P}_t\}^{-1} \\ &= (\mathbf{P}_t'\mathbf{P}_t)^{-1} \otimes r = \mathbf{I}_{t-1} \otimes r \end{aligned}$$

$$\begin{aligned} \mathbf{I}_t \otimes \mathbf{I}_r - \mathbf{X}(\mathbf{X}'\mathbf{X})^{-1}\mathbf{X}' &= \mathbf{I}_t \otimes \mathbf{I}_r \\ &\quad - \{(\mathbf{I}_t \otimes \mathbf{1}_r)[(\mathbf{I}_t \otimes \mathbf{1}_r)'(\mathbf{I}_t \otimes \mathbf{1}_r)]^{-1}(\mathbf{I}_t \otimes \mathbf{1}_r)'\} \\ &= \mathbf{I}_t \otimes \mathbf{I}_r - \left(\mathbf{I}_t \otimes \frac{1}{r}\mathbf{J}_r\right) \\ &= \mathbf{I}_t \otimes \left(\mathbf{I}_r - \frac{1}{r}\mathbf{J}_r\right). \end{aligned}$$

Therefore, the third form of the likelihood ratio statistic is given by

$$V = \frac{\mathbf{Y}'\left[\left(\mathbf{P}_t' \otimes \frac{1}{r}\mathbf{1}_r'\right)'(\mathbf{I}_{t-1} \otimes r)\left(\mathbf{P}_t' \otimes \frac{1}{r}\mathbf{1}_r'\right)\right]\mathbf{Y}/(t-1)}{\mathbf{Y}'\mathbf{I}_t \otimes \left(\mathbf{I}_r - \frac{1}{r}\mathbf{J}_r\right)\mathbf{Y}/(tr-t)}$$

$$= \frac{\mathbf{Y}'\left[\left(\mathbf{I}_t - \frac{1}{t}\mathbf{J}_t\right) \otimes \frac{1}{r}\mathbf{J}_r\right]\mathbf{Y}/(t-1)}{\mathbf{Y}'\left[\mathbf{I}_t \otimes \left(\mathbf{I}_r - \frac{1}{r}\mathbf{J}_r\right)\right]\mathbf{Y}/[t(r-1)]} \sim F_{t-1,t(r-1)}(\lambda).$$

Therefore, V equals the mean square for the treatments divided by the mean square for the nested replicates, which is the usual ANOVA test for equality of the treatment means.

6.5 CONFIDENCE BANDS ON LINEAR COMBINATIONS OF β

The likelihood ratio statistic is now used to construct confidence bands on individual linear combinations of β. Assume the model $\mathbf{Y} = \mathbf{X}\beta + \mathbf{E}$ where $\mathbf{E} \sim \mathrm{N}_n(\mathbf{0}, \sigma^2\mathbf{I}_n)$. Let the $q \times p$ matrix $\mathbf{H}$ and the $q \times 1$ vector $\mathbf{h}$ from Section 6.4 equal a $1 \times p$ row vector $\mathbf{g}'$ and a scalar g_0, respectively. The hypothesis $\mathrm{H}_0 : \mathbf{H}\beta = \mathbf{h}$ versus $\mathrm{H}_1 : \mathbf{H}\beta \neq \mathbf{h}$ becomes $\mathrm{H}_0 : \mathbf{g}'\beta = g_0$ versus $\mathrm{H}_1 : \mathbf{g}'\beta \neq g_0$. The third form of the likelihood ratio statistic equals

$$V = \frac{(\mathbf{g}'\hat{\beta} - g_0)^2}{[\mathbf{Y}'(\mathbf{I}_n - \mathbf{X}(\mathbf{X}'\mathbf{X})^{-1}\mathbf{X}')\mathbf{Y}/(n-p)][\mathbf{g}'(\mathbf{X}'\mathbf{X})^{-1}\mathbf{g}]} \sim F_{1,n-p}(\lambda)$$

or

$$\sqrt{V} = \frac{(\mathbf{g}'\hat{\beta} - g_0)}{\{[\mathbf{Y}'(\mathbf{I}_n - \mathbf{X}(\mathbf{X}'\mathbf{X})^{-1}\mathbf{X}')\mathbf{Y}/(n-p)][\mathbf{g}'(\mathbf{X}'\mathbf{X})^{-1}\mathbf{g}]\}^{1/2}} \sim t_{n-p}(\lambda).$$

Therefore,

$$1 - \gamma = P\left(-t_{n-p}^{\gamma/2} \leq \sqrt{V} \leq t_{n-p}^{\gamma/2}\right)$$

where $t_{n-p}^{\gamma/2}$ is the $100(1-\gamma/2)^{\text{th}}$ percentile point of a central t distribution with $n-p$ degrees of freedom. Substitute for $\sqrt{V}$ and solve for $g_0 = \mathbf{g}'\beta$ in the preceding probability statement. The resulting $100(1-\gamma)\%$ confidence band on $\mathbf{g}'\beta$ is

$$\mathbf{g}'\hat{\beta} \pm t_{n-p}^{\gamma/2}\{[\mathbf{Y}'(\mathbf{I}_n - \mathbf{X}(\mathbf{X}'\mathbf{X})^{-1}\mathbf{X}')\mathbf{Y}/(n-p)][\mathbf{g}'(\mathbf{X}'\mathbf{X})^{-1}\mathbf{g}]\}^{1/2}.$$

In the next example, confidence bands are placed on certain linear combinations of the treatment means in a one-way ANOVA problem.

Example 6.5.1 Consider the one-way ANOVA problem from Example 6.4.3. The objective is to construct individual confidence bands on $\beta_1 - \beta_2$, and on $\bar{\beta}. = \sum_{i=1}^{t} \beta_i/t = \mathbf{1}_t'\boldsymbol{\beta}/t$. For the one-way ANOVA problem, $n = tr$, $p = t$,

$$\begin{aligned}\hat{\boldsymbol{\beta}} &= (\mathbf{X}'\mathbf{X})^{-1}\mathbf{X}'\mathbf{Y} \\ &= [(\mathbf{I}_t \otimes \mathbf{1}_r)'(\mathbf{I}_t \otimes \mathbf{1}_r)]^{-1}(\mathbf{I}_t \otimes \mathbf{1}_r)'\mathbf{Y} \\ &= (\bar{Y}_{1.}, \ldots, \bar{Y}_{t.})'\end{aligned}$$

and

$$\begin{aligned}\mathbf{Y}'[\mathbf{I}_n - \mathbf{X}(\mathbf{X}'\mathbf{X})^{-1}\mathbf{X}']\mathbf{Y} &= \mathbf{Y}'\{\mathbf{I}_t \otimes \mathbf{I}_r - (\mathbf{I}_t \otimes \mathbf{1}_r)[(\mathbf{I}_t \otimes \mathbf{1}_r)' \\ &\quad (\mathbf{I}_t \otimes \mathbf{1}_r)]^{-1}(\mathbf{I}_t \otimes \mathbf{1}_r)'\}\mathbf{Y} \\ &= \mathbf{Y}'\left[\mathbf{I}_t \otimes \left(\mathbf{I}_r - \frac{1}{r}\mathbf{J}_r\right)\right]\mathbf{Y}.\end{aligned}$$

For $\beta_1 - \beta_2$, the $t \times 1$ vector $\mathbf{g} = (1, -1, 0, \ldots, 0)'$, $\mathbf{g}'\hat{\boldsymbol{\beta}} = \bar{Y}_{1.} - \bar{Y}_{2.}$ and

$$\begin{aligned}\mathbf{g}'(\mathbf{X}'\mathbf{X})^{-1}\mathbf{g} &= (1, -1, 0, \ldots, 0)[(\mathbf{I}_t \otimes \mathbf{1}_r)'(\mathbf{I}_t \otimes \mathbf{1}_r)]^{-1} \\ &\quad (1, -1, 0, \ldots, 0)' \\ &= 2/r.\end{aligned}$$

For $\bar{\beta}. = (1/t)\mathbf{1}_t'\boldsymbol{\beta}$, the $t \times 1$ vector $\mathbf{g} = (1/t)\mathbf{1}_t$, $\mathbf{g}'\hat{\boldsymbol{\beta}} = \bar{Y}..$ and

$$\begin{aligned}\mathbf{g}'(\mathbf{X}'\mathbf{X})^{-1}\mathbf{g} &= [(1/t)\mathbf{1}_t]'[(\mathbf{I}_t \otimes \mathbf{1}_r)'(\mathbf{I}_t \otimes \mathbf{1}_r)]^{-1}[(1/t)\mathbf{1}_t] \\ &= (rt)^{-1}.\end{aligned}$$

Therefore, $100(1 - \gamma)\%$ confidence bands on $\beta_1 - \beta_2$ are

$$\bar{Y}_{1.} - \bar{Y}_{2.} \pm t_{t(r-1)}^{\gamma/2}\left\{2\mathbf{Y}'\left[\mathbf{I}_t \otimes \left(\mathbf{I}_r - \frac{1}{r}\mathbf{J}_r\right)\right]\mathbf{Y}/[tr(r-1)]\right\}^{1/2}$$

and $100(1 - \gamma)\%$ confidence bands on $\bar{\beta}.$ are

$$\bar{Y}.. \pm t_{t(r-1)}^{\gamma/2}\left\{\mathbf{Y}'\left[\mathbf{I}_t \otimes \left(\mathbf{I}_r - \frac{1}{r}\mathbf{J}_r\right)\right]\mathbf{Y}/[t^2 r(r-1)]\right\}^{1/2}.$$

EXERCISES

1. Let $Y_i = \beta_0 + \beta_1 x_i + E_i$ for $i = 1, \ldots, n$ where $E_i \sim iid\ \mathrm{N}_1(0, \sigma^2)$. Construct individual confidence bands on β_0 and β_1 and write the answer in terms of $\sum x_i$, $\sum Y_i$, $\sum x_i^2$, $\sum Y_i^2$, $\Sigma x_i Y_i$, and n.

2. Consider the one-way classification from Example 6.4.3. Find the minimum variance unbiased estimator of $\sum_{i=1}^{t} \mu_i/\sigma^2_{R(T)}$.

3. Let $Y_1, Y_2, \ldots, Y_n$ be a random sample of size n from a $N_1(\mu, \sigma^2)$ distribution where $-\infty < \mu < \infty$ and $\sigma^2 > 0$. Consider the γ level test $H_0 : \mu = \mu_0$ versus $H_1 : \mu \neq \mu_0$ where μ_0 is given. Show that the likelihood ratio test is equivalent to the two-sided test, reject H_0 if $|T| > t^{\gamma/2}_{n-1}$ where $T = (\bar{Y} - \mu_0)/[s/\sqrt{n}]$, $\bar{Y} = \sum_{i=1}^{n} Y_i$ and $s^2 = \sum_{i=1}^{n} (Y_i - \bar{Y})^2/(n-1)$.

4. Let $\mathbf{Y} = \mathbf{X}\boldsymbol{\beta} + \mathbf{E}$ where $\mathbf{E} \sim N_n(0, \sigma^2 \mathbf{I}_n)$. A $100(1-\gamma)\%$ confidence interval on $\mathbf{h}'\beta$ is $\mathbf{h}'\hat{\beta} \pm t^{\gamma/2}_{n-p}[\widehat{\text{std}}(\mathbf{h}'\hat{\beta})]$ where $\widehat{\text{std}}(\mathbf{h}'\hat{\beta})$ is the estimated standard deviation of $\mathbf{h}'\hat{\beta}$.

   (a) Find the distribution of L^2 where L is the length of the interval.

   (b) If $p = 3, n = 20, \boldsymbol{\beta} = (\beta_1, \beta_2, \beta_3)'$, $\mathbf{Y}'\mathbf{Y} = 418$, $\mathbf{X}'\mathbf{X} = \mathbf{I}_3$, and $\mathbf{X}'\mathbf{Y} = (5, 10, 15)'$, construct individual 95% confidence bands on $\beta_1, \beta_1 + \beta_2$, and $\beta_1 - \beta_3$.

5. Let $\mathbf{Y}_1 = \mu_1 + \delta_1$, $\mathbf{Y}_2 = \mu_1 + \mu_2 + \delta_1 + \delta_2$, $\mathbf{Y}_3 = \mu_1 + \delta_3$, and $\mathbf{Y}_4 = \mu_1 + \mu_2 + \delta_3 + \delta_4$ where μ_1 and μ_2 are unknown parameters and $\delta = (\delta_1, \delta_2, \delta_3, \delta_4)' \sim N_4(\mathbf{0}, \sigma^2_\delta \mathbf{I}_4)$.

   (a) Write the model as $\mathbf{Y} = \mathbf{X}\boldsymbol{\beta} + \mathbf{E}$ where $\mathbf{E} = (E_1, E_2, E_3, E_4)'$. Define each term and distribution explicitly.

   (b) Find the MLE of $\boldsymbol{\mu} = (\mu_1, \mu_2)'$.

   (c) Find the distribution of the MLE of $\boldsymbol{\mu}$.

6. Let $Y_{ij} = \mu_i + E_{ij}$ where the E_{ij}'s are distributed as independent $N_1(0, j\sigma^2)$ random variables. Let $\mathbf{Y} = (Y_{11}, Y_{12}, Y_{21}, Y_{22})'$. Find the likelihood ratio statistic for testing the hypothesis $H_0 : \mu_1 = \mu_2$ versus $H_1 : \mu_1 \neq \mu_2$.

7. Consider the paired t-test problem of Exercise 5 in Chapter 4.

   (a) Calculate the MLEs of μ_1, μ_2, σ^2_B, and σ^2_{BT}.

   (b) Construct the likelihood ratio statistic for testing the hypothesis $H_0 = \mu_1 = \mu_2$ versus $H_1 : \mu_1 \neq \mu_2$.

8. Consider the experiment from Exercise 6 in Chapter 4. Calculate the MLEs of $\mu_{11}, \ldots, \mu_{53}, \sigma^2_{P(F)}, \sigma^2_{C(FP)}, \sigma^2_{PT(F)}$, and $\sigma^2_{CT(FP)}$.

9. Consider the experiment from Exercise 7 in Chapter 4.

   (a) Calculate the MLEs of $\mu_{111}, \ldots, \mu_{222}, \sigma^2_{P(S)}, \sigma^2_{PM(S)}$, and $\sigma^2_{PO(MS)}$.

   (b) Write the model as $\mathbf{Y} = \mathbf{X}\boldsymbol{\beta} + \mathbf{E}$. Define the vector $\mathbf{p}$ in Kronecker product form such that $\mathbf{p}'\boldsymbol{\beta} = \bar{\mu}_{1..} - \bar{\mu}_{2..} = (\mu_{111} + \mu_{112} + \mu_{121} + \mu_{122} - \mu_{211} - \mu_{212} - \mu_{221} - \mu_{222})/4$.

(c) Find the MLE of $\mathbf{p}'\boldsymbol{\beta}$ from part b.

(d) Find the standard error of the MLE of $\mathbf{p}'\boldsymbol{\beta}$.

10. Consider the experiment from Example 4.5.4.

(a) Find Σ^{-1} in terms of σ_B^2 and σ_{BV}^2.

(b) Write the model as $\mathbf{Y} = \mathbf{X}\boldsymbol{\beta} + \mathbf{E}$ and find the MLEs of $\boldsymbol{\beta}$, σ_B^2, and σ_{BV}^2.

11. Under the conditions of Theorem 6.3.1, prove that the eigenvalues of Σ are $a_1, \ldots, a_m$ with multiplicities $p_1 + r_1, \ldots, p_m + r_m$, respectively.

7 Unbalanced Designs and Missing Data

In complete, balanced factorial experiments, the same number of replicate values is observed within each combination of the factors. Kronecker products may be used in such experiments to construct covariance and sum of squares matrices. However, in other types of experiments, the number of replicates per combination of the factors varies, or certain factor combinations may have no observations at all. Kronecker products are often not useful for constructing covariance and sums of squares matrices in such unbalanced and missing data experiments. To accommodate these unbalanced and missing data situations, replication and pattern matrices are introduced.

7.1 REPLICATION MATRICES

This section beings with a reexamination of the data in Table 5.1.1. This data set contains four distinct combinations of speed and grade; {(speed, grade)} {(20, 0), (20, 6), (50, 0), (50, 6)}. Four replicates $(Y_1, \ldots, Y_4)$ were observed in the (20, 0) speed $\times$ grade combination, one replicate (Y_5) in the (20, 6) combination, two replicates (Y_6, Y_7) in the (50, 0) combination, and three replicates (Y_8, Y_9, Y_{10}) in the (50, 6) combination. Thus, the observations Y_i appear in $k = 4$ distinct groups with $r_1 = 4, r_2 = 1, r_3 = 2$, and $r_4 = 3$ observations per group.

Assume the model $\mathbf{Y} = \mathbf{X}\boldsymbol{\beta} + \mathbf{E}$ where the 10×1 random vector $\mathbf{Y} = (Y_1, \ldots, Y_{10})'$, the 3×1 vector $\boldsymbol{\beta} = (\beta_1, \beta_2, \beta_3)'$, the 10×1 error vector $\mathbf{E} \sim \mathrm{N}_{10}(\mathbf{0}, \sigma^2\mathbf{I}_{10})$, and the 10×3 matrix $\mathbf{X}$ is given by

$$\mathbf{X} = \begin{bmatrix} \mathbf{1}_4 \otimes (1, 20, \quad 0) \\ \mathbf{1}_1 \otimes (1, 20, 120) \\ \mathbf{1}_2 \otimes (1, 50, \quad 0) \\ \mathbf{1}_3 \otimes (1, 50, 300) \end{bmatrix}.$$

where the three columns of $\mathbf{X}$ correspond to an intercept, a speed variable, and a speed $\times$ grade variable. Note that the rows of $\mathbf{X}$ appear in four distinct groups with r_j rows per group for $j = 1, \ldots, 4$. Therefore, the 10×3 matrix $\mathbf{X}$ can be rewritten as $\mathbf{X} = \mathbf{R}\mathbf{X}_\mathrm{d}$ where the 10×4 replication matrix $\mathbf{R}$ is given by

$$\mathbf{R} = \begin{bmatrix} \mathbf{1}_4 & & & \\ & \mathbf{1}_1 & & \mathbf{0} \\ & \mathbf{0} & \mathbf{1}_2 & \\ & & & \mathbf{1}_3 \end{bmatrix}$$

and the 4×3 matrix $\mathbf{X}_\mathrm{d}$ defines the distinct speed and speed $\times$ grade combinations where

$$\mathbf{X}_\mathrm{d} = \begin{bmatrix} 1 & 20 & 0 \\ 1 & 20 & 120 \\ 1 & 50 & 0 \\ 1 & 50 & 300 \end{bmatrix}.$$

In general, the model $\mathbf{Y} = \mathbf{X}\boldsymbol{\beta} + \mathbf{E}$ can always be rewritten as $\mathbf{Y} = \mathbf{R}\mathbf{X}_\mathrm{d}\boldsymbol{\beta} + \mathbf{E}$ where $r_j \geq 1$ is the number of replicate observations for the jth distinct set of $x_1, \ldots, x_{p-1}$ values for $j = 1, \ldots, k \geq p$; the replication matrix $\mathbf{R}$ is an $n \times k$ block diagonal matrix with $\mathbf{1}_{r_j}$ on the diagonal; $\mathbf{X}_\mathrm{d}$ is a $k \times p$ matrix of distinct $x_1, \ldots, x_{p-1}$ values with $\mathbf{1}_k$ in the first column; and $n = \sum_{j=1}^{k} r_j$.

The ordinary least-squares estimate of $\boldsymbol{\beta}$ is

$$\begin{aligned} \hat{\boldsymbol{\beta}} &= (\mathbf{X}_\mathrm{d}'\mathbf{R}'\mathbf{R}\mathbf{X}_\mathrm{d})^{-1}\mathbf{X}_\mathrm{d}'\mathbf{R}'\mathbf{Y} \\ &= (\mathbf{X}_\mathrm{d}'\mathbf{D}\mathbf{X}_\mathrm{d})^{-1}\mathbf{X}_\mathrm{d}'\{\mathbf{D}\mathbf{D}^{-1}\}\mathbf{R}'\mathbf{Y} \\ &= (\mathbf{X}_\mathrm{d}'\mathbf{D}\mathbf{X}_\mathrm{d})^{-1}\mathbf{X}_\mathrm{d}'\mathbf{D}\bar{\mathbf{Y}} \end{aligned}$$

where the $k \times k$ diagonal matrix

$$\mathbf{D} = \mathbf{R}'\mathbf{R} = \begin{bmatrix} r_1 & & & \\ & r_2 & \mathbf{0} & \\ & \mathbf{0} & \ddots & \\ & & & r_k \end{bmatrix}$$

and the $k \times 1$ vector $\bar{\mathbf{Y}} = \mathbf{D}^{-1}\mathbf{R}'\mathbf{Y} = (\bar{Y}_1, \ldots, \bar{Y}_k)'$. The random variable $\bar{Y}_j$ is the average of the observed Y values in the jth group identified by distinct $x_1, \ldots, x_{p-1}$ values. The least-squares estimate of σ^2 is

$$\begin{aligned} \hat{\sigma}^2 &= \mathbf{Y}'[\mathbf{I}_n - \mathbf{R}\mathbf{X}_\text{d}(\mathbf{X}_\text{d}'\mathbf{R}'\mathbf{R}\mathbf{X}_\text{d})^{-1}\mathbf{X}_\text{d}'\mathbf{R}']\mathbf{Y}/(n-p) \\ &= \mathbf{Y}'[\mathbf{I}_n - \mathbf{R}\mathbf{X}_\text{d}(\mathbf{X}_\text{d}'\mathbf{D}\mathbf{X}_\text{d})^{-1}\mathbf{X}_\text{d}'\mathbf{R}']\mathbf{Y}/(n-p). \end{aligned}$$

The least-squares estimator of σ^2 is the mean square residual and this estimator is unbiased for σ^2 provided $\text{E}(\mathbf{Y}) = \mathbf{X}\boldsymbol{\beta} = \mathbf{R}\mathbf{X}_\text{d}\boldsymbol{\beta}$.

In Table 5.5.1 the sum of squares for the mean, regression, lack of fit, and pure error were presented for the model $\mathbf{Y} = \mathbf{X}\boldsymbol{\beta} + \mathbf{E}$. These sums of squares are now reconstructed for the equivalent model $\mathbf{Y} = \mathbf{R}\mathbf{X}_\text{d}\boldsymbol{\beta} + \mathbf{E}$. First, substitute $\mathbf{R}\mathbf{X}_\text{d}$ for $\mathbf{X}$ in the regression sum of squares $\mathbf{Y}'[\mathbf{X}(\mathbf{X}'\mathbf{X})^{-1}\mathbf{X}' - \frac{1}{n}\mathbf{J}_n]\mathbf{Y}$ and into the residual sum of squares $\mathbf{Y}'[\mathbf{I}_n - \mathbf{X}(\mathbf{X}'\mathbf{X})^{-1}\mathbf{X}']\mathbf{Y}$ to obtain

$$\begin{aligned} \text{SSReg} &= \mathbf{Y}'\left[\mathbf{R}\mathbf{X}_\text{d}(\mathbf{X}_\text{d}'\mathbf{D}\mathbf{X}_\text{d})^{-1}\mathbf{X}_\text{d}'\mathbf{R}' - \frac{1}{n}\mathbf{J}_n\right]\mathbf{Y} \\ \text{SSRes} &= \mathbf{Y}'[\mathbf{I}_n - \mathbf{R}\mathbf{X}_\text{d}(\mathbf{X}_\text{d}'\mathbf{D}\mathbf{X}_\text{d})^{-1}\mathbf{X}_\text{d}'\mathbf{R}']\mathbf{Y}. \end{aligned}$$

Observe that the pure error sum of squares can be rewritten as

$$\begin{aligned} \mathbf{Y}'\mathbf{A}_\text{pe}\mathbf{Y} &= \mathbf{Y}'\begin{bmatrix} \mathbf{I}_{r_1} - \frac{1}{r_1}\mathbf{J}_{r_1} & & \mathbf{0} \\ & \ddots & \\ \mathbf{0} & & \mathbf{I}_{r_k} - \frac{1}{r_k}\mathbf{J}_{r_k} \end{bmatrix}\mathbf{Y} \\ &= \mathbf{Y}'\left\{\mathbf{I}_n - \begin{bmatrix} \mathbf{1}_{r_1} & & \mathbf{0} \\ & \ddots & \\ \mathbf{0} & & \mathbf{1}_{r_k} \end{bmatrix}\begin{bmatrix} r_1^{-1} & & \mathbf{0} \\ & \ddots & \\ \mathbf{0} & & r_k^{-1} \end{bmatrix}\begin{bmatrix} \mathbf{1}_{r_1}' & & \mathbf{0} \\ & \ddots & \\ \mathbf{0} & & \mathbf{1}_{r_k}' \end{bmatrix}\right\}\mathbf{Y} \\ &= \mathbf{Y}'[\mathbf{I}_n - \mathbf{R}\mathbf{D}^{-1}\mathbf{R}']\mathbf{Y} \end{aligned}$$

Table 7.1.1
ANOVA Table with Pure Error and Lack of Fit

Source	df	SS
Overall mean	1	$\mathbf{Y}'\frac{1}{n}\mathbf{J}_n\mathbf{Y}$
Regression ($\beta_1, \ldots, \beta_{p-1}$)	$p-1$	$\mathbf{Y}'[\mathbf{R}\mathbf{X}_d(\mathbf{X}_d'\mathbf{D}\mathbf{X}_d)^{-1}\mathbf{X}_d'\mathbf{R}' - \frac{1}{n}\mathbf{J}_n]\mathbf{Y}$
Lack of fit	$k-p$	$\mathbf{Y}'\{\mathbf{R}[\mathbf{D}^{-1} - \mathbf{X}_d(\mathbf{X}_d'\mathbf{D}\mathbf{X}_d)^{-1}\mathbf{X}_d']\mathbf{R}'\}\mathbf{Y}$
Pure error	$n-k$	$\mathbf{Y}'[\mathbf{I}_n - \mathbf{R}\mathbf{D}^{-1}\mathbf{R}']\mathbf{Y}$
Total	n	$\mathbf{Y}'\mathbf{Y}$

where $\mathbf{A}_{pe}$ is the pure error sum of squares matrix defined in Section 5.5. Therefore, by subtraction the lack of fit sum of squares is

$$\begin{aligned}\text{SSLack of Fit} &= \text{SSRes} - \text{SSPure Error}\\ &= \mathbf{Y}'\{[\mathbf{I}_n - \mathbf{R}\mathbf{X}_d(\mathbf{X}_d'\mathbf{D}\mathbf{X}_d)^{-1}\mathbf{X}_d'\mathbf{R}'] - [\mathbf{I}_n - \mathbf{R}\mathbf{D}^{-1}\mathbf{R}']\}\mathbf{Y}\\ &= \mathbf{Y}'\{\mathbf{R}[\mathbf{D}^{-1} - \mathbf{X}_d(\mathbf{X}_d'\mathbf{D}\mathbf{X}_d)^{-1}\mathbf{X}_d']\mathbf{R}'\}\mathbf{Y}.\end{aligned}$$

Table 7.1.1 gives the ANOVA table for the model $\mathbf{Y} = \mathbf{R}\mathbf{X}_d\boldsymbol{\beta} + \mathbf{E}$.

In the next example problem, a replication matrix is used in a one-way classification problem when the number of observations in each level is unequal.

Example 7.1.1 Consider a one-way classification problem with three fixed treatment levels and $r_1 = 3, r_2 = 4$, and $r_3 = 2$ observations in each level, respectively. Let Y_{ij} represent the jth numbered observation in the ith fixed treatment level for $i = 1, 2, 3$ and $j = 1, \ldots, r_i$. The 9×1 vector of observations is $\mathbf{Y} = (Y_{11}, Y_{12}, Y_{13}, Y_{21}, Y_{22}, Y_{23}, Y_{24}, Y_{31}, Y_{32})'$. Assume the model $\mathbf{Y} = \mathbf{R}\mathbf{X}_d\boldsymbol{\beta} + \mathbf{E}$ where the 9×1 random error vector $\mathbf{E} \sim \text{N}_9(\mathbf{0}, \sigma^2_{R(T)}\mathbf{I}_9)$, the 9×3 replication matrix

$$\mathbf{R} = \begin{bmatrix} \mathbf{1}_3 & & \mathbf{0} \\ & \mathbf{1}_4 & \\ \mathbf{0} & & \mathbf{1}_2 \end{bmatrix},$$

the 3×3 matrix $\mathbf{X}_d = \mathbf{I}_3$, and the 3×1 vector $\boldsymbol{\beta} = (\beta_1, \beta_2, \beta_3)'$. For this model, $k = p = 3$ and $n = 9$. It is not possible to compute a lack of fit sum of squares since $k - p = 0$. Furthermore, β_i represents the mean effect of the ith treatment level for $i = 1, 2, 3$. The value $k = 3$ indicates that there are three distinct treatment levels and each treatment level is defined by a row of the $\mathbf{X}_d$ matrix. A 1 in the first row of the $\mathbf{X}_d$ matrix indicates treatment level one, a 1 in the second row indicates treatment level two, and a 1 in the third row indicates treatment level three.

For the example, the sum of squares due to the mean is $\mathbf{Y}'\frac{1}{9}\mathbf{J}_9\mathbf{Y}$ with 1 degree of freedom. The sum of squares for regression is equivalent to the sum of squares for treatments with $p-1 = 3-1 = 2$ degrees of freedom. The matrix for the sum of squares for lack of fit has zero rank. That is, $\mathbf{R}[\mathbf{D}^{-1} - \mathbf{X}_\mathrm{d}(\mathbf{X}_\mathrm{d}'\mathbf{D}\mathbf{X}_\mathrm{d})^{-1}\mathbf{X}_\mathrm{d}']\mathbf{R}' = \mathbf{0}$ or $\mathbf{R}\mathbf{D}^{-1}\mathbf{R}' = \mathbf{R}\mathbf{X}_\mathrm{d}(\mathbf{X}_\mathrm{d}'\mathbf{D}\mathbf{X}_\mathrm{d})^{-1}\mathbf{X}_\mathrm{d}'\mathbf{R}'$. Therefore, the sum of squares for treatments equals

$$\mathbf{Y}'\left[\mathbf{R}\mathbf{X}_\mathrm{d}(\mathbf{X}_\mathrm{d}'\mathbf{D}\mathbf{X}_\mathrm{d})^{-1}\mathbf{X}_\mathrm{d}'\mathbf{R}' - \frac{1}{9}\mathbf{J}_9\right]\mathbf{Y} = \mathbf{Y}'\left[\mathbf{R}\mathbf{D}^{-1}\mathbf{R}' - \frac{1}{9}\mathbf{J}_9\right]\mathbf{Y}$$

$$= \mathbf{Y}'\left\{\begin{bmatrix} \frac{1}{3}\mathbf{J}_3 & & \mathbf{0} \\ & \frac{1}{4}\mathbf{J}_4 & \\ \mathbf{0} & & \frac{1}{2}\mathbf{J}_2 \end{bmatrix} - \frac{1}{9}\mathbf{J}_9\right\}\mathbf{Y}.$$

Finally the sum of squares for pure error equals

$$\mathbf{Y}'[\mathbf{I}_9 - \mathbf{R}\mathbf{D}^{-1}\mathbf{R}']\mathbf{Y} = \mathbf{Y}'\begin{bmatrix} \mathbf{I}_3 - \frac{1}{3}\mathbf{J}_3 & & \mathbf{0} \\ & \mathbf{I}_4 - \frac{1}{4}\mathbf{J}_4 & \\ \mathbf{0} & & \mathbf{I}_2 - \frac{1}{2}\mathbf{J}_2 \end{bmatrix}\mathbf{Y}.$$

Replication matrices can also be used when it is of interest to partition the regression sum of squares into Type I sums of squares. Assume that there are k distinct groups of x values with r_j replicate observations in each group for $j = 1, \ldots, k$. The $n \times p$ matrix $\mathbf{X}$ can be written as

$$\mathbf{X} = \mathbf{R}\mathbf{X}_\mathrm{d} = \mathbf{R}[\mathbf{X}_{1\mathrm{d}}|\mathbf{X}_{2\mathrm{d}}|\cdots|\mathbf{X}_{m\mathrm{d}}]$$

where $\mathbf{R}$ is the $n \times k$ block diagonal replication matrix with $\mathbf{1}_{r_j}$ on the diagonal for $j = 1, \ldots, k$ and $\mathbf{X}_\mathrm{d} = [\mathbf{X}_{1\mathrm{d}}|\mathbf{X}_{2\mathrm{d}}|\cdots|\mathbf{X}_{m\mathrm{d}}]$ is a $k \times p$ matrix. Each $k \times p_s$ matrix $\mathbf{X}_{s\mathrm{d}}$ contains the k distinct values of p_s x variables. Furthermore, $p = \sum_{s=1}^{m} p_s$. Let $\mathbf{X}_{1\mathrm{d}} = \mathbf{1}_k$, $\mathbf{S}_1 = \mathbf{R}\mathbf{X}_{1\mathrm{d}}$, $\mathbf{S}_2 = \mathbf{R}[\mathbf{X}_{1\mathrm{d}}|\mathbf{X}_{2\mathrm{d}}], \ldots, \mathbf{S}_{m-1} = \mathbf{R}[\mathbf{X}_{1\mathrm{d}}|\mathbf{X}_{2\mathrm{d}}|\cdots|\mathbf{X}_{m-1,\mathrm{d}}]$, and $\mathbf{S}_m = \mathbf{R}\mathbf{X}_\mathrm{d}$. The Type I sums of squares using $\mathbf{R}, \mathbf{S}_1, \ldots, \mathbf{S}_m$ and $\mathbf{Y}$ are given in Table 7.1.2.

In the next example, a replication matrix is used to generate Type I sums of squares when the model has a covariance matrix with several variance components.

Example 7.1.2 Consider a two-way cross classification with b random blocks, t fixed treatment levels, and $r_{ij} \geq 1$ replicate observations per block treatment combination for $i = 1, \ldots, b$ and $j = 1, \ldots, t$. The total number of observations $n = \sum_{i=1}^{b}\sum_{j=1}^{t} r_{ij}$. Use the model $\mathbf{Y} = \mathbf{R}\mathbf{X}_\mathrm{d}\boldsymbol{\beta} + \mathbf{E}$ where the $n \times 1$ random vector $\mathbf{Y} = (Y_{111}, \ldots, Y_{11r_{11}}, \ldots, Y_{bt1}, \ldots, Y_{btr_{bt}})'$; the $n \times bt$ replication matrix $\mathbf{R}$ is a block diagonal matrix with $\mathbf{1}_{r_{ij}}$ on the diagonal; the $bt \times t$ matrix $\mathbf{X}_\mathrm{d} =$

Table 7.1.2
Type I Sums of Squares with Replication

Source	df	Type I SS
Overall mean $\mathbf{X}_1$	$p_1 = 1$	$\mathbf{Y}'\frac{1}{n}\mathbf{J}_n\mathbf{Y}$
$\mathbf{X}_2\|\mathbf{X}_1$	p_2	$\mathbf{Y}'[\mathbf{S}_2(\mathbf{S}_2'\mathbf{S}_2)^{-1}\mathbf{S}_2' - \frac{1}{n}\mathbf{J}_n]\mathbf{Y}$
$\mathbf{X}_3\|\mathbf{X}_1, \mathbf{X}_2$	p_3	$\mathbf{Y}'[\mathbf{S}_3(\mathbf{S}_3'\mathbf{S}_3)^{-1}\mathbf{S}_3' - \mathbf{S}_2(\mathbf{S}_2'\mathbf{S}_2)^{-1}\mathbf{S}_2']\mathbf{Y}$
$\vdots$	$\vdots$	$\vdots$
$\mathbf{X}_m\|\mathbf{X}_1, \ldots, \mathbf{X}_{m-1}$	p_m	$\mathbf{Y}'[\mathbf{S}_m(\mathbf{S}_m'\mathbf{S}_m)^{-1}\mathbf{S}_m' - \mathbf{S}_{m-1}(\mathbf{S}_{m-1}'\mathbf{S}_{m-1})^{-1}\mathbf{S}_{m-1}']\mathbf{Y}$
Lack of fit	$k - p$	$\mathbf{Y}'\{\mathbf{R}[\mathbf{D}^{-1} - \mathbf{X}_{\mathrm{d}}(\mathbf{X}_{\mathrm{d}}'\mathbf{D}\mathbf{X}_{\mathrm{d}})^{-1}\mathbf{X}_{\mathrm{d}}']\mathbf{R}'\}\mathbf{Y}$
Pure error	$n - k$	$\mathbf{Y}'[\mathbf{I}_n - \mathbf{R}\mathbf{D}^{-1}\mathbf{R}']\mathbf{Y}$
Total	n	$\mathbf{Y}'\mathbf{Y}.$

$(\mathbf{X}_{1\mathrm{d}}|\mathbf{X}_{2\mathrm{d}}) = (\mathbf{1}_b \otimes \mathbf{1}_t|\mathbf{1}_b \otimes \mathbf{P}_t)$ where $\mathbf{P}_t'$ is the $(t-1) \times t$ lower portion of a t-dimensional Helmert matrix; the $t \times 1$ vector $\boldsymbol{\beta} = (\beta_1|\beta_2, \ldots, \beta_t)'$; and the $n \times 1$ random vector $\mathbf{E} = (E_{111}, \ldots, E_{11r_{11}}, \ldots, E_{bt1}, \ldots, E_{btr_{bt}})' \sim \mathrm{N}_n(\mathbf{0}, \Sigma)$. To construct the $n \times n$ covariance matrix Σ, redefine the random variable E_{ijv} as

$$E_{ijv} = B_i + BT_{ij} + R(BT)_{(ij)v}$$

for $v = 1, \ldots, r_{ij}$ where the random variables B_i represent the random block effect such that $B_i \sim iid\, \mathrm{N}_1(0, \sigma_B^2)$; the random variables BT_{ij} represent the random block treatment interaction such that the $b(t-1) \times 1$ vector $(\mathbf{I}_b \otimes \mathbf{P}_t')\ (BT_{11}, \ldots, BT_{bt})' \sim \mathrm{N}_{b(t-1)}(\mathbf{0}, \sigma_{BT}^2\mathbf{I}_b \otimes \mathbf{I}_{t-1})$; and the random variables $R(BT)_{(ij)v}$ represent the random nested replicates such that $R(BT)_{(ij)v} \sim iid\, \mathrm{N}_1(0, \sigma_{R(BT)}^2)$. Furthermore, assume that B_i, $(BT_{11}, \ldots, BT_{bt})'$, and $R(BT)_{(ij)v}$ are uncorrelated. Next, construct the covariance matrix when there is one replicate observation per block treatment combination. If $r_{ij} = 1$ for all i, j then the $bt \times bt$ covariance matrix is given by

$$\Sigma_{\mathrm{d}} = \sigma_B^2[\mathbf{I}_b \otimes \mathbf{J}_t] + \sigma_{BT}^2\left[\mathbf{I}_b \otimes \left(\mathbf{I}_t - \frac{1}{t}\mathbf{J}_t\right)\right]$$

where the subscript d on Σ_{d} denotes a covariance matrix for one replicate observation in each of the bt distinct block treatment combinations. Note that the variance of $R(BT)_{(ij)v}$ is nonestimable when $r_{ij} = 1$ for all i, j. Therefore, $\sigma_{R(BT)}^2$ does not appear in Σ_{d}. Now expand the covariance structure Σ_{d} to include all $n = \sum_{i=1}^{b}\sum_{j=1}^{t} r_{ij}$ observations by premultiplying Σ_{d} by $\mathbf{R}$, postmultiplying Σ_{d} and $\mathbf{R}'$, and adding a variance component that will account for the estimable variance of the $R(BT)_{(ij)k}$ variables when $r_{ij} \geq 1$. Therefore, the $n \times n$ covariance

Table 7.1.3
Type I Sums of Squares for Example 7.1.2

Source	df	Type I SS
Overall mean μ	1	$\mathbf{Y}'\mathbf{S}_1(\mathbf{S}_1'\mathbf{S}_1)^{-1}\mathbf{S}_1'\mathbf{Y} = \mathbf{Y}'\frac{1}{n}\mathbf{J}_n\mathbf{Y}$
Block $(B)\vert\mu$	$b-1$	$\mathbf{Y}'[\mathbf{T}_1(\mathbf{T}_1'\mathbf{T}_1)^{-1}\mathbf{T}_1' - \frac{1}{n}\mathbf{J}_n]\mathbf{Y}$
Treatment $(T)\vert\mu, B$	$t-1$	$\mathbf{Y}'[\mathbf{S}_2(\mathbf{S}_2'\mathbf{S}_2)^{-1}\mathbf{S}_2' - \mathbf{T}_1(\mathbf{T}_1'\mathbf{T}_1)^{-1}\mathbf{T}_1']\mathbf{Y}$
$BT\vert\mu, B, T$	$(b-1)(t-1)$	$\mathbf{Y}'[\mathbf{T}_2(\mathbf{T}_2'\mathbf{T}_2)^{-1}\mathbf{T}_2' - \mathbf{S}_2(\mathbf{S}_2'\mathbf{S}_2)^{-1}\mathbf{S}_2']\mathbf{Y}$
Rep (BT)	$n-bt$	$\mathbf{Y}'[\mathbf{I}_n - \mathbf{R}\mathbf{D}^{-1}\mathbf{R}']\mathbf{Y}$
Total	n	$\mathbf{Y}'\mathbf{Y}$

matrix Σ is given by

$$\Sigma = \mathbf{R}\Sigma_d\mathbf{R}' + \sigma^2_{R(BT)}\mathbf{I}_n.$$

The ANOVA table with Type I sums of squares can also be constructed for this example. First, consider what the sums of squares would be if there was one replicate observation per block treatment combination. If $r_{ij} = 1$ for all i, j, the matrices for the sums of squares due to the overall mean, blocks, treatments, and the block by treatment interaction are given by

$$\mathbf{B}_1 = \frac{1}{b}\mathbf{J}_b \otimes \frac{1}{t}\mathbf{J}_t = \mathbf{X}_{1d}(\mathbf{X}_{1d}'\mathbf{X}_{1d})^{-1}\mathbf{X}_{1d}'$$

$$\mathbf{B}_2 = \left(\mathbf{I}_b - \frac{1}{b}\mathbf{J}_b\right) \otimes \frac{1}{t}\mathbf{J}_t = \mathbf{Z}_{1d}(\mathbf{Z}_{1d}'\mathbf{Z}_{1d})^{-1}\mathbf{Z}_{1d}'$$

$$\mathbf{B}_3 = \frac{1}{b}\mathbf{J}_b \otimes \left(\mathbf{I}_t - \frac{1}{t}\mathbf{J}_t\right) = \mathbf{X}_{2d}(\mathbf{X}_{2d}'\mathbf{X}_{2d})^{-1}\mathbf{X}_{2d}'$$

$$\mathbf{B}_4 = \left(\mathbf{I}_b - \frac{1}{b}\mathbf{J}_b\right) \otimes \left(\mathbf{I}_t - \frac{1}{t}\mathbf{J}_t\right) = \mathbf{Z}_{2d}(\mathbf{Z}_{2d}'\mathbf{Z}_{2d})^{-1}\mathbf{Z}_{2d}'$$

respectively, where $\mathbf{X}_{1d} = \mathbf{1}_b \otimes \mathbf{1}_t$, $\mathbf{Z}_{1d} = \mathbf{P}_b \otimes \mathbf{1}_t$, $\mathbf{X}_{2d} = \mathbf{1}_b \otimes \mathbf{P}_t$, $\mathbf{Z}_{2d} = \mathbf{P}_b \otimes \mathbf{P}_t$, and $\mathbf{P}_\ell'$ is the $(\ell - 1) \times \ell$ lower portion of an ℓ-dimensional Helmert matrix. Let $\mathbf{S}_1 = \mathbf{R}\mathbf{X}_{1d} = \mathbf{1}_n, \mathbf{T}_1 = \mathbf{R}(\mathbf{X}_{1d}|\mathbf{Z}_{1d}), \mathbf{S}_2 = \mathbf{R}(\mathbf{X}_{1d}|\mathbf{Z}_{1d}|\mathbf{X}_{2d}), \mathbf{T}_2 = \mathbf{R}(\mathbf{X}_{1d}|\mathbf{Z}_{1d}|\mathbf{X}_{2d}|\mathbf{Z}_{2d})$, and $\mathbf{D} = \mathbf{R}'\mathbf{R}$. Matrices $\mathbf{S}_1, \mathbf{T}_1, \mathbf{S}_2, \mathbf{T}_2$, and $\mathbf{R}$ are used to construct Type I sums of squares in Table 7.1.3.

In the next section pattern matrices are used in data structures with crossed factors and missing data.

		Random blocks		
		1	2	3
Fixed treatments	1		Y_{21}	Y_{31}
	2	Y_{12}		Y_{32}
	3	Y_{13}	Y_{23}	

Figure 7.2.1 Missing Data Example.

7.2 PATTERN MATRICES AND MISSING DATA

In some data sets, certain data points accidentally end up missing. In other data sets, data points are intentionally not observed in certain cells. For example, in fractional factorials or incomplete block designs, data are not observed in certain cells. In either case, the overall structure of such experiments follows the complete, balanced factorial form except that the actual observed data set contains "holes" where no data are observed. These holes are located in patterns in fractional factorial experiments and incomplete block designs. However, in other experiments, the holes appear irregularly. Such experiments with missing data can be examined using pattern matrices. The following example introduces the topic.

Consider the two-way cross classification described in Figure 7.2.1. The experiment contains three random blocks and three fixed treatment levels. However, the observed data set contains no observations in the (1, 1), (2, 2), and (3, 3) block treatment combinations and one observation in each of the other six block treatment combinations. This may have arisen from a balanced incomplete block design.

We begin our discussion by first examining the experiment when one observation is present in all nine block treatment combinations. In this complete, balanced design, let the 9×1 random vector $\mathbf{Y}^* = (Y_{11}, Y_{12}, Y_{13}, Y_{21}, Y_{22}, Y_{23}, Y_{31}, Y_{32}, Y_{33})'$. Write the model $\mathbf{Y}^* = \mathbf{X}^*\boldsymbol{\beta}+\mathbf{E}^*$ where the 9×3 matrix $\mathbf{X}^* = \mathbf{1}_3\otimes\mathbf{I}_3$, the 3×1 vector $\boldsymbol{\beta} = (\beta_1, \beta_2, \beta_3)'$, the 9×1 error vector $\mathbf{E}^* = (E_{11}, E_{12}, E_{13}, E_{21}, E_{22}, E_{23}, E_{31}, E_{32}, E_{33})' \sim \mathrm{N}_9(\mathbf{0}, \Sigma^*)$, and

$$\Sigma^* = \sigma_B^2[\mathbf{I}_3 \otimes \mathbf{J}_3] + \sigma_{BT}^2 \left[\mathbf{I}_3 \otimes \left(\mathbf{I}_3 - \frac{1}{3}\mathbf{J}_3\right)\right].$$

The 9×9 covariance matrix Σ^* is built by setting $E_{ij} = B_i + (BT)_{ij}$ and applying the covariance matrix algorithm from Chapter 4. For this complete, balanced data set, the sums of squares matrices for the mean, blocks, treatments, and the block by treatment interaction are given by

$$\mathbf{B}_1 = \frac{1}{3}\mathbf{J}_3 \otimes \frac{1}{3}\mathbf{J}_3 = \mathbf{X}_1(\mathbf{X}_1'\mathbf{X}_1)^{-1}\mathbf{X}_1'$$

$$\mathbf{B}_2 = \left(\mathbf{I}_3 - \frac{1}{3}\mathbf{J}_3\right) \otimes \frac{1}{3}\mathbf{J}_3 = \mathbf{Z}_1(\mathbf{Z}_1'\mathbf{Z}_1)^{-1}\mathbf{Z}_1'$$

$$\mathbf{B}_3 = \frac{1}{3}\mathbf{J}_3 \otimes \left(\mathbf{I}_3 - \frac{1}{3}\mathbf{J}_3\right) = \mathbf{X}_2(\mathbf{X}_2'\mathbf{X}_2)^{-1}\mathbf{X}_2'$$

$$\mathbf{B}_4 = \left(\mathbf{I}_3 - \frac{1}{3}\mathbf{J}_3\right) \otimes \left(\mathbf{I}_3 - \frac{1}{3}\mathbf{J}_3\right) = \mathbf{Z}_2(\mathbf{Z}_2'\mathbf{Z}_2)^{-1}\mathbf{Z}_2'$$

respectively, where $\mathbf{X}_1 = \mathbf{1}_3 \otimes \mathbf{1}_3$, $\mathbf{Z}_1 = \mathbf{Q}_3 \otimes \mathbf{1}_3$, $\mathbf{X}_2 = \mathbf{1}_3 \otimes \mathbf{Q}_3$, $\mathbf{Z}_2 = \mathbf{Q}_3 \otimes \mathbf{Q}_3$, and

$$\mathbf{Q}_3' = \begin{bmatrix} 1 & -1 & 0 \\ 1 & 1 & -2 \end{bmatrix}.$$

The actual data set only contains six observations since Y_{11}, Y_{22}, and Y_{33} are missing. Let the 6×1 vector $\mathbf{Y} = (Y_{12}, Y_{13}, Y_{21}, Y_{23}, Y_{31}, Y_{32})'$ depict the actual observed data set. Note that $\mathbf{Y} = \mathbf{MY}^*$ where the 6×9 pattern matrix $\mathbf{M}$ is given by

$$\mathbf{M} = \begin{bmatrix} 0 & 1 & 0 & 0 & 0 & 0 & 0 & 0 & 0 \\ 0 & 0 & 1 & 0 & 0 & 0 & 0 & 0 & 0 \\ 0 & 0 & 0 & 1 & 0 & 0 & 0 & 0 & 0 \\ 0 & 0 & 0 & 0 & 0 & 1 & 0 & 0 & 0 \\ 0 & 0 & 0 & 0 & 0 & 0 & 1 & 0 & 0 \\ 0 & 0 & 0 & 0 & 0 & 0 & 0 & 1 & 0 \end{bmatrix}.$$

Furthermore, note $\mathbf{MM}' = \mathbf{I}_6$ and the 9×9 matrix

$$\mathbf{M}'\mathbf{M} = \begin{bmatrix} 0 & & & & & & & & \\ & 1 & & & & & & & \\ & & 1 & & & & \mathbf{0} & & \\ & & & 1 & & & & & \\ & & & & 0 & & & & \\ & & & & & 1 & & & \\ & & & \mathbf{0} & & & 1 & & \\ & & & & & & & 1 & \\ & & & & & & & & 0 \end{bmatrix}.$$

The vector of actual observations $\mathbf{Y}$ contains the second, third, fourth, sixth, seventh, and eighth elements of the complete data vector $\mathbf{Y}^*$. Therefore, the second, third, fourth, sixth, seventh, and eighth diagonal elements of $\mathbf{M}'\mathbf{M}$ are ones and all other elements of $\mathbf{M}'\mathbf{M}$ are zero. Furthermore, $\mathbf{M}$ is a 6×9 matrix of zeros and ones, with a one placed in the second, third, fourth, sixth, seventh, and eighth columns of rows one through six, respectively.

Since the 9×1 complete vector $\mathbf{Y}^* \sim \mathrm{N}_9(\mathbf{X}^*\boldsymbol{\beta}, \Sigma^*)$, the 6×1 vector of actual observations $\mathbf{Y} = \mathbf{M}\mathbf{Y}^* \sim \mathrm{N}_6(\mathbf{X}\boldsymbol{\beta}, \Sigma)$ where

$$\mathbf{X}\boldsymbol{\beta} = \mathbf{M}\mathbf{X}^*\boldsymbol{\beta} = \begin{bmatrix} 0 & 1 & 0 & 0 & 0 & 0 & 0 & 0 & 0 \\ 0 & 0 & 1 & 0 & 0 & 0 & 0 & 0 & 0 \\ 0 & 0 & 0 & 1 & 0 & 0 & 0 & 0 & 0 \\ 0 & 0 & 0 & 0 & 0 & 1 & 0 & 0 & 0 \\ 0 & 0 & 0 & 0 & 0 & 0 & 1 & 0 & 0 \\ 0 & 0 & 0 & 0 & 0 & 0 & 0 & 1 & 0 \end{bmatrix} \begin{bmatrix} \mathbf{I}_3 \\ \mathbf{I}_3 \\ \mathbf{I}_3 \end{bmatrix} \begin{bmatrix} \beta_1 \\ \beta_2 \\ \beta_3 \end{bmatrix}$$

$$= (\beta_2, \beta_3, \beta_1, \beta_3, \beta_1, \beta_2)'$$

and

$$\Sigma = \mathbf{M}\Sigma^*\mathbf{M}' = \mathbf{M}\left\{\sigma_B^2[\mathbf{I}_3 \otimes \mathbf{J}_3] + \sigma_{BT}^2\left[\mathbf{I}_3 \otimes \left(\mathbf{I}_3 - \frac{1}{3}\mathbf{J}_3\right)\right]\right\}\mathbf{M}'$$

$$= \sigma_B^2[\mathbf{I}_3 \otimes \mathbf{J}_2] + \sigma_{BT}^2\left[\mathbf{I}_3 \otimes \left(\mathbf{I}_2 - \frac{1}{3}\mathbf{J}_2\right)\right].$$

The Type I sums of squares for this problem are presented in Table 7.2.1 using matrices $\mathbf{S}_1 = \mathbf{M}\mathbf{X}_1, \mathbf{T}_1 = \mathbf{M}[\mathbf{X}_1|\mathbf{Z}_1]$, and $\mathbf{S}_2 = \mathbf{M}[\mathbf{X}_1|\mathbf{Z}_1|\mathbf{X}_2]$. The sum of squares matrices $\mathbf{A}_1, \ldots, \mathbf{A}_4$ is Table 7.2.1 were calculated numerically using PROC IML in SAS. The PROC IML output for this section is presented in Section A2.1 of Appendix 2. The resulting idempotent matrices are as follows:

$$\mathbf{A}_1 = \frac{1}{3}\mathbf{J}_3 \otimes \frac{1}{2}\mathbf{J}_2$$

$$\mathbf{A}_2 = \left(\mathbf{I}_3 - \frac{1}{3}\mathbf{J}_3\right) \otimes \frac{1}{2}\mathbf{J}_2$$

$$\mathbf{A}_3 = \frac{1}{3}\begin{bmatrix} 2 & 1 & -1 \\ 1 & 2 & 1 \\ -1 & 1 & 2 \end{bmatrix} \otimes \left(\mathbf{I}_2 - \frac{1}{2}\mathbf{J}_2\right)$$

$$\mathbf{A}_4 = \frac{1}{3}\begin{bmatrix} 1 & -1 & 1 \\ -1 & 1 & -1 \\ 1 & -1 & 1 \end{bmatrix} \otimes \left(\mathbf{I}_2 - \frac{1}{2}\mathbf{J}_2\right).$$

Depending on the pattern of the missing data, some Type I sum of squares matrices may have zero rank. In the example data set, the Type I sums of squares

Table 7.2.1
Type I Sums of Squares for the Missing Data Example

Source	df	Type I SS
Overall mean μ	1	$\mathbf{Y}'\mathbf{S}_1(\mathbf{S}_1'\mathbf{S}_1)^{-1}\mathbf{S}_1'\mathbf{Y} = \mathbf{Y}'\mathbf{A}_1\mathbf{Y}$
Block $(B)\vert\mu$	2	$\mathbf{Y}'[\mathbf{T}_1(\mathbf{T}_1'\mathbf{T}_1)^{-1}\mathbf{T}_1' - \mathbf{S}_1(\mathbf{S}_1'\mathbf{S}_1)^{-1}\mathbf{S}_1']\mathbf{Y} = \mathbf{Y}'\mathbf{A}_2\mathbf{Y}$
Treatment $(T)\vert\mu, B$	2	$\mathbf{Y}'[\mathbf{S}_2(\mathbf{S}_2'\mathbf{S}_2)^{-1}\mathbf{S}_2' - \mathbf{T}_1(\mathbf{T}_1'\mathbf{T}_1)^{-1}\mathbf{T}_1']\mathbf{Y} = \mathbf{Y}'\mathbf{A}_3\mathbf{Y}$
$BT\vert\mu, B, T$	1	$\mathbf{Y}'[\mathbf{I}_6 - \mathbf{S}_2)(\mathbf{S}_2'\mathbf{S}_2)^{-1}\mathbf{S}_2']\mathbf{Y} = \mathbf{Y}'\mathbf{A}_4\mathbf{Y}$
Total	6	$\mathbf{Y}'\mathbf{Y}$

matrices for all factors have ranks greater than zero. However, because there are only $n = 6$ observations, there is only one degree of freedom for the $BT|\mu, B, T$ effect. Furthermore, the sum of squares for $BT|\mu, B, T$ is calculated as the residual effect after μ, B, and T have been removed. Therefore, the sum of squares matrix for $BT|\mu, B, T$ is $\mathbf{A}_4 = \mathbf{I}_6 - \mathbf{S}_2(\mathbf{S}_2'\mathbf{S}_2)^{-1}\mathbf{S}_2'$.

Since the matrices $\mathbf{A}_1, \ldots, \mathbf{A}_4, \mathbf{M}$, and Σ have been determined for the example problem, the distributions of $\mathbf{Y}'\mathbf{A}_s\mathbf{Y}$ for $s = 1, \ldots, 4$ can be derived. From the PROC IML output note that $\mathbf{A}_1\Sigma = (2\sigma_B^2 + \frac{1}{3}\sigma_{BT}^2)\mathbf{A}_1$, $\mathbf{A}_2\Sigma = (2\sigma_B^2 + \frac{1}{3}\sigma_{BT}^2)\mathbf{A}_2$, $\mathbf{A}_3\Sigma = \sigma_{BT}^2\mathbf{A}_3$, and $\mathbf{A}_4\Sigma = \sigma_{BT}^2\mathbf{A}_4$. Therefore, by Corollary 3.1.2(a), $\mathbf{Y}'\mathbf{A}_1\mathbf{Y} \sim (2\sigma_B^2 + \frac{1}{3}\sigma_{BT}^2)\chi_1^2(\lambda_1)$, $\mathbf{Y}'\mathbf{A}_2\mathbf{Y} \sim (2\sigma_B^2 + \frac{1}{3}\sigma_{BT}^2)\chi_2^2(\lambda_2)$, $\mathbf{Y}'\mathbf{A}_3\mathbf{Y} \sim \sigma_{BT}^2\chi_2^2(\lambda_3)$, and $\mathbf{Y}'\mathbf{A}_4\mathbf{Y} \sim \sigma_{BT}^2\chi_1^2(\lambda_4)$ where

$$
\begin{aligned}
\lambda_1 &= (\mathbf{X}\beta)'\mathbf{A}_1(\mathbf{X}\beta)/\left[2\left(2\sigma_B^2 + \frac{1}{3}\sigma_{BT}^2\right)\right] \\
&= (\beta_1 + \beta_2 + \beta_3)^2/(6\sigma_B^2 + \sigma_{BT}^2) \\
\lambda_2 &= (\mathbf{X}\beta)'\mathbf{A}_2(\mathbf{X}\beta)/\left[2\left(2\sigma_B^2 + \frac{1}{3}\sigma_{BT}^2\right)\right] \\
&= (\beta_1^2 + \beta_2^2 + \beta_3^2 - \beta_1\beta_2 - \beta_1\beta_3 - \beta_2\beta_3)/\left[2\left(6\sigma_B^2 + \sigma_{BT}^2\right)\right] \\
\lambda_3 &= (\mathbf{X}\beta)'\mathbf{A}_3(\mathbf{X}\beta)/(2\sigma_{BT}^2) \\
&= (\beta_1^2 + \beta_2^2 + \beta_3^2 - \beta_1\beta_2 - \beta_1\beta_3 - \beta_2\beta_3)/(2\sigma_{BT}^2) \\
\lambda_4 &= (\mathbf{X}\beta)'\mathbf{A}_4(\mathbf{X}\beta)/(2\sigma_{BT}^2) \\
&= 0.
\end{aligned}
$$

Furthermore, $\mathbf{A}_s\Sigma\mathbf{A}_t = \mathbf{0}_{6\times 6}$ for all $s \neq t$, $s, t = 1, \ldots, 4$. By Theorem 3.2.1, the four sums of squares $\mathbf{Y}'\mathbf{A}_1\mathbf{Y}, \ldots, \mathbf{Y}'\mathbf{A}_4\mathbf{Y}$ and mutually independent. Therefore,

$$F^* = \frac{\mathbf{Y}'\mathbf{A}_3\mathbf{Y}/2}{\mathbf{Y}'\mathbf{A}_4\mathbf{Y}/1} \sim F_{2,1}(\lambda_3)$$

where $\lambda_3 = 0$ under the hypothesis $\mathrm{H}_0 : \beta_1 = \beta_2 = \beta_3$. A γ level rejection region for the hypothesis $\mathrm{H}_0 : \beta_1 = \beta_2 = \beta_3$ versus H_1: not all β's equal is to reject H_0 if $F^* > F^{\gamma}_{2,1}$ where $F^{\gamma}_{2,1}$ is the $100(1-\gamma)$ percentile point of a central F distribution with 2 and 1 degrees of freedom. Note that H_0 is equivalent to hypothesis that there is no treatment effect.

The Type I sums of squares can also be used to provide unbiased estimators of the variance components σ^2_B and σ^2_{BT}. The mean square for $BT|\mu, B, T$ provides an unbiased estimator of σ^2_{BT} since $\lambda_4 = 0$ and $\mathrm{E}(\mathbf{Y}'\mathbf{A}_4\mathbf{Y}/1) = \mathrm{E}(\sigma^2_{BT}\chi^2_1(0)) = \sigma^2_{BT}$.

Constructing an unbiased estimator for σ^2_B involves a little more work. In complete balanced designs, the sum of squares for blocks can be used to find an unbiased estimator for σ^2_B. However, in this balanced, incomplete design problem, the block effect is confounded with the treatment effect. Therefore, the Type I sum of squares for Block$(B)|\mu$ has a noncentrality parameter $\lambda_2 > 0$ and cannot be used directly with $\mathbf{Y}'\mathbf{A}_4\mathbf{Y}$ to form an unbiased estimator of σ^2_B. One solution to this problem is to calculate the sum of squares due to blocks after the overall mean and the treatment effect have been removed. After doing so, the block effect does not contain any treatment effects. As a result, the Type I sum of squares due to Block $(B)|\mu, T$ has a zero noncentrality parameter and can be used with $\mathbf{Y}'\mathbf{A}_4\mathbf{Y}$ to construct an unbiased estimator of σ^2_B. The Type I sum of squares due to Block $(B)|\mu, T$ is given by

$$\text{SS Block}(B)|\mu, T = \mathbf{Y}'\mathbf{A}^*_3\mathbf{Y}$$

where $\mathbf{A}^*_3 = \mathbf{T}^*_1(\mathbf{T}^{*\prime}_1\mathbf{T}^*_1)^{-1}\mathbf{T}^{*\prime}_1 - \mathbf{S}^*_2(\mathbf{S}^{*\prime}_2\mathbf{S}^*_2)^{-1}\mathbf{S}^{*\prime}_2$ with $\mathbf{S}^*_2 = \mathbf{M}[\mathbf{X}_1|\mathbf{X}_2]$ and $\mathbf{T}^*_1 = \mathbf{M}[\mathbf{X}_1|\mathbf{X}_2|\mathbf{Z}_1]$. Note that the matrices $\mathbf{S}^*_2$ and $\mathbf{T}^*_1$ now order the overall mean matrix $\mathbf{X}_1$ first, the treatment matrix $\mathbf{X}_2$ second, and the block matrix $\mathbf{Z}_1$ third. From the PROC IML output, the 6×6 matrix $\mathbf{A}^*_3$ for the example data set equals

$$\mathbf{A}^*_3 = \frac{1}{6}\begin{bmatrix} 2 & 1 & 1 & -1 & -1 & -2 \\ 1 & 2 & -1 & -2 & 1 & -1 \\ 1 & -1 & 2 & 1 & -2 & -1 \\ -1 & -2 & 1 & 2 & -1 & 1 \\ -1 & 1 & -2 & -1 & 2 & 1 \\ -2 & -1 & -1 & 1 & 1 & 2 \end{bmatrix}.$$

Furthermore, $tr(\mathbf{A}^*_3\mathbf{\Sigma}) = (3\sigma^2_B + \sigma^2_{BT})$ and

$$(\mathbf{X}\boldsymbol{\beta})'\mathbf{A}^*_3(\mathbf{X}\boldsymbol{\beta}) = (\beta_2, \beta_3, \beta_1, \beta_3, \beta_1, \beta_2)\mathbf{A}^*_3(\beta_2, \beta_3, \beta_1, \beta_3, \beta_1, \beta_2)' = 0.$$

Therefore, an unbiased estimator of σ_B^2 is provided by the quadratic form $\frac{1}{3}\mathbf{Y}'[\mathbf{A}_3^* - \mathbf{A}_4]\mathbf{Y}$ since

$$\mathrm{E}\left\{\frac{1}{3}\mathbf{Y}'[\mathbf{A}_3^* - \mathbf{A}_4]\mathbf{Y}\right\} = \frac{1}{3}\{(3\sigma_B^2 + \sigma_{BT}^2) - \sigma_{BT}^2\} = \sigma_B^2.$$

The procedure just described is now generalized. Let the $n^* \times 1$ vector $\mathbf{Y}^*$ represent the observations from a complete, balanced factorial experiment with model $\mathbf{Y}^* = \mathbf{X}^*\boldsymbol{\beta} + \mathbf{E}^*$ where $\mathbf{X}^*$ is an $n^* \times p$ matrix of constants, $\boldsymbol{\beta}$ is a $p \times 1$ vector of unknown parameters, and the $n^* \times 1$ random vector $\mathbf{E}^* \sim \mathrm{N}_n(\mathbf{0}, \Sigma^*)$ where Σ^* can be expressed as a function of one or more unknown parameters. Suppose the $n \times 1$ vector $\mathbf{Y}$ represents the actual observed data with $n \leq n^*$ where n^* is the number of observations in the complete data set, n is the number of actual observations, and $n^* - n$ is the number of missing observations. The $n \times 1$ random vector $\mathbf{Y} = \mathbf{M}\mathbf{Y}^* \sim \mathrm{N}_n(\mathbf{X}\boldsymbol{\beta}, \Sigma)$ where $\mathbf{M}$ is an $n \times n^*$ pattern matrix of zeros and ones, $\mathbf{X} = \mathbf{M}\mathbf{X}^*$, and $\Sigma = \mathbf{M}\Sigma^*\mathbf{M}'$. Each of the n rows of $\mathbf{M}$ has a single value of one and $(n^* - 1)$ zeros. The ij^{th} element of $\mathbf{M}$ is a 1 when the ith element in the actual data vector $\mathbf{Y}$ matches the jth element in the complete data vector $\mathbf{Y}^*$ for $i = 1, \ldots, n$ and $j = 1, \ldots, n^*$. Furthermore, the $n \times n$ matrix $\mathbf{M}\mathbf{M}' = \mathbf{I}_n$ and the $n^* \times n^*$ matrix $\mathbf{M}'\mathbf{M}$ is an idempotent, diagonal matrix of rank n with n ones and $n^* - n$ zeros on the diagonal. The ones on the diagonal of $\mathbf{M}'\mathbf{M}$ correspond to the ordered location of the actual data points in the complete data vector $\mathbf{Y}^*$ and the zeros on the diagonal of $\mathbf{M}'\mathbf{M}$ correspond to the ordered location of the missing data in the complete data vector $\mathbf{Y}^*$.

Finally, let $\mathbf{X}_s(\mathbf{X}_s'\mathbf{X}_s)^{-1}\mathbf{X}_s'$ and $\mathbf{Z}_s(\mathbf{Z}_s'\mathbf{Z}_s)^{-1}\mathbf{Z}_s'$ be the sum of squares matrices for the fixed and random effects in the complete data set for $s = 1, \ldots, m$ where $\operatorname{rank}(\mathbf{X}_s) = p_s \geq 0$, $\operatorname{rank}(\mathbf{Z}_s) = q_s \geq 0$, and $\mathbf{X}_1 = \mathbf{1}_n^*$. Let $\mathbf{S}_s = \mathbf{M}[\mathbf{X}_1|\mathbf{Z}_1|\mathbf{X}_2|\cdots|\mathbf{Z}_{s-1}|\mathbf{X}_s]$ and $\mathbf{T}_s = \mathbf{M}[\mathbf{X}_1|\mathbf{Z}_1|\mathbf{X}_2|\cdots|\mathbf{X}_s|\mathbf{Z}_s]$ for $s = 1, \ldots, m$. The Type I sum of squares for the mean is $\mathbf{Y}'\mathbf{S}_1(\mathbf{S}_1'\mathbf{S}_1)^{-1}\mathbf{S}_1'\mathbf{Y} = \mathbf{Y}'\frac{1}{n}\mathbf{J}_n\mathbf{Y}$. The Type I sums of squares for the intermediate fixed effects take the form

$$\mathbf{Y}'\mathbf{S}_s(\mathbf{S}_s'\mathbf{S}_s)^{-1}\mathbf{S}_s'\mathbf{Y} - \mathbf{Y}'\mathbf{T}_{s-1}(\mathbf{T}_{s-1}'\mathbf{T}_{s-1})^{-1}\mathbf{T}_{s-1}'\mathbf{Y}.$$

The Type I sum of squares for the intermediate random effects take the form

$$\mathbf{Y}'\mathbf{T}_s(\mathbf{T}_s'\mathbf{T}_s)^{-1}\mathbf{T}_s'\mathbf{Y} - \mathbf{Y}'\mathbf{S}_s(\mathbf{S}_s'\mathbf{S}_s)^{-1}\mathbf{S}_s'\mathbf{Y}$$

for $s = 2, \ldots, \leq m$. However, the missing data may cause some of these Type I sum of squares matrices to have zero rank. Furthermore, it may be necessary to calculate Type I sums of squares in various orders to obtain unbiased estimators of the variance components. The estimation of variance components with Type I sums of squares is discussed in detail in Chapter 10.

7.3 USING REPLICATION AND PATTERN MATRICES TOGETHER

Replication and pattern matrices can be used together in factorial experiments where certain combinations of the factors are missing and other combinations of the factors contain an unequal number of replicate observations. For example, consider the two-way cross classification described in Figure 7.3.1. The experiment contains three random blocks and three fixed treatment levels. The data set contains no observations in the (1, 1), (2, 2), and (3, 3) block treatment combinations and either one or two observations in the other six combinations.

As in Section 7.2, begin by examining an experiment with exactly one observation in each of the nine block treatment combinations. In this complete, balanced design the 9×1 random vector $\mathbf{Y}^* = (Y_{111}, Y_{121}, Y_{131}, Y_{211}, Y_{221}, Y_{231}, Y_{311}, Y_{321}, Y_{331})'$. Use the model $\mathbf{Y}^* = \mathbf{X}^*\boldsymbol{\beta} + \mathbf{E}$ where $\mathbf{E} \sim \mathrm{N}_9(\mathbf{0}, \Sigma^*)$. The matrices $\mathbf{X}^*, \boldsymbol{\beta}, \mathbf{E}, \Sigma^*, \mathbf{X}_1, \mathbf{Z}_1, \mathbf{X}_2, \mathbf{Z}_2$, and $\mathbf{M}$ are defined as in Section 7.2.

For the data set in Figure 7.3.1, let the 9×6 replication matrix $\mathbf{R}$ identify the nine replicate observations in the six block treatment combinations that contain data. The replication matrix $\mathbf{R}$ is given by

$$\mathbf{R} = \begin{bmatrix} 1 & & & & & \\ & \mathbf{1}_2 & & & \mathbf{0} & \\ & & \mathbf{1}_2 & & & \\ & \mathbf{0} & & 1 & & \\ & & & & 1 & \\ & & & & & \mathbf{1}_2 \end{bmatrix}.$$

Finally, let the 9×1 random vector of actual observations $\mathbf{Y} = (Y_{121}, Y_{131}, Y_{132}, Y_{211}, Y_{212}, Y_{231}, Y_{311}, Y_{321}, Y_{322})'$. Therefore, $\mathbf{Y} \sim \mathrm{N}_9(\mathbf{X}\boldsymbol{\beta}, \Sigma)$ where

$$\begin{aligned} \mathbf{X}\boldsymbol{\beta} = \mathbf{RMX}^*\boldsymbol{\beta} &= \mathbf{R}(\beta_2, \beta_3, \beta_1, \beta_3, \beta_1, \beta_2)' \\ &= (\beta_2, \beta_3\mathbf{1}_2', \beta_1\mathbf{1}_2', \beta_3, \beta_1, \beta_2\mathbf{1}_2')' \end{aligned}$$

Fixed treatments (j)	Random blocks (i): 1	2	3
1		Y_{211} Y_{212}	Y_{311}
2	Y_{121}		Y_{321} Y_{322}
3	Y_{131} Y_{132}	Y_{231}	

Figure 7.3.1 Missing Data Example with Unequal Replication.

Table 7.3.1
Type I Sums of Squares for the Missing Data Example with Unequal Replication

Source	df	Type I SS	
Overall mean μ	1	$\mathbf{Y}'\mathbf{S}_1(\mathbf{S}_1'\mathbf{S}_1)^{-1}\mathbf{S}_1'\mathbf{Y}$	$= \mathbf{Y}'\mathbf{A}_1\mathbf{Y}$
Block $(B)\vert\mu$	2	$\mathbf{Y}'[\mathbf{T}_1(\mathbf{T}_1'\mathbf{T}_1)^{-1}\mathbf{T}_1' - \mathbf{S}_1(\mathbf{S}_1'\mathbf{S}_1)^{-1}\mathbf{S}_1']\mathbf{Y}$	$= \mathbf{Y}'\mathbf{A}_2\mathbf{Y}$
Treatment $(T)\vert\mu, B$	2	$\mathbf{Y}'[\mathbf{S}_2(\mathbf{S}_2'\mathbf{S}_2)^{-1}\mathbf{S}_2' - \mathbf{T}_1(\mathbf{T}_1'\mathbf{T}_1)^{-1}\mathbf{T}_1']\mathbf{Y}$	$= \mathbf{Y}'\mathbf{A}_3\mathbf{Y}$
$BT\vert\mu, B, T$	1	$\mathbf{Y}'[\mathbf{R}\mathbf{D}^{-1}\mathbf{R}' - \mathbf{S}_2(\mathbf{S}_2'\mathbf{S}_2)^{-1}\mathbf{S}_2']\mathbf{Y}$	$= \mathbf{Y}'\mathbf{A}_4\mathbf{Y}$
Pure error	3	$\mathbf{Y}'[\mathbf{I}_9 - \mathbf{R}\mathbf{D}^{-1}\mathbf{R}']\mathbf{Y}$	$= \mathbf{Y}'\mathbf{A}_5\mathbf{Y}$
Total	9	$\mathbf{Y}'\mathbf{Y}$	

and

$$\begin{aligned}\Sigma &= \mathbf{RM}\Sigma^*\mathbf{M}'\mathbf{R}' + \sigma^2_{R(BT)}\mathbf{I}_9\\ &= \mathbf{R}\left\{\sigma^2_B[\mathbf{I}_3 \otimes \mathbf{J}_2] + \sigma^2_{BT}\left[\mathbf{I}_3 \otimes \left(\mathbf{I}_2 - \frac{1}{3}\mathbf{J}_2\right)\right]\right\}\mathbf{R}' + \sigma^2_{R(BT)}\mathbf{I}_9.\end{aligned}$$

The Type I sums of squares for this problem are presented in Table 7.3.1 using matrices $\mathbf{R}$, $\mathbf{D}$, $\mathbf{S}_1 = \mathbf{RMX}_1$, $\mathbf{T}_1 = \mathbf{RM}[\mathbf{X}_1|\mathbf{Z}_1]$, and $\mathbf{S}_2 = \mathbf{RM}[\mathbf{X}_1|\mathbf{Z}_1|\mathbf{X}_2]$.

The sums of squares matrices $\mathbf{A}_1, \ldots, \mathbf{A}_5$ in Table 7.3.1, the matrices $\mathbf{A}_1 \Sigma\mathbf{A}_1, \ldots, \mathbf{A}_5\Sigma\mathbf{A}_5$, $\mathbf{A}_3^*$, $\mathbf{A}_3^*\Sigma\mathbf{A}_3^*$, and the noncentrality parameters $\lambda_1, \ldots, \lambda_5, \lambda_3^*$ were calculated numerically using PROC IML in SAS. The PROC IML output for this section is presented in Section A2.2 of Appendix 2.

From the PROC IML output note that

$$\begin{aligned}\mathbf{A}_1\Sigma\mathbf{A}_1 &= \left(3\sigma^2_B + \frac{2}{3}\sigma^2_{BT} + \sigma^2_{R(BT)}\right)\mathbf{A}_1\\ \mathbf{A}_2\Sigma\mathbf{A}_2 &= \left(3\sigma^2_B + \frac{2}{3}\sigma^2_{BT} + \sigma^2_{R(BT)}\right)\mathbf{A}_2\\ \mathbf{A}_3\Sigma\mathbf{A}_3 &= \left(\frac{4}{3}\sigma^2_{BT} + \sigma^2_{R(BT)}\right)\mathbf{A}_3\\ \mathbf{A}_4\Sigma\mathbf{A}_4 &= \left(\frac{4}{3}\sigma^2_{BT} + \sigma^2_{R(BT)}\right)\mathbf{A}_4\\ \mathbf{A}_5\Sigma\mathbf{A}_5 &= \left(\sigma^2_{R(BT)}\right)\mathbf{A}_5.\end{aligned}$$

Therefore, by Corollary 3.1.2(a),

$$\mathbf{Y}'\mathbf{A}_1\mathbf{Y} \sim \left(3\sigma_B^2 + \frac{2}{3}\sigma_{BT}^2 + \sigma_{R(BT)}^2\right)\chi_1^2(\lambda_1)$$

$$\mathbf{Y}'\mathbf{A}_2\mathbf{Y} \sim \left(3\sigma_B^2 + \frac{2}{3}\sigma_{BT}^2 + \sigma_{R(BT)}^2\right)\chi_2^2(\lambda_2)$$

$$\mathbf{Y}'\mathbf{A}_3\mathbf{Y} \sim \left(\frac{4}{3}\sigma_{BT}^2 + \sigma_{R(BT)}^2\right)\chi_2^2(\lambda_3)$$

$$\mathbf{Y}'\mathbf{A}_4\mathbf{Y} \sim \left(\frac{4}{3}\sigma_{BT}^2 + \sigma_{R(BT)}^2\right)\chi_1^2(\lambda_4)$$

$$\mathbf{Y}'\mathbf{A}_5\mathbf{Y} \sim \sigma_{R(BT)}^2\chi_3^2(\lambda_5)$$

where

$$\begin{aligned}
\lambda_1 &= (\mathbf{X}\boldsymbol{\beta})'\mathbf{A}_1(\mathbf{X}\boldsymbol{\beta})\Big/\left[2\left(3\sigma_B^2 + \frac{2}{3}\sigma_{BT}^2 + \sigma_{R(BT)}^2\right)\right]\\
&= (\beta_1 + \beta_2 + \beta_3)^2\Big/\left[2\left(3\sigma_B^2 + \frac{2}{3}\sigma_{BT}^2 + \sigma_{R(BT)}^2\right)\right]\\
\lambda_2 &= (\mathbf{X}\boldsymbol{\beta})'\mathbf{A}_2(\mathbf{X}\boldsymbol{\beta})\Big/\left[2\left(3\sigma_B^2 + \frac{2}{3}\sigma_{BT}^2 + \sigma_{R(BT)}^2\right)\right]\\
&= \left(\beta_1^2 + \beta_2^2 + \beta_3^2 - \beta_1\beta_2 - \beta_1\beta_3 - \beta_2\beta_3\right)\Big/\left[3\left(3\sigma_B^2 + \frac{2}{3}\sigma_{BT}^2 + \sigma_{R(BT)}^2\right)\right]\\
\lambda_3 &= (\mathbf{X}\boldsymbol{\beta})'\mathbf{A}_3(\mathbf{X}\boldsymbol{\beta})\Big/\left[2\left(\frac{4}{3}\sigma_{BT}^2 + \sigma_{R(BT)}^2\right)\right]\\
&= 2\left(\beta_1^2 + \beta_2^2 + \beta_3^2 - \beta_1\beta_2 - \beta_1\beta_3 - \beta_2\beta_3\right)\big/\left[\left(4\sigma_{BT}^2 + 3\sigma_{R(BT)}^2\right)\right]\\
\lambda_4 &= (\mathbf{X}\boldsymbol{\beta})'\mathbf{A}_4(\mathbf{X}\boldsymbol{\beta})\Big/\left[2\left(\frac{4}{3}\sigma_{BT}^2 + \sigma_{R(BT)}^2\right)\right]\\
&= 0.\\
\lambda_5 &= (\mathbf{X}\boldsymbol{\beta})'\mathbf{A}_5(\mathbf{X}\boldsymbol{\beta})\big/\left(2\sigma_{R(BT)}^2\right)\\
&= 0.
\end{aligned}$$

The quadratic forms $\frac{3}{4}[\mathbf{Y}'\mathbf{A}_4\mathbf{Y} - (\mathbf{Y}'\mathbf{A}_5\mathbf{Y}/3)]$ and $\mathbf{Y}'\mathbf{A}_5\mathbf{Y}/3$ are unbiased estimators of σ_{BT}^2 and $\sigma_{R(BT)}^2$, respectively, since

$$\mathrm{E}\left\{\frac{3}{4}[\mathbf{Y}'\mathbf{A}_4\mathbf{Y} - (\mathbf{Y}'\mathbf{A}_5\mathbf{Y}/3)]\right\} = \mathrm{E}\left\{\frac{3}{4}\left[\left(\frac{4}{3}\sigma_{BT}^2 + \sigma_{R(BT)}^2\right)\chi_1^2(0) - \sigma_{R(BT)}^2\left(\chi_3^2(0)/3\right)\right]\right\} = \sigma_{BT}^2$$

and

$$\mathrm{E}[\mathbf{Y}'\mathbf{A}_5\mathbf{Y}/3] = \mathrm{E}[\sigma^2_{R(BT)}\chi^2_3(0)/3] = \sigma^2_{R(BT)}.$$

Furthermore, $\mathbf{A}_s\Sigma\mathbf{A}_t = \mathbf{0}_{6\times 6}$ for all $s \neq t$, $s, t = 1, \ldots, 5$. By Theorem 3.2.1, the five sums of squares $\mathbf{Y}'\mathbf{A}_1\mathbf{Y}, \ldots, \mathbf{Y}'\mathbf{A}_5\mathbf{Y}$, are mutually independent. Therefore,

$$F^* = \frac{\mathbf{Y}'\mathbf{A}_3\mathbf{Y}/2}{\mathbf{Y}'\mathbf{A}_4\mathbf{Y}/1} \sim F_{2,1}(\lambda_3)$$

where $\lambda_3 = 0$ under the hypothesis $\mathrm{H}_0 : \beta_1 = \beta_2 = \beta_3$. A γ level rejection region for the hypothesis $\mathrm{H}_0 : \beta_1 = \beta_2 = \beta_3$ versus H_1: not all β's equal is to reject H_0 if $F^* > F^{\gamma}_{2,1}$ where $F^{\gamma}_{2,1}$ is the $100(1-\gamma)$ percentile point of a central F distribution with 2 and 1 degrees of freedom.

The Type I sum of squares due to Block $(B)|\mu, T$ is given by

$$\text{SS Block } (B)|\mu, T = \mathbf{Y}'\mathbf{A}^*_3\mathbf{Y}$$

where $\mathbf{A}^*_3 = \mathbf{T}^*_1(\mathbf{T}^{*\prime}_1\mathbf{T}^*_1)^{-1}\mathbf{T}^*_1, -\mathbf{S}^*_2(\mathbf{S}^{*\prime}_2\mathbf{S}^*_2)^{-1}\mathbf{S}^*_2$, with $\mathbf{S}^*_2 = \mathbf{RM}[\mathbf{X}_1|\mathbf{X}_2]$ and $\mathbf{T}^*_1 = \mathbf{RM}[\mathbf{X}_1|\mathbf{X}_2|\mathbf{Z}_1]$. Furthermore, $\beta'\mathbf{X}'\mathbf{A}^*_3\mathbf{X}\beta = 0$, $tr(\mathbf{A}^*_3\Sigma) = 4\sigma^2_B + \frac{4}{3}\sigma^2_{BT} + 2\sigma^2_{R(BT)}$, and $\mathrm{E}[\mathbf{Y}'\mathbf{A}^*_3\mathbf{Y}] = 4\sigma^2_B + \frac{4}{3}\sigma^2_{BT} + 2\sigma^2_{R(BT)}$. Therefore, an unbiased estimator of σ^2_B is provided by $\frac{1}{4}\{[\mathbf{Y}'\mathbf{A}^*_3\mathbf{Y}] - [\mathbf{Y}'\mathbf{A}_4\mathbf{Y} + (\mathbf{Y}'\mathbf{A}_5\mathbf{Y}/3)]\}$ since

$$\begin{aligned}
&\frac{1}{4}\mathrm{E}\{[\mathbf{Y}'\mathbf{A}^*_3\mathbf{Y}] - [\mathbf{Y}'\mathbf{A}_4\mathbf{Y} + (\mathbf{Y}'\mathbf{A}_5\mathbf{Y}/3)\}\\
&= \frac{1}{4}\left\{\left(4\sigma^2_B + \frac{4}{3}\sigma^2_{BT} + 2\sigma^2_{R(BT)}\right) - \mathrm{E}\left[\left(\frac{4}{3}\sigma^2_{BT} + \sigma^2_{R(BT)}\right)\chi^2_1(0)\right.\right.\\
&\quad \left.\left. + \sigma^2_{R(BT)}(\chi^2_3(0)/3)\right]\right\} = \sigma^2_B.
\end{aligned}$$

EXERCISES

1. If $\mathbf{B}_1$, $\mathbf{B}_2$, $\mathbf{B}_3$, and $\mathbf{B}_4$ are the $n \times n$ sum of squares matrices in Table 7.1.1, prove $\mathbf{B}^2_r = \mathbf{B}_r$ for $r = 1, \ldots, 4$ and $\mathbf{B}_r\mathbf{B}_s = \mathbf{0}_{n\times n}$ for $r \neq s$.

2. In Example 7.1.1 prove that the sums of squares due to the mean, regression, and pure error are distributed as multiples of chi-square random variables. Find the three noncentrality parameters in terms of β_1, β_2, and β_3.

3. In Example 7.1.2 let $b = t = 3$, $r_{11} = r_{22} = r_{33} = 2$, $r_{12}, r_{23} = r_{31} = 1$, $r_{13} = r_{21} = r_{32} = 3$, and thus $n = 18$. Let $\mathbf{B}_1, \mathbf{B}_2, \mathbf{B}_3, \mathbf{B}_4, \mathbf{B}_5$ be the Type I

sum of squares matrices for the overall mean μ, $B|\mu$, $T|\mu,B$, $BT|\mu,B,T$, and $R(BT)|\mu$, B,T,BT, respectively. Construct and $\mathbf{B}_1$, $\mathbf{B}_2$, $\mathbf{B}_3$, $\mathbf{B}_4$, $\mathbf{B}_5$.

4. From Exercise 3, construct the $n \times n$ covariance matrix Σ.

5. From Exercises 3 and 4, calculate $tr(\mathbf{B_r}\Sigma)$ and $\beta'\mathbf{X}_\mathrm{d}'\mathbf{R}'\mathbf{B}_r\mathbf{R}\mathbf{X}_\mathrm{d}\beta$ for $r = 1, \ldots, 5$.

6. From Exercise 3, find the distributions of $\mathbf{Y}'\mathbf{B}_r\mathbf{Y}$ for $r = 1, \ldots, 5$. Are these five variables mutually independent?

8 Balanced Incomplete Block Designs

The analysis of any balanced incomplete block design (BIBD) is developed in this chapter.

8.1 GENERAL BALANCED INCOMPLETE BLOCK DESIGN

In Section 7.2 a special case of a balanced incomplete block design was discussed. As shown in Figure 7.2.1, the special case has three random blocks, three fixed treatments, two observations in every block, and two replicate observations per treatment, with six of the nine block treatment combinations containing data.

We adopt Yates's/Kempthorne's notation to characterize the general class of BIBDs. Let

b = the number of blocks

t = the number of treatments

Fixed treatments (j)	Random blocks (i)					
	1	2	3	4	5	6
1	Y_{11}		Y_{31}		Y_{51}	
2	Y_{12}			Y_{42}		Y_{62}
3		Y_{23}	Y_{33}			Y_{63}
4		Y_{24}		Y_{44}	Y_{54}	

Figure 8.1.1 Balanced Incomplete Block Example.

k = the number of observations per block

r = the number of replicate observations per treatment.

The general BIBD has b random blocks, t fixed treatments, k observations per block, and r replicate observations per treatment, with bk (or tr) of the bt block treatment combinations containing data. Furthermore, the number of times any two treatments occur together in a block is $\lambda = r(k-1)/(t-1)$. In the Figure 7.2.1 example, $b = 3, t = 3, k = 2$, and $r = 2$.

The total number of block treatment combinations containing data is bk or tr, establishing the relationship $bk = tr$. To obtain a design where each block contains k treatments, the number of blocks equals all combinations of treatments taken k at a time or $b = t!/[k!(t-k)!]$.

A second example of a BIBD is depicted in Figure 8.1.1. This design has $b = 6$ random blocks, $t = 4$ fixed treatments, $k = 2$ observations in every block, and $r = 3$ replicate observations per treatment, with $bk = 12$ of the $bt = 24$ block treatment combinations containing data.

Next we seek a model for the general balanced incomplete block design. To this end, begin with a model for the complete balanced design with b blocks, t treatments, and one observation per block/treatment combination. The model is

$$\mathbf{Y}^* = \mathbf{X}^* \boldsymbol{\tau} + \mathbf{E}^*$$

where the $bt \times 1$ vector $\mathbf{Y}^* = (\mathbf{Y}_{11}, \ldots, \mathbf{Y}_{1t}, \ldots, \mathbf{Y}_{b1}, \ldots, \mathbf{Y}_{bt})'$, the $bt \times t$ matrix $\mathbf{X}^* = \mathbf{1}_b \otimes \mathbf{I}_t$, the $t \times 1$ vector of unknown treatments means $\boldsymbol{\tau} = (\tau_1, \ldots, \tau_t)'$ and the $bt \times 1$ random vector $\mathbf{E}^* = (E_{11}, \ldots, E_{1t}, \ldots, E_{b1}, \ldots, E_{bt})' \sim \mathrm{N}_{bt}(\mathbf{0}, \Sigma^*)$ where

$$\boldsymbol{\Sigma}^* = \sigma_B^2[\mathbf{I}_b \otimes \mathbf{J}_t] + \sigma_{BT}^2 \left[\mathbf{I}_b \otimes \left(\mathbf{I}_t - \frac{1}{t}\mathbf{J}_t\right)\right].$$

Let the $bk \times 1$ vector $\mathbf{Y}$ represent the actual observed data for the balanced incomplete block design. Note that $\mathbf{Y}$ can be represented by $\mathbf{Y} = \mathbf{M}\mathbf{Y}^*$ where $\mathbf{M}$

is a $bk \times bt$ pattern matrix $\mathbf{M}$. The matrix $\mathbf{M}$ takes the form

$$\mathbf{M} = \begin{bmatrix} \mathbf{M}_1 & & & \\ & \mathbf{M}_2 & \mathbf{0} & \\ & \mathbf{0} & \ddots & \\ & & & \mathbf{M}_b \end{bmatrix}$$

where $\mathbf{M}_i$ is a $k \times t$ matrix for $i = 1, \ldots, b$. Furthermore, $\mathbf{MM}' = \mathbf{I}_b \otimes \mathbf{I}_k$ and

$$\mathbf{M}(\mathbf{I}_b \otimes \mathbf{J}_t)\mathbf{M}' = \begin{bmatrix} \mathbf{M}_1\mathbf{J}_t\mathbf{M}'_1 & \mathbf{0} \\ & \ddots & \\ \mathbf{0} & \mathbf{M}_b\mathbf{J}_t\mathbf{M}'_b \end{bmatrix} = \mathbf{I}_b \otimes \mathbf{J}_k.$$

Each of the k rows of $\mathbf{M}_i$ has $t - 1$ zeros and a single value of one where the one indicates the treatment level of the jth observation in the ith block for $j = 1, \ldots, k$. Therefore, the model used for the balanced incomplete block design is $\mathbf{Y} = \mathbf{X}\boldsymbol{\tau} + \mathbf{E}$ where the $bk \times t$ matrix $\mathbf{X} = \mathbf{MX}^*$, $\mathbf{E} \sim \mathrm{N}_{bk}(\mathbf{0}, \Sigma)$ and

$$\begin{aligned} \Sigma &= \mathbf{M}\Sigma^*\mathbf{M}' \\ &= \mathbf{M}\left\{\sigma_B^2[\mathbf{I}_b \otimes \mathbf{J}_t] + \sigma_{BT}^2\left[\mathbf{I}_b \otimes \left(\mathbf{I}_t - \frac{1}{t}\mathbf{J}_t\right)\right]\right\}\mathbf{M}' \\ &= \sigma_B^2\mathbf{M}[\mathbf{I}_b \otimes \mathbf{J}_t]\mathbf{M}' + \sigma_{BT}^2\left[\mathbf{M}(\mathbf{I}_b \otimes \mathbf{I}_t)\mathbf{M}' - \frac{1}{t}\mathbf{M}(\mathbf{I}_b \otimes \mathbf{J}_t)\mathbf{M}'\right] \\ &= \sigma_B^2[\mathbf{I}_b \otimes \mathbf{J}_k] + \sigma_{BT}^2\left[\mathbf{I}_b \otimes \left(\mathbf{I}_k - \frac{1}{t}\mathbf{J}_k\right)\right]. \end{aligned}$$

Rearranging terms, Σ can also be written as

$$\Sigma = \left(\sigma_B^2 - \frac{1}{t}\sigma_{BT}^2\right)[\mathbf{I}_b \otimes \mathbf{J}_k] + \sigma_{BT}^2[\mathbf{I}_b \otimes \mathbf{I}_k]$$

or

$$\Sigma = \left(k\sigma_B^2 + \frac{t-k}{t}\sigma_{BT}^2\right)\left[\mathbf{I}_b \otimes \frac{1}{k}\mathbf{J}_k\right] + \sigma_{BT}^2\left[\mathbf{I}_b \otimes \left(\mathbf{I}_k - \frac{1}{k}\mathbf{J}_k\right)\right].$$

This last form of Σ can be used to find Σ^{-1}. Note $\left[\mathbf{I}_b \otimes \frac{1}{k}\mathbf{J}_k\right]$ and $\left[\mathbf{I}_b \otimes \left(\mathbf{I}_k - \frac{1}{k}\mathbf{J}_k\right)\right]$ are idempotent matrices where $\left[\mathbf{I}_b \otimes \left(\frac{1}{k}\mathbf{J}_k\right)\right]\left[\mathbf{I}_b \otimes \left(\mathbf{I}_k - \frac{1}{k}\mathbf{J}_k\right)\right] = \mathbf{0}$. Therefore,

$$\Sigma^{-1} = \left(k\sigma_B^2 + \frac{t-k}{t}\sigma_{BT}^2\right)^{-1}\left[\mathbf{I}_b \otimes \frac{1}{k}\mathbf{J}_k\right] + \sigma_{BT}^{-2}\left[\mathbf{I}_b \otimes \left(\mathbf{I}_k - \frac{1}{k}\mathbf{J}_k\right)\right].$$

Table 8.2.1
First ANOVA Table with Type I Sum of Squares for BIBD

Source	df	Type I SS
Overall mean μ	1	$\mathbf{Y}'\mathbf{A}_1\mathbf{Y} = \mathbf{Y}'\frac{1}{b}\mathbf{J}_b \otimes \frac{1}{k}\mathbf{J}_k\mathbf{Y}$
Block $(B)\|\mu$	$b-1$	$\mathbf{Y}'\mathbf{A}_2\mathbf{Y} = \mathbf{Y}'(\mathbf{I}_b - \frac{1}{b}\mathbf{J}_b) \otimes \frac{1}{k}\mathbf{J}_k\mathbf{Y}$
Treatment $(T)\|\mu, B$	$t-1$	$\mathbf{Y}'\mathbf{A}_3\mathbf{Y}$
$BT\|\mu, B, T$	$bk-b-t+1$	$\mathbf{Y}'\mathbf{A}_4\mathbf{Y}$
Total	bk	$\mathbf{Y}'\mathbf{Y}$

8.2 ANALYSIS OF THE GENERAL CASE

The treatment differences can be examined and the variance parameters σ_B^2 and σ_{BT}^2 can be estimated by calculating the Type I sums of squares from two ANOVA tables. Some notation is necessary to describe these calculations.

Let the $bt \times 1$ matrix $\mathbf{X}_1 = \mathbf{1}_b \otimes \mathbf{1}_t$, the $bt \times (b-1)$ matrix $\mathbf{Z}_1 = \mathbf{Q}_b \otimes \mathbf{1}_t$, the $bt \times (t-1)$ matrix $\mathbf{X}_2 = \mathbf{I}_b \otimes \mathbf{Q}_t$ and the $bt \times (b-1)(t-1)$ matrix $\mathbf{Z}_2 = \mathbf{Q}_b \otimes \mathbf{Q}_t$ where $\mathbf{Q}_b$ and $\mathbf{Q}_t$ are $b \times (b-1)$ and $t \times (t-1)$ matrices defined as in Section 5.7. Furthermore, let $\mathbf{S}_1 = \mathbf{M}\mathbf{X}_1$, $\mathbf{T}_1 = \mathbf{M}[\mathbf{X}_1|\mathbf{Z}_1]$, $\mathbf{S}_2 = \mathbf{M}[\mathbf{X}_1|\mathbf{Z}_1|\mathbf{X}_2]$, and $\mathbf{T}_2 = \mathbf{M}[\mathbf{X}_1|\mathbf{Z}_1|\mathbf{X}_2|\mathbf{Z}_2]$. The first ANOVA table is presented in Table 8.2.1.

The Type I sums of squares due to the overall mean μ, due to Blocks $(B)|\mu$, Treatments $(T)|\mu, B$, and $BT|\mu, B, T$ are represented by $\mathbf{Y}'A_1\mathbf{Y}$, $\mathbf{Y}'\mathbf{A}_2\mathbf{Y}$, $\mathbf{Y}'A_3\mathbf{Y}$, and $\mathbf{Y}'\mathbf{A}_4\mathbf{Y}$, respectively, where

$$\mathbf{A}_1 = \mathbf{S}_1(\mathbf{S}_1'\mathbf{S}_1)^{-1}\mathbf{S}_1' = \frac{1}{b}\mathbf{J}_b \otimes \frac{1}{k}\mathbf{J}_k$$

$$\mathbf{A}_2 = \mathbf{T}_1(\mathbf{T}_1'\mathbf{T}_1)^{-1}\mathbf{T}_1' - \mathbf{S}_1(\mathbf{S}_1'\mathbf{S}_1)^{-1}\mathbf{S}_1' = \left(\mathbf{I}_b - \frac{1}{b}\mathbf{J}_b\right) \otimes \frac{1}{k}\mathbf{J}_k$$

$$\mathbf{A}_3 = \mathbf{S}_2(\mathbf{S}_2'\mathbf{S}_2)^{-1}\mathbf{S}_2' - \mathbf{T}_1(\mathbf{T}_1'\mathbf{T}_1)^{-1}\mathbf{T}_1'$$

$$\mathbf{A}_4 = (\mathbf{I}_b \otimes \mathbf{I}_k) - \mathbf{A}_1 - \mathbf{A}_2 - \mathbf{A}_3.$$

Note $\mathbf{A}_1 + \mathbf{A}_2 = \mathbf{I}_b \otimes \frac{1}{k}\mathbf{J}_k$, $\sum_{u=1}^4 \mathbf{A}_u = \mathbf{I}_b \otimes \mathbf{I}_k$, and $\sum_{u=1}^4 \text{rank}(\mathbf{A}_u) = bk$. By Theorem 1.1.7, $\mathbf{A}_u^2 = \mathbf{A}_u$ for $u = 1, \ldots, 4$ and $\mathbf{A}_u\mathbf{A}_v = \mathbf{0}$ for $u \neq v$. Also $\mathbf{A}_u(\mathbf{A}_1 + \mathbf{A}_2) = \mathbf{A}_u\left(\mathbf{I}_b \otimes \frac{1}{k}\mathbf{J}_k\right) = \mathbf{0}$ or $\mathbf{A}_u(\mathbf{I}_b \otimes \mathbf{J}_k) = \mathbf{0}$ for $u = 3, 4$. Therefore,

$$\mathbf{A}_u\Sigma = \sigma_B^2\mathbf{A}_u[\mathbf{I}_b \otimes \mathbf{J}_k] + \sigma_{BT}^2\mathbf{A}_u\left[\mathbf{I}_b \otimes \left(\mathbf{I}_k - \frac{1}{t}\mathbf{J}_k\right)\right] = \sigma_{BT}^2\mathbf{A}_u$$

for $u = 3, 4$. Furthermore, $\mathbf{A}_u\Sigma = \left(k\sigma_B^2 + \frac{t-k}{t}\sigma_{BT}^2\right)\mathbf{A}_u$ for $u = 1, 2$. Therefore, by Corollary 3.1.2(a), $\mathbf{Y}'\mathbf{A}_1\mathbf{Y} \sim a_1\chi_1^2(\lambda_1)$, $\mathbf{Y}'\mathbf{A}_2\mathbf{Y} \sim a_2\chi_{b-1}^2(\lambda_2)$, $\mathbf{Y}'\mathbf{A}_3\mathbf{Y} \sim$

Table 8.2.2
Second ANOVA Table with Type I Sum of Squares for BIBD

Source	df	Type I SS
Overall mean μ	1	$\mathbf{Y}'\mathbf{A}_1\mathbf{Y} = \mathbf{Y}'\frac{1}{b}\mathbf{J}_b \otimes \frac{1}{k}\mathbf{J}_k\mathbf{Y}$
Treatments $(T)\vert\mu$	$t-1$	$\mathbf{Y}'\mathbf{A}_2^*\mathbf{Y}$
Block $(B)\vert\mu, T$	$b-1$	$\mathbf{Y}'\mathbf{A}_3^*\mathbf{Y}$
$BT\vert\mu, B, T$	$bk-b-t+1$	$\mathbf{Y}'\mathbf{A}_4\mathbf{Y}$
Total	bk	$\mathbf{Y}'\mathbf{Y}$

$a_3\chi^2_{t-1}(\lambda_3)$, and $\mathbf{Y}'\mathbf{A}_4\mathbf{Y} \sim a_4\chi^2_{bk-t-b+1}(\lambda_4)$ where $a_1 = a_2 = \left(k\sigma_B^2 + \frac{t-k}{t}\sigma^2_{BT}\right)$, $a_3 = a_4 = \sigma^2_{BT}$, $\lambda_u = (\mathbf{X}\boldsymbol{\tau})'\mathbf{A}_u(\mathbf{X}\boldsymbol{\tau})/(2a_u)$ for $u = 1, 2, 3$ and $\lambda_4 = 0$. Finally, by Theorem 3.2.1, $\mathbf{Y}'\mathbf{A}_1\mathbf{Y}$, $\mathbf{Y}'\mathbf{A}_2\mathbf{Y}$, $\mathbf{Y}'\mathbf{A}_3\mathbf{Y}$, and $\mathbf{Y}'\mathbf{A}_4\mathbf{Y}$ are mutually independent.

A test on the treatment means can be constructed by noting that $\tau_1 = \cdots = \tau_t$ implies $\lambda_3 = 0$. Therefore, a γ level rejection region for the hypothesis $\mathrm{H}_0 : \tau_1 = \cdots = \tau_t$ versus H_1 : not all τ_j's equal is to reject H_0 if $F^* > F^{\gamma}_{t-1,\ bk-t-b+1}$ where

$$F^* = \frac{\mathbf{Y}'\mathbf{A}_3\mathbf{Y}/(t-1)}{\mathbf{Y}'\mathbf{A}_4\mathbf{Y}/(bk-t-b+1)}.$$

Unbiased estimators of σ_B^2 and σ^2_{BT} can also be constructed. An unbiased estimator of σ^2_{BT} is provided by

$$\hat{\sigma}^2_{BT} = \mathbf{Y}'\mathbf{A}_4\mathbf{Y}/(bk-t-b+1)$$

since $\mathbf{Y}'\mathbf{A}_4\mathbf{Y} \sim \sigma^2_{BT}\chi^2_{bk-t-b+1}(\lambda_4 = 0)$. An unbiased estimator of σ_B^2 can be developed using a second ANOVA table. Let the Type I sums of squares due to the overall mean μ and due to $BT|\mu, B, T$ be represented by $\mathbf{Y}'\mathbf{A}_1\mathbf{Y}$ and $\mathbf{Y}'\mathbf{A}_4\mathbf{Y}$, as before. Let the Type I sums of squares due to Treatments $(T)|\mu$ and due to Blocks $(B)|\mu, T$ be represented by $\mathbf{Y}'\mathbf{A}_2^*\mathbf{Y}$ and $\mathbf{Y}'\mathbf{A}_3^*\mathbf{Y}$, respectively. The matrices $\mathbf{A}_1$ and $\mathbf{A}_4$ are defined in Table 8.2.1. Matrices $\mathbf{A}_2^*$ and $\mathbf{A}_3^*$ can be constructed by setting $\mathbf{S}_2^* = \mathbf{M}[\mathbf{X}_1|\mathbf{X}_2]$ and $\mathbf{T}_1^* = \mathbf{M}[\mathbf{X}_1|\mathbf{X}_2|\mathbf{Z}_1]$. Then

$$\mathbf{A}_2^* = \mathbf{S}_2^*(\mathbf{S}_2^{*\prime}\mathbf{S}_2^*)^{-1}\mathbf{S}_2^{*\prime} - \mathbf{A}_1$$

$$\mathbf{A}_3^* = \mathbf{T}_1^*(\mathbf{T}_1^{*\prime}\mathbf{T}_1^*)^{-1}\mathbf{T}_1^{*\prime} - \mathbf{S}_2^*(\mathbf{S}_2^{*\prime}\mathbf{S}_2^*)^{-1}\mathbf{S}_2^{*\prime}.$$

The second ANOVA table is presented in Table 8.2.2.

The expected mean square for Blocks $(B)|\mu, T$ is now derived and then used to generate an unbiased estimator of σ_B^2. First, let $\mathbf{G}$ be the $t \times bk$ matrix such that $\mathbf{GY} = (\bar{Y}_{.1}, \ldots, \bar{Y}_{.t})'$. The $t \times t$ covariance matrix $\mathbf{G\Sigma G}'$ has diagonal elements equal to $\left(\sigma_B^2 + \frac{t-1}{t}\sigma^2_{BT}\right)/r$ and off-diagonals equal to $(k-1)\left(\sigma_B^2 - \frac{1}{t}\sigma^2_{BT}\right)/[r(t-1)]$.

The sum of squares due to Treatments $(T)|\mu$ can be reexpressed as $r\mathbf{Y}'\mathbf{G}'\big(\mathbf{I}_t - \frac{1}{t}\mathbf{J}_t\big)\mathbf{G}\mathbf{Y}$. Therefore, $\mathbf{A}_2^* = r\mathbf{G}'\big(\mathbf{I}_t - \frac{1}{t}\mathbf{J}_t\big)\mathbf{G}$ and

$$\begin{aligned}
tr[\mathbf{A}_2^*\Sigma] &= r\ \mathrm{tr}\left[\mathbf{G}'\left(\mathbf{I}_t - \frac{1}{t}\mathbf{J}_t\right)\mathbf{G}\Sigma\right] \\
&= r\ \mathrm{tr}\left[\left(\mathbf{I}_t - \frac{1}{t}\mathbf{J}_t\right)\mathbf{G}\Sigma\mathbf{G}'\right] \\
&= r\left\{\frac{t}{r}\left(\sigma_B^2 + \frac{t-1}{t}\sigma_{BT}^2\right) - \frac{1}{t}\left[\frac{t}{r}\left(\sigma_B^2 + \frac{t-1}{t}\sigma_{BT}^2\right)\right.\right. \\
&\quad \left.\left. + \frac{t(k-1)}{r}\left(\sigma_B^2 - \frac{1}{t}\sigma_{BT}^2\right)\right]\right\} \\
&= (t-k)\sigma_B^2 + \frac{1}{t}[(t-1)^2 + (k-1)]\sigma_{BT}^2.
\end{aligned}$$

But $\mathbf{A}_2 + \mathbf{A}_3 = \mathbf{A}_2^* + \mathbf{A}_3^*$, therefore

$$\begin{aligned}
\mathrm{tr}[\mathbf{A}_3^*\Sigma] &= \mathrm{tr}[(\mathbf{A}_2 + \mathbf{A}_3)\Sigma] - \mathrm{tr}[\mathbf{A}_2^*\Sigma] \\
&= (b-1)\left[k\sigma_B^2 + \frac{(t-k)}{t}\sigma_{BT}^2\right] + (t-1)\sigma_{BT}^2 \\
&\quad - \left\{(t-k)\sigma_B^2 + \frac{1}{t}[(t-1)^2 + (k-1)]\sigma_{BT}^2\right\} \\
&= (bk-t)\sigma_B^2 + \frac{b(t-k)}{t}\sigma_{BT}^2.
\end{aligned}$$

Furthermore,

$$\begin{aligned}
\mathrm{E}[\mathbf{Y}'\mathbf{A}_3^*\mathbf{Y}/(b-1)] &= \mathrm{tr}[\mathbf{A}_3^*\Sigma]/(b-1) \\
&= \frac{(bk-t)}{(b-1)}\sigma_B^2 + \frac{b(t-k)}{t(b-1)}\sigma_{BT}^2
\end{aligned}$$

since $\boldsymbol{\tau}'\mathbf{X}'\mathbf{A}_3^*\mathbf{X}\boldsymbol{\tau} = 0$. Therefore, an unbiased estimator of σ_B^2 is provided by

$$\hat{\sigma}_B^2 = \frac{1}{(bk-t)}\left[\mathbf{Y}'\mathbf{A}_3^*\mathbf{Y} - \frac{b(t-k)}{t(bk-b-t+1)}\mathbf{Y}'\mathbf{A}_4\mathbf{Y}\right]$$

because

$$\begin{aligned}
&\mathrm{E}\ \left\{\frac{1}{(bk-t)}\left[\mathbf{Y}'\mathbf{A}_3^*\mathbf{Y} - \frac{b(t-k)}{t(bk-b-t+1)}\mathbf{Y}'\mathbf{A}_4\mathbf{Y}\right]\right\} = \\
&\frac{1}{(bk-t)}\left[(bk-t)\sigma_B^2 + \frac{b(t-k)}{t}\sigma_{BT}^2 - \frac{b(t-k)}{t}\sigma_{BT}^2\right] = \sigma_B^2.
\end{aligned}$$

Treatment comparisons are of particular interest in balanced incomplete block designs. Treatment comparisons can be described as $\mathbf{h}'\boldsymbol{\tau}$ where $\mathbf{h}$ is a $t \times 1$ vector of constants. By the Gauss–Markov theorem, the best linear unbiased estimator of $\mathbf{h}'\boldsymbol{\tau}$ is given by $\mathbf{h}'[\mathbf{X}'\Sigma^{-1}\mathbf{X}]^{-1}\mathbf{X}'\Sigma^{-1}\mathbf{Y}$ where Σ^{-1} is given in Section 8.1. Therefore, the BLUE of $\mathbf{h}'\boldsymbol{\tau}$ is a function of the unknown parameters σ_B^2 and σ_{BT}^2. However, an estimated covariance matrix, $\hat{\Sigma}$, can be constructed by using Σ with σ_B^2 and σ_{BT}^2 replaced by $\hat{\sigma}_B^2$ and $\hat{\sigma}_{BT}^2$, respectively. An estimator of $\mathbf{h}'\boldsymbol{\tau}$ is provided by $\mathbf{h}'[\mathbf{X}'\hat{\Sigma}^{-1}\mathbf{X}]^{-1}\mathbf{X}'\hat{\Sigma}^{-1}\mathbf{Y}$.

8.3 MATRIX DERIVATIONS OF KEMPTHORNE'S INTERBLOCK AND INTRABLOCK TREATMENT DIFFERENCE ESTIMATORS

In Section 26.4 of Kempthorne's (1952) *Design and Analysis of Experiments* text, he develops two types of treatment comparison estimators for balanced incomplete block designs. The first estimator is derived from intrablock information within the blocks. The second estimator is derived from interblock information between the blocks. The purpose of this section is to develop Kempthorne's estimators in matrix form and to relate these estimators to the discussion presented in Sections 8.1 and 8.2 of this text.

Within this section we adopt Kempthorne's notation, namely:

$$V_j = \text{the total of all observations in treatment } j$$

$$T_j = \text{the total of all observations in blocks containing treatment } j$$

$$Q_j = V_j - T_j/k.$$

Kempthorne indicates that differences between treatments j and j' (i.e., $\tau_j - \tau_{j'}$) can be estimated by

$$\hat{\theta}_1 = k(t-1)[Q_j - Q_{j'}]/[tr(k-1)] \tag{1}$$

$$\hat{\theta}_2 = (t-1)[T_j - T_{j'}]/[r(t-k)]. \tag{2}$$

The statistics $\hat{\theta}_1$ and $\hat{\theta}_2$ are unbiased estimators of $\tau_j - \tau_{j'}$ where $\hat{\theta}_1$ is derived from intrablock information and $\hat{\theta}_2$ is derived from interblock information. The following summary provides matrix procedures for calculating estimators (1) and (2).

Estimator 1

Let $\mathbf{A}_1$ and $\mathbf{A}_3$ be the $bk \times bk$ idempotent matrices constructed in Section 8.2 where rank$(\mathbf{A}_1) = 1$ and rank$(\mathbf{A}_3) = t - 1$. Let $\mathbf{A}_1 = \mathbf{R}_1\mathbf{R}_1'$ and $\mathbf{A}_3 = \mathbf{R}_3\mathbf{R}_3'$

where $\mathbf{R}_1 = (1/\sqrt{bk})\mathbf{1}_b \otimes \mathbf{1}_k$ and $\mathbf{R}_3$ is the $bk \times (t-1)$ matrix whose columns are the eigenvectors of $\mathbf{A}_3$ that correspond to the $t-1$ eigenvalues equal to 1. The estimator $\hat{\theta}_1$ is given by

$$\hat{\theta}_1 = \mathbf{g}'(\mathbf{R}'\mathbf{X})^{-1}\mathbf{R}'\mathbf{Y}$$

where the $bk \times t$ matrix $\mathbf{X}$ and the $bk \times 1$ vector $\mathbf{Y}$ are defined in Section 8.1, the $bk \times t$ matrix $\mathbf{R} = [\mathbf{R}_1|\mathbf{R}_3]$, and $\mathbf{g}$ is a $t \times 1$ vector with a one in row j, a minus one in row j' and zeros elsewhere.

A second form of the estimator $\hat{\theta}_1$ can be developed by a different procedure. First, construct a $t \times b$ matrix $\mathbf{N}$. The ij^{th} element of $\mathbf{N}$ equals 1 if the ij^{th} block treatment combination in the BIBD factorial contains an observation and the ij^{th} element of $\mathbf{N}$ equals 0 if the ij^{th} block treatment combination is empty for $i = 1, \ldots, b$ and $j = 1, \ldots, t$. For example, the 3×3 matrix $\mathbf{N}$ corresponding to the data in Figure 7.2.1 is given by

$$\mathbf{N} = \begin{bmatrix} 0 & 1 & 1 \\ 1 & 0 & 1 \\ 1 & 1 & 0 \end{bmatrix}$$

and the 4×6 matrix $\mathbf{N}$ corresponding to the data in Figure 8.1.1 is given by

$$\mathbf{N} = \begin{bmatrix} 1 & 0 & 1 & 0 & 1 & 0 \\ 1 & 0 & 0 & 1 & 0 & 1 \\ 0 & 1 & 1 & 0 & 0 & 1 \\ 0 & 1 & 0 & 1 & 1 & 0 \end{bmatrix}.$$

The matrix $\mathbf{N}$ is sometimes called the *incidence matrix*. The estimator $\hat{\theta}_1$ is given by

$$\hat{\theta}_1 = \{k(t-1)/[tr(k-1)]\}\mathbf{g}'\left[\mathbf{N} \,\square \left(\mathbf{I}_k - \frac{1}{k}\mathbf{J}_k\right)\right]\mathbf{Y}$$

where $\mathbf{N}$ is the $t \times b$ matrix constructed above and $\mathbf{N} \,\square (\mathbf{I}_k - \frac{1}{k}\mathbf{J}_k)$ is a $t \times bk$ BIB product matrix. For example, the 3×6 matrix $[\mathbf{N} \,\square \left(\mathbf{I}_k - \frac{1}{k}\mathbf{J}_k\right)]$ corresponding to the data in Figure 7.2.1 is given by

$$\left[\mathbf{N} \,\square \left(\mathbf{I}_2 - \frac{1}{2}\mathbf{J}_2\right)\right] = \frac{1}{2}\begin{bmatrix} 0 & 0 & 1 & -1 & 1 & -1 \\ 1 & -1 & 0 & 0 & -1 & 1 \\ -1 & 1 & -1 & 1 & 0 & 0 \end{bmatrix}$$

and the 4×12 matrix $[\mathbf{N} \,\square (\mathbf{I}_k - \frac{1}{k}\mathbf{J}_k)]$ corresponding to the data in Figure 8.1.1

is given by

$$\left[\mathbf{N} \square \left(\mathbf{I}_2 - \frac{1}{2}\mathbf{J}_2\right)\right] = 1/2 \begin{bmatrix} 1 & -1 & 0 & 0 & 1 & -1 & 0 & 0 & 1 & -1 & 0 & 0 \\ -1 & 1 & 0 & 0 & 0 & 0 & 1 & -1 & 0 & 0 & 1 & -1 \\ 0 & 0 & 1 & -1 & -1 & 1 & 0 & 0 & 0 & 0 & -1 & 1 \\ 0 & 0 & -1 & 1 & 0 & 0 & -1 & 1 & -1 & 1 & 0 & 0 \end{bmatrix}.$$

The following relationships are used to derive the variance of $\hat{\theta}_1$.

$$\mathbf{g}'\mathbf{g} = 2$$

$$\mathbf{g}'\mathbf{J}_t\mathbf{g} = 0$$

$$\left[\mathbf{N} \square \left(\mathbf{I}_k - \frac{1}{k}\mathbf{J}_k\right)\right][\mathbf{I}_b \otimes \mathbf{1}_k] = \mathbf{0}_{t \times b}$$

$$\left[\mathbf{N} \square \left(\mathbf{I}_k - \frac{1}{k}\mathbf{J}_k\right)\right][\mathbf{I}_b \otimes \mathbf{J}_k] = \mathbf{0}_{t \times bk}$$

$$\left[\mathbf{N} \square \left(\mathbf{I}_k - \frac{1}{k}\mathbf{J}_k\right)\right]\left[\mathbf{N} \square \left(\mathbf{I}_k - \frac{1}{k}\mathbf{J}_k\right)\right]' = \frac{tr(k-1)}{k(t-1)}\left[\mathbf{I}_t - \frac{1}{t}\mathbf{J}_t\right].$$

Therefore,

$$\begin{aligned} \text{var}(\hat{\theta}_1) &= \{k(t-1)/[tr(k-1)]\}^2\mathbf{g}'\left[\mathbf{N} \square \left(\mathbf{I}_k - \frac{1}{k}\mathbf{J}_k\right)\right] \\ &\quad \times \left\{\sigma_B^2[\mathbf{I}_b \otimes \mathbf{J}_k] + \sigma_{BT}^2\left[\mathbf{I}_b \otimes \left(\mathbf{I}_k - \frac{1}{t}\mathbf{J}_k\right)\right]\right\}\left[\mathbf{N} \square \left(\mathbf{I}_k - \frac{1}{k}\mathbf{J}_k\right)\right]'\mathbf{g} \\ &= \sigma_{BT}^2\{k(t-1)/[tr(k-1)]\}^2\mathbf{g}'\left[\mathbf{N} \square \left(\mathbf{I}_k - \frac{1}{k}\mathbf{J}_k\right)\right] \\ &\quad \times \left[\mathbf{N} \square \left(\mathbf{I}_k - \frac{1}{k}\mathbf{J}_k\right)\right]'\mathbf{g} \\ &= \sigma_{BT}^2\{k(t-1)/[tr(k-1)]\}\mathbf{g}'\left(\mathbf{I}_t - \frac{1}{t}\mathbf{J}_t\right)\mathbf{g} \\ &= 2k(t-1)\sigma_{BT}^2/[tr(k-1)]. \end{aligned}$$

The matrix $\left[\mathbf{N} \square \left(\mathbf{I}_k - \frac{1}{k}\mathbf{J}_k\right)\right]$ is used to create the estimator $\hat{\theta}_1$. This estimator is constructed from the treatment effect after the effects due to the overall mean and blocks have been removed. Similarly, $\mathbf{Y}'\mathbf{A}_3\mathbf{Y}$ is the sum of squares due to treatments after the overall mean and blocks have been removed. Therefore, $\left[\mathbf{N} \square \left(\mathbf{I}_k - \frac{1}{k}\mathbf{J}_k\right)\right]$ and $\mathbf{A}_3$ are related and can be shown to satisfy the relationship

$$\mathbf{A}_3 = \frac{k(t-1)}{tr(k-1)}\left[\mathbf{N} \square \left(\mathbf{I}_k - \frac{1}{k}\mathbf{J}_k\right)\right]'\left[\mathbf{N} \square \left(\mathbf{I}_k - \frac{1}{k}\mathbf{J}_k\right)\right].$$

There is a one-to-one relationship between the $t \times b$ incidence matrix $\mathbf{N}$ and the $bk \times bt$ pattern matrix $\mathbf{M}$. In Appendix 3 a SAS computer program generates the $bk \times bt$ pattern matrix $\mathbf{M}$ for a balanced incomplete design, when the dimensions b, t, k, and the $t \times b$ incidence matrix $\mathbf{N}$ are supplied as inputs. A second SAS program generates the incidence matrix $\mathbf{N}$, when the dimensions b, t, k, and the pattern matrix $\mathbf{M}$ are supplied as inputs.

Estimator 2

The estimator $\hat{\theta}_2$ is given by

$$\hat{\theta}_2 = (t-1)\mathbf{g}'[\mathbf{N} \otimes \mathbf{1}_k']\mathbf{Y}/[r(t-k)]$$

where $\mathbf{g}$, $\mathbf{N}$, and $\mathbf{Y}$ are defined as in estimator 1. The following relationships are used to derive the variance of $\hat{\theta}_2$:

$$\mathbf{NN}' = \frac{r(t-k)}{(t-1)}\left[\mathbf{I}_t + \frac{(k-1)}{(t-k)}\mathbf{J}_t\right] \qquad \text{and} \qquad \mathbf{g}'\mathbf{NN}'\mathbf{g} = 2r(t-k)/(t-1).$$

Therefore,

$$\begin{aligned}
\operatorname{var}(\hat{\theta}_2) &= \{(t-1)^2/[r^2(t-k)^2]\}\mathbf{g}'[\mathbf{N} \otimes \mathbf{1}_k'] \\
&\quad \times \left\{\sigma_B^2[\mathbf{I}_b \otimes \mathbf{J}_k] + \sigma_{BT}^2\left[\mathbf{I}_b \otimes \left(\mathbf{I}_k - \frac{1}{t}\mathbf{J}_k\right)\right]\right\}[\mathbf{N}' \otimes \mathbf{1}_k]\mathbf{g} \\
&= \{(t-1)^2k/[r^2(t-k)^2]\}\left(k\sigma_B^2 + \frac{t-k}{t}\sigma_{BT}^2\right)\mathbf{g}'\mathbf{NN}'\mathbf{g} \\
&= \frac{2k(t-1)}{r(t-k)}\left[k\sigma_B^2 + \frac{t-k}{t}\sigma_{BT}^2\right].
\end{aligned}$$

Finally, Kempthorne suggests constructing the best linear unbiased estimator of the treatment differences by combining $\hat{\theta}_1$ and $\hat{\theta}_2$, weighting inversely as their variances. That is, the BLUE of $\tau_j - \tau_{j'}$ is given by

$$\hat{\theta}_3 = \frac{\hat{\theta}_1/\operatorname{var}(\hat{\theta}_1) + \hat{\theta}_2/\operatorname{var}(\hat{\theta}_2)}{[1/\operatorname{var}(\hat{\theta}_1) + 1/\operatorname{var}(\hat{\theta}_2)]}.$$

Note that $\hat{\theta}_3$ is a function of σ_B^2 and σ_{BT}^2 since $\operatorname{var}(\hat{\theta}_1)$ and $\operatorname{var}(\hat{\theta}_2)$ are functions of σ_B^2 and σ_{BT}^2. However, as discussed in Section 8.2, σ_B^2 and σ_{BT}^2 can be estimated by $\hat{\sigma}_B^2$ and $\hat{\sigma}_{BT}^2$, respectively.

Therefore, $\hat{\theta}_3$ can be estimated by

$$\hat{\theta}_3^* = \frac{\hat{\theta}_1/\widehat{\operatorname{var}}(\hat{\theta}_1) + \hat{\theta}_2/\widehat{\operatorname{var}}(\hat{\theta}_2)}{[1/\widehat{\operatorname{var}}(\hat{\theta}_1) + 1/\widehat{\operatorname{var}}(\hat{\theta}_2)]}$$

where $\widehat{\mathrm{var}}(\hat{\theta}_1)$ and $\widehat{\mathrm{var}}(\hat{\theta}_2)$ equal $\mathrm{var}(\hat{\theta}_1)$ and $\mathrm{var}(\hat{\theta}_2)$, respectively, with σ_B^2 and σ_{BT}^2 replaced by $\hat{\sigma}_B^2$ and $\hat{\sigma}_{BT}^2$.

It should be noted that the estimator $\hat{\theta}_3^*$ given above equals the estimator $\mathbf{h}'[\mathbf{X}'\hat{\Sigma}^{-1}\mathbf{X}]^{-1}\mathbf{X}'\hat{\Sigma}^{-1}\mathbf{Y}$ given in Section 8.2 when $\mathbf{h} = \mathbf{g}$.

EXERCISES

Use the example design given in Figure 8.1.1 to answer Exercises 1–11.

1. Define the $bk \times bt$ pattern matrix $\mathbf{M}$, identifying the $t \times t$ matrices $\mathbf{M}_1, \ldots, \mathbf{M}_b$ explicitly.
2. Construct the matrices $\mathbf{A}_1$, $\mathbf{A}_2$, $\mathbf{A}_3$, $\mathbf{A}_4$, $\mathbf{A}_2^*$, and $\mathbf{A}_3^*$.
3. Verify that $\mathbf{A}_r\Sigma = \left(k\sigma_B^2 + \frac{t-k}{t}\sigma_{BT}^2\right)\mathbf{A}_r$ for $r = 1, 2$ and $\mathbf{A}_r\Sigma = \sigma_{BT}^2\mathbf{A}_r$ for $r = 3, 4$.
4. Verify that $\lambda_4 = 0$.
5. Verify that $\lambda_3 = 0$ under $\mathrm{H}_0 : \tau_1 = \cdots = \tau_t$.
6. Construct the $t \times bk$ matrix $\mathbf{G}$ such that $\mathbf{GY} = (\bar{Y}_{.1}, \ldots, \bar{Y}_{.t})'$.
7. Verify that $\mathbf{G}\Sigma\mathbf{G}'$ is a $t \times t$ matrix with $\left(\sigma_B^2 + \frac{t-1}{t}\sigma_{BT}^2\right)/r$ on the diagonal and $(k-1)\left(\sigma_B^2 - \frac{1}{t}\sigma_{BT}^2\right)/[r(t-1)]$ on the off-diagonal.
8. Verify that $\mathbf{A}_2^* = r\mathbf{G}'\left(\mathbf{I}_t - \frac{1}{t}\mathbf{J}_t\right)\mathbf{G}$.
9. Verify that $\mathrm{tr}(\mathbf{A}_2^*\Sigma) = (t-k)\sigma_B^2 + \frac{1}{t}[(t-1)^2 + (k-1)]\sigma_{BT}^2$.
10. Verify that $\mathrm{tr}(\mathbf{A}_3^*\Sigma) = (bk - t)\sigma_B^2 + \frac{b(t-k)}{t}\sigma_{BT}^2$.
11. Verify that $\boldsymbol{\tau}'\mathbf{X}'\mathbf{A}_3^*\mathbf{X}\boldsymbol{\tau} = 0$.

Use the following data set of answer Exercises 12–16.

		Random blocks		
		1	2	3
Fixed treatments	1		7	12
	2	8		16
	3	15	20	

12. Calculate the Type I sum of squares for Tables 8.2.1 and 8.2.2.
13. Compute unbiased estimates of σ_B^2 and σ_{BT}^2.

14. Test the hypothesis $H_0 : \tau_1 = \tau_2 = \tau_3$ versus H_1 : not all τ_j's are equal. Use $\gamma = 0.05$.

15. Compute Kempthorne's estimates of θ_1 and θ_2 for the differences $\tau_1 - \tau_2$, $\tau_1 - \tau_3$, and $\tau_2 - \tau_3$.

16. Compute Kempthorne's estimates of θ_3 for the differences $\tau_1 - \tau_2$, $\tau_1 - \tau_3$, and $\tau_2 - \tau_3$ and then verify that these three estimates equal $\mathbf{h}'(\mathbf{X}'\hat{\Sigma}^{-1}\mathbf{X})^{-1}\mathbf{X}'\hat{\Sigma}^{-1}\mathbf{Y}$ when $\mathbf{h} = (1, -1, 0)'$, $(1, 0, -1)'$ and $(0, 1, -1)'$, respectively.

9 Less Than Full Rank Models

In Chapter 7 models of the form $\mathbf{Y} = \mathbf{R}\mathbf{X}_d\boldsymbol{\beta} + \mathbf{E}$ were discussed when the $k \times p$ matrix $\mathbf{X}_d$ had full column rank p. In this chapter, models of the same form are developed when the matrix $\mathbf{X}_d$ does not have full column rank.

9.1 MODEL ASSUMPTIONS AND EXAMPLES

Consider the model

$$\mathbf{Y} = \mathbf{R}\mathbf{X}_d\boldsymbol{\beta} + \mathbf{E}$$

where $\mathbf{Y}$ is an $n \times 1$ random vector of observations, $\mathbf{R}$ is an $n \times k$ replication matrix, $\mathbf{X}_d$ is a $k \times p$ known matrix of rank $k < p$, and $\mathbf{E}$ is an $n \times 1$ vector of random errors. For the present assume $\mathrm{E}(\mathbf{E}) = \mathbf{0}$ and $\mathrm{cov}(\mathbf{E}) = \sigma^2\mathbf{I}_n$. The assumption $\mathbf{E} \sim \mathrm{N}_n(\mathbf{0}, \sigma^2\mathbf{I}_n)$ will be added in Section 9.3 when hypothesis testing and confidence intervals are discussed. In Section 9.5 the analysis is described when $\mathbf{E} \sim \mathrm{N}_n(\mathbf{0}, \sigma^2\mathbf{V})$ and $\mathbf{V}$ is an $n \times n$ positive definite matrix.

The least squares estimators of $\boldsymbol{\beta}$ are obtained by minimizing the quadratic form $(\mathbf{Y} - \mathbf{R}\mathbf{X}_d\boldsymbol{\beta})'(\mathbf{Y} - \mathbf{R}\mathbf{X}_d\boldsymbol{\beta})$ with respect to $\boldsymbol{\beta}$. The solution $\hat{\boldsymbol{\beta}}$, satisfies the normal equations

$$\mathbf{X}_d'\mathbf{D}\mathbf{X}_d\hat{\boldsymbol{\beta}} = \mathbf{X}_d'\mathbf{R}'\mathbf{Y}$$

where the nonsingular $k \times k$ matrix $\mathbf{D} = \mathbf{R}'\mathbf{R}$. Since the $p \times p$ matrix $\mathbf{X}_d'\mathbf{D}\mathbf{X}_d$ has rank $k < p$, $\mathbf{X}_d'\mathbf{D}\mathbf{X}_d$ is singular and the usual least-squares solution $\hat{\boldsymbol{\beta}} = (\mathbf{X}_d'\mathbf{D}\mathbf{X}_d)^{-1}\mathbf{X}_d'\mathbf{R}'\mathbf{Y}$ does not exist. Therefore, the analysis approach described in Section 7.1 is not appropriate and an alternative solution must be found. In Section 9.2 the mean model solution is developed. Before we proceed with the mean model, a few examples of less than full rank models are presented.

Example 9.1.1 Searle (1971, p.165) discussed an experiment introduced by Federer (1955). In the experiment, a fixed treatment **A** with three levels has $r_1 = 3, r_2 = 2$, and $r_3 = 1$ replicate observations per level. Let Y_{ij} represent the jth observation in the ith treatment level for $i = 1, 2, 3$ and $j = 1, \ldots r_i$ with the 6×1 random vector $\mathbf{Y} = (Y_{11}, Y_{12}, Y_{13}, Y_{21}, Y_{22}, Y_{31})'$. The experimental layout is presented in Figure 9.1.1.

Searle (1971) employs the less than full rank model

$$Y_{ij} = \alpha + \alpha_i + E_{ij}$$

where α is the overall mean, α_i is the effect of the ith treatment level, and E_{ij} is a random error term particular to observation Y_{ij}. The model can be rewritten in matrix form as

$$\mathbf{Y} = \mathbf{R}\mathbf{X}_{\text{d}}\boldsymbol{\beta} = \mathbf{E}$$

where the 6×3 replication matrix $\mathbf{R}$, the 3×4 matrix $\mathbf{X}_{\text{d}}$, the 4×1 vector $\boldsymbol{\beta}$ and the 6×1 error vector $\mathbf{E}$ are given by

$$\mathbf{R} = \begin{bmatrix} \mathbf{1}_3 & & \mathbf{0} \\ & \mathbf{1}_2 & \\ \mathbf{0} & & 1 \end{bmatrix}, \quad \mathbf{X}_{\text{d}} = \begin{bmatrix} 1 & 1 & 0 & 0 \\ 1 & 0 & 1 & 0 \\ 1 & 0 & 0 & 1 \end{bmatrix}, \quad \boldsymbol{\beta} = \begin{bmatrix} \alpha \\ \alpha_1 \\ \alpha_2 \\ \alpha_3 \end{bmatrix}$$

and $\mathbf{E} = (E_{11}, E_{12}, E_{13}, E_{21}, E_{22}, E_{31})'$. The 3×4 matrix $\mathbf{X}_{\text{d}}$ has rank 3 and therefore $\mathbf{X}_{\text{d}}'\mathbf{D}\mathbf{X}_{\text{d}}$ is singular. In this case $n = 6$, $p = 4$, and $k = 3$.

	Fixed factor $A(i)$		
	1	2	3
Replicates (j)	Y_{11}	Y_{21}	Y_{31}
	Y_{12}	Y_{22}	
	Y_{13}		

Figure 9.1.1 Searle's (1971) Less than Full Rank Example.

The main difficulty with the model in Example 9.1.1 is that $p = 4$ fixed parameters ($\alpha, \alpha_1, \alpha_2$, and α_3) are being used to depict $k = 3$ distinct fixed treatment levels. Therefore $k < p$ and the less than full rank model overparameterizes the problem.

Less than full rank models can also originate in experiments with missing data. In Chapter 7 pattern matrices proved very useful when analyzing missing data experiments. However, as shown in the next example, pattern matrices do not in general solve the difficulties associated with the less than full rank model.

Example 9.1.2 Consider the two-way layout described in Figure 9.1.2. Fixed factor A has three levels, fixed factor B has two levels, and there are $r_{11} = r_{32} = 2$ and $r_{12} = r_{21} = r_{31} = 1$ replicate observations per A, B combination. Note there are no observations in the $(i, j) = (2, 2)$ A, B combination.

As in Section 7.3, we develop the model for this experiment using a pattern matrix. Let $\mathbf{Y}^* = \mathbf{X}^*\boldsymbol{\beta} + \mathbf{E}^*$ describe an experiment with one observation in each of the six distinct combinations of factors A and B where the 6×1 random vector $\mathbf{Y}^* = (Y_{111}, Y_{121}, Y_{211}, Y_{221}, Y_{311}, Y_{321})'$, the 6×6 matrix $\mathbf{X}^* = [\mathbf{X}_1|\mathbf{X}_2|\mathbf{X}_3|\mathbf{X}_4]$, the 6×1 vector $\boldsymbol{\beta} = (\beta_1, \ldots, \beta_6)'$ and the 6×1 error vector $\mathbf{E}^* = (E_{111}, E_{121}, E_{211}, E_{221}, E_{311}, E_{321})'$ with $\mathbf{X}_1 = \mathbf{1}_3 \otimes \mathbf{1}_2$, $\mathbf{X}_2 = \mathbf{Q}_3 \otimes \mathbf{1}_2$, $\mathbf{X}_3 = \mathbf{1}_3 \otimes \mathbf{Q}_2$, $\mathbf{X}_4 = \mathbf{Q}_3 \otimes \mathbf{Q}_2$, $\mathbf{Q}_2 = (1, -1)'$, and

$$\mathbf{Q}_3 = \begin{bmatrix} 1 & 1 \\ -1 & 1 \\ 0 & -2 \end{bmatrix}$$

Let the 7×1 random vector of actual observations $\mathbf{Y} = (Y_{111}, Y_{112}, Y_{121}, Y_{211}, Y_{311}, Y_{321}, Y_{322})'$. Therefore, the model for the actual data set is

$$\mathbf{Y} = \mathbf{RMX}^*\boldsymbol{\beta} + \mathbf{E}$$

where the 7×5 replication matrix $\mathbf{R}$, the 5×6 pattern matrix $\mathbf{M}$, and the 7×1

		Fixed factor $A(i)$		
		1	2	3
Fixed factor B (j)	1	Y_{111} Y_{112}	Y_{211}	Y_{311}
	2	Y_{121}		Y_{321} Y_{322}

Figure 9.1.2 Less than Full Rank Example Using a Pattern Matrix.

vector $\mathbf{E}$ are given by

$$\mathbf{R} = \begin{bmatrix} \mathbf{1}_2 & & & & \\ & 1 & & \mathbf{0} & \\ & & 1 & & \\ & \mathbf{0} & & 1 & \\ & & & & \mathbf{1}_2 \end{bmatrix}, \mathbf{M} = \begin{bmatrix} 1 & 0 & 0 & 0 & 0 & 0 \\ 0 & 1 & 0 & 0 & 0 & 0 \\ 0 & 0 & 1 & 0 & 0 & 0 \\ 0 & 0 & 0 & 0 & 1 & 0 \\ 0 & 0 & 0 & 0 & 0 & 1 \end{bmatrix}$$

and $\mathbf{E} = (E_{111}, E_{112}, E_{121}, E_{211}, E_{311}, E_{321}, E_{322})'$. The preceding model can be rewritten as

$$\mathbf{Y} = \mathbf{R}\mathbf{X}_d\boldsymbol{\beta} + \mathbf{E}$$

where the 5×6 matrix $\mathbf{X}_d = \mathbf{M}\mathbf{X}^*$. In this problem, $n = 7$, $p = 6$, $k = 5$, with $\mathbf{X}_d'\mathbf{D}\mathbf{X}_d$ is a 6×6 singular matrix of rank 5.

The main difficulty with Example 9.1.2 is that $p = 6$ fixed parameters are used to depict the $k = 5$ distinct fixed treatment combinations that contain data. Note that the use of a pattern matrix did not solve the overparameterization problem.

In the next section the mean model is introduced to solve this overparameterization problem.

9.2 THE MEAN MODEL SOLUTION

In less than full rank models, the number of fixed parameters is greater than the number of distinct fixed treatment combinations that contain data. As a consequence, the least-squares estimator of $\boldsymbol{\beta}$ does not exist and the analysis cannot be carried out as before. One solution to the problem is to use a mean model where the number of fixed parameters equals the number of distinct fixed treatment combinations that contain data. Examples 9.1.1 and 9.1.2 are now reintroduced to illustrate how the mean model is formulated.

Example 9.2.1 Reconsider the experiment described in Example 9.1.1. Let $\mathrm{E}(Y_{ij}) = \mu_i$ represent the expected value of the jth observation in the ith fixed treatment level. Use the mean model

$$Y_{ij} = \mu_i + E_{ij}.$$

In matrix form the mean model is given by

$$\mathbf{Y} = \mathbf{R}\boldsymbol{\mu} + \mathbf{E}$$

where the 3×1 vector $\boldsymbol{\mu} = (\mu_1, \mu_2, \mu_3)'$ and where $\mathbf{Y}$, $\mathbf{R}$, and $\mathbf{E}$ are defined as in Example 9.1.1. Note the 6×3 replication matrix $\mathbf{R}$ has full column rank $k = 3$.

Example 9.2.2 Reconsider the experiment described in Example 9.1.2. Let $\mathrm{E}(Y_{ijk}) = \mu_{ij}$ represent the expected value of the kth observation in the ij^{th} combination of fixed factors A and B. Use the mean model

$$\mathbf{Y} = \mathbf{R}\boldsymbol{\mu} + \mathbf{E}$$

where the 5×1 vector $\boldsymbol{\mu} = (\mu_{11}, \mu_{12}, \mu_{21}, \mu_{31}, \mu_{32})'$ and where $\mathbf{Y}$, $\mathbf{R}$, and $\mathbf{E}$ are defined as in Example 9.1.2. Note the 7×5 replication matrix $\mathbf{R}$ has full column rank $k = 5$.

In general, the less than full rank model is given by

$$\mathbf{Y} = \mathbf{R}\mathbf{X}_{\mathrm{d}}\boldsymbol{\beta} + \mathbf{E}$$

where the $k \times p$ matrix $\mathbf{X}_{\mathrm{d}}$ had rank $k < p$. The equivalent mean model is

$$\mathbf{Y} = \mathbf{R}\boldsymbol{\mu} + \mathbf{E}$$

where the $n \times k$ replication matrix $\mathbf{R}$ has full column rank k and the elements of the $k \times 1$ mean vector $\boldsymbol{\mu}$ are the expected values of the observations in the k fixed factor combinations that contain data. Since the two models are equivalent $\mathbf{R}\boldsymbol{\mu} = \mathbf{R}\mathbf{X}_{\mathrm{d}}\boldsymbol{\beta}$. Premultiplying each side of this relationship by $(\mathbf{R}'\mathbf{R})^{-1}\mathbf{R}'$ produces

$$\boldsymbol{\mu} = \mathbf{X}_{\mathrm{d}}\boldsymbol{\beta}.$$

This equation defines the relationship between the vector $\boldsymbol{\mu}$ from the mean model and the vector $\boldsymbol{\beta}$ from the overparameterized model.

9.3 MEAN MODEL ANALYSIS WHEN COV(E) = $\sigma^2\mathbf{I}_n$

The analysis of the mean model follows the same analysis sequence provided in Chapter 5. Since the $n \times k$ replication matrix $\mathbf{R}$ has full column rank, the ordinary least-squares estimator of the $k \times 1$ vector $\boldsymbol{\mu}$ is given by

$$\begin{aligned}\hat{\boldsymbol{\mu}} &= (\mathbf{R}'\mathbf{R})^{-1}\mathbf{R}'\mathbf{Y}\\ &= \mathbf{D}^{-1}\mathbf{R}'\mathbf{Y} = \bar{\mathbf{Y}}\end{aligned}$$

where $\bar{\mathbf{Y}} = \mathbf{D}^{-1}\mathbf{R}'\mathbf{Y}$ is the $k \times 1$ vector whose elements are the averages of the observations in the k distinct fixed factor combinations that contain data. The least-squares estimator $\hat{\boldsymbol{\mu}}$ is an unbiased estimator of $\boldsymbol{\mu}$ since

$$\begin{aligned}\mathrm{E}(\hat{\boldsymbol{\mu}}) &= \mathrm{E}[(\mathbf{R}'\mathbf{R})^{-1}\mathbf{R}'\mathbf{Y}]\\ &= (\mathbf{R}'\mathbf{R})^{-1}\mathbf{R}'\mathbf{R}\boldsymbol{\mu} = \boldsymbol{\mu}.\end{aligned}$$

Table 9.3.1
Mean Model ANOVA Table

Source	df	SS	
Overall mean	1	$\mathbf{Y}'\frac{1}{n}\mathbf{J}_n\mathbf{Y}$	$= \mathbf{Y}'\mathbf{A}_1\mathbf{Y}$
Treatment combinations	$k-1$	$\mathbf{Y}'[\mathbf{R}\mathbf{D}^{-1}\mathbf{R}' - \frac{1}{n}\mathbf{J}_n]\mathbf{Y}$	$= \mathbf{Y}'\mathbf{A}_2\mathbf{Y}$
Residual	$n-k$	$\mathbf{Y}'[\mathbf{I}_n - \mathbf{R}\mathbf{D}^{-1}\mathbf{R}']\mathbf{Y}$	$= \mathbf{Y}'\mathbf{A}_{\text{pe}}\mathbf{Y}$
Total	n	$\mathbf{Y}'\mathbf{Y}$	

The $k \times k$ covariance matrix of $\hat{\boldsymbol{\mu}}$ is given by

$$\text{cov}(\hat{\boldsymbol{\mu}}) = (\mathbf{R}'\mathbf{R})^{-1}\mathbf{R}'(\sigma^2\mathbf{I}_n)\mathbf{R}(\mathbf{R}'\mathbf{R})^{-1} = \sigma^2\mathbf{D}^{-1}$$

and the least-squares estimator of σ^2 is

$$\begin{aligned}\hat{\sigma}^2 &= \mathbf{Y}'[\mathbf{I}_n - \mathbf{R}(\mathbf{R}'\mathbf{R})^{-1}\mathbf{R}']\mathbf{Y}/(n-k)\\ &= \mathbf{Y}'[\mathbf{I}_n - \mathbf{R}\mathbf{D}^{-1}\mathbf{R}']\mathbf{Y}/(n-k)\\ &= \mathbf{Y}'\mathbf{A}_{\text{pe}}\mathbf{Y}/(n-k)\end{aligned}$$

where $\mathbf{A}_{\text{pe}}$ is the $n \times n$ pure error sum of squares matrix originally defined in Section 5.5. The quadratic form $\hat{\sigma}^2$ provides an unbiased estimator of σ^2 since

$$\begin{aligned}\text{E}(\hat{\sigma}^2) &= \text{E}[\mathbf{Y}'\mathbf{A}_{\text{pe}}\mathbf{Y}/(n-k)]\\ &= \{tr[\mathbf{A}_{\text{pe}}(\sigma^2\mathbf{I}_n)] + \boldsymbol{\mu}'\mathbf{R}'\mathbf{A}_{\text{pe}}\mathbf{R}\boldsymbol{\mu}\}/(n-k)\\ &= \sigma^2\end{aligned}$$

where $\text{tr}(\mathbf{A}_{\text{pe}}) = n - k$ and $\mathbf{A}_{\text{pe}}\mathbf{R} = \mathbf{0}_{n\times k}$. Furthermore, by Theorem 5.2.1, the least-squares estimator $\mathbf{t}'\mu = \mathbf{t}'\bar{\mathbf{Y}}$ is the BLUE of $\mathbf{t}'\boldsymbol{\mu}$ for any $k \times 1$ nonzero vector $\mathbf{t}$.

An ANOVA table that partitions the total sum of squares for the mean model is presented in Table 9.3.1.

The expected mean squares are calculated below using Theorem 1.3.2 with the $k \times 1$ mean vector $\boldsymbol{\mu} = (\mu_1, \mu_2, \ldots, \mu_k)'$.

$$\begin{aligned}\text{EMS (overall mean)} &= \text{E}\left[\mathbf{Y}'\frac{1}{n}\mathbf{J}_n\mathbf{Y}\right]\\ &= tr[(1/n)\sigma^2\mathbf{J}_n] + \boldsymbol{\mu}'\mathbf{R}'\frac{1}{n}\mathbf{J}_n\mathbf{R}\boldsymbol{\mu}\\ &= \sigma^2 + \frac{1}{n}\sum_{u=1}^{k}\sum_{v=1}^{k} r_u r_v \mu_u \mu_v\end{aligned}$$

$$
\begin{aligned}
\text{EMS (treatment combinations)} &= \mathrm{E}\left[\mathbf{Y}'\left(\mathbf{RD}^{-1}\mathbf{R}' - \frac{1}{n}\mathbf{J}_n\right)\mathbf{Y}\right]/(k-1) \\
&= \left\{tr[\sigma^2\mathbf{RD}^{-1}\mathbf{R}'] = \boldsymbol{\mu}'\mathbf{R}'\mathbf{RD}^{-1}\mathbf{R}'\mathbf{R}\boldsymbol{\mu}\right. \\
&\quad \left. - \mathrm{E}\left[\mathbf{Y}'\frac{1}{n}\mathbf{J}_n\mathbf{Y}\right]\right\}/(k-1) \\
&= \left[(n-1)\sigma^2 + \sum_{u=1}^{k} r_u(r_u - 1)\mu_u^2\right. \\
&\quad \left. - (2/n)\sum_{u=1,u<v}^{k} r_u r_v \mu_u \mu_v\right]/(k-1)
\end{aligned}
$$

and the EMS (residual) $= \mathrm{E}[\mathbf{Y}'\mathbf{A}_{\mathrm{pe}}\mathbf{Y}]/(n-k)] = \sigma^2$, as derived earlier.

Hypothesis tests and confidence bands can be constructed if it is assumed that the $n \times 1$ random vector $\mathbf{E} \sim \mathrm{N}_n(\mathbf{0}, \sigma^2\mathbf{I}_n)$. In Section 5.5 it was shown that $\mathbf{Y}'\mathbf{A}_{\mathrm{pe}}\mathbf{Y} \sim \sigma^2\chi^2_{n-k}(0)$. Furthermore,

$$
\begin{aligned}
\mathbf{A}_2^2 &= \left[\mathbf{RD}^{-1}\mathbf{R}' - \frac{1}{n}\mathbf{J}_n\right]^2 \\
&= \mathbf{RD}^{-1}\mathbf{R}'\mathbf{RD}^{-1}\mathbf{R}' + \frac{1}{n}\mathbf{J}_n - \mathbf{RD}^{-1}\mathbf{R}'\frac{1}{n}\mathbf{J}_n - \frac{1}{n}\mathbf{J}_n\mathbf{RD}^{-1}\mathbf{R}' \\
&= \mathbf{RD}^{-1}\mathbf{R}'\mathbf{RD}^{-1}\mathbf{R}' + \frac{1}{n}\mathbf{J}_n - \frac{1}{n}\mathbf{J}_n - \frac{1}{n}\mathbf{J}_n \\
&= \mathbf{RD}^{-1}\mathbf{R}' - \frac{1}{n}\mathbf{J}_n \\
&= \mathbf{A}_2
\end{aligned}
$$

and

$$
\begin{aligned}
\mathbf{A}_2(\sigma^2\mathbf{I}_n)\mathbf{A}_{\mathrm{pe}} &= \sigma^2\left[\mathbf{RD}^{-1}\mathbf{R}' - \frac{1}{n}\mathbf{J}_n\right][\mathbf{I}_n - \mathbf{RD}^{-1}\mathbf{R}'] \\
&= \sigma^2\left[\mathbf{RD}^{-1}\mathbf{R}' - \mathbf{RD}^{-1}\mathbf{R}'\mathbf{RD}^{-1}\mathbf{R}' - \frac{1}{n}\mathbf{J}_n + \frac{1}{n}\mathbf{J}_n\mathbf{RD}^{-1}\mathbf{R}'\right] \\
&= \sigma^2\left[\mathbf{RD}^{-1}\mathbf{R}' - \mathbf{RD}^{-1}\mathbf{R}' - \frac{1}{n}\mathbf{J}_n + \frac{1}{n}\mathbf{J}_n\right] \\
&= \mathbf{0}_{n\times n}.
\end{aligned}
$$

since $\mathbf{J}_n\mathbf{RD}^{-1}\mathbf{R}' = \mathbf{J}_n$. Therefore, by Corollary 3.1.2(a), $\mathbf{Y}'\mathbf{A}_2\mathbf{Y} \sim \sigma^2\chi^2_{k-1}(\lambda_2)$

where

$$\lambda_2 = \boldsymbol{\mu}'\mathbf{R}'\mathbf{A}_2\mathbf{R}\boldsymbol{\mu}/(2\sigma^2)$$
$$= \boldsymbol{\mu}'\left[\mathbf{R}'\mathbf{R}\mathbf{D}^{-1}\mathbf{R}'\mathbf{R} - \mathbf{R}'\frac{1}{n}\mathbf{J}_n\mathbf{R}\right]\boldsymbol{\mu}/(2\sigma^2)$$
$$= \boldsymbol{\mu}'\mathbf{R}'\left(\mathbf{I}_n - \frac{1}{n}\mathbf{J}_n\right)\mathbf{R}\boldsymbol{\mu}/(2\sigma^2).$$

By Theorem 3.2.1, $\mathbf{Y}'\mathbf{A}_2\mathbf{Y}$ and $\mathbf{Y}'\mathbf{A}_{\text{pe}}\mathbf{Y}$ are independent and therefore

$$F^* = \frac{\mathbf{Y}'\mathbf{A}_2\mathbf{Y}/(k-1)}{\mathbf{Y}'\mathbf{A}_{\text{pe}}\mathbf{Y}/(n-k)} \sim F_{k-1,n-k}(\lambda_2).$$

Furthermore, if the k elements of $\boldsymbol{\mu}$ are equal then $\lambda_2 = 0$. Therefore, a γ level test for $\text{H}_0 : \boldsymbol{\mu} = \alpha\mathbf{1}_k$ versus $\text{H}_1 : \boldsymbol{\mu} \neq \alpha\mathbf{1}_k$ is to reject H_0 if $F^* > F^{\gamma}_{k-1,n-k}$ where α is the overall mean.

Confidence bands can be constructed on the linear combinations $\mathbf{t}'\boldsymbol{\mu}$ where $\mathbf{t}$ is a $k \times 1$ nonzero vector of constants. Under the normality assumption $\mathbf{t}'\hat{\boldsymbol{\mu}} \sim \text{N}_1(\mathbf{t}'\boldsymbol{\mu}, \sigma^2\mathbf{t}'\mathbf{D}^{-1}\mathbf{t})$ since

$$\text{E}[\mathbf{t}'\hat{\boldsymbol{\mu}}] = \text{E}[\mathbf{t}'\mathbf{D}^{-1}\mathbf{R}'\mathbf{Y}] = \mathbf{t}'\mathbf{D}^{-1}\mathbf{R}'\mathbf{R}\boldsymbol{\mu} = \mathbf{t}'\boldsymbol{\mu}$$

and

$$\text{cov}[\mathbf{t}'\hat{\boldsymbol{\mu}}] = \mathbf{t}'\mathbf{D}^{-1}\mathbf{R}'(\sigma^2\mathbf{I}_n)\mathbf{R}\mathbf{D}^{-1}\mathbf{t} = \sigma^2\mathbf{t}'\mathbf{D}^{-1}\mathbf{t}.$$

By Theorem 3.2.2, $\mathbf{t}'\hat{\boldsymbol{\mu}}$ and $\mathbf{Y}'\mathbf{A}_{\text{pe}}\mathbf{Y}$ are independent since

$$\mathbf{t}'\mathbf{D}^{-1}\mathbf{R}'(\sigma^2\mathbf{I}_n)[\mathbf{I}_n - \mathbf{R}\mathbf{D}^{-1}\mathbf{R}'] = \sigma^2\mathbf{t}'[\mathbf{D}^{-1}\mathbf{R}' - \mathbf{D}^{-1}\mathbf{R}'\mathbf{R}\mathbf{D}^{-1}\mathbf{R}'] = \mathbf{0}_{1\times n}.$$

Therefore,

$$t^* = \frac{\mathbf{t}'\hat{\boldsymbol{\mu}} - \mathbf{t}'\boldsymbol{\mu}}{[\mathbf{Y}'\mathbf{A}_{\text{pe}}\mathbf{Y}/(n-k)]^{1/2}} \sim t_{n-k}(0).$$

A $100(1-\gamma)\%$ confidence band on $\mathbf{t}'\boldsymbol{\mu}$ is given by

$$\mathbf{t}'\hat{\boldsymbol{\mu}} \pm t^{\gamma/2}_{n-k}\{[\mathbf{Y}'\mathbf{A}_{\text{pe}}\mathbf{Y}/(n-k)][\mathbf{t}'\mathbf{D}^{-1}\mathbf{t}]\}^{1/2}.$$

9.4 ESTIMABLE FUNCTIONS

In Section 9.1 the less than full rank model $\mathbf{Y} = \mathbf{R}\mathbf{X}_{\text{d}}\boldsymbol{\beta} + \mathbf{E}$ was introduced. The mean model $\mathbf{Y} = \mathbf{R}\boldsymbol{\mu} + \mathbf{E}$ was developed in Sections 9.2 and 9.3 to solve the difficulties caused by the less than full rank model. Arguably, there is no need to develop the less than full rank model since the mean model solved the overparameterization problem. However, less than full rank models are used (in

SAS, for example), so it seems worthwhile to explore them and their relationship to the mean model.

For the less than full rank model, the least-squares estimator $\hat{\boldsymbol{\beta}}$ satisfies the system of normal equations

$$\mathbf{X}_d'\mathbf{D}\mathbf{X}_d\hat{\boldsymbol{\beta}} = \mathbf{X}_d'\mathbf{R}'\mathbf{Y}.$$

However, since $\mathbf{X}_d'\mathbf{D}\mathbf{X}_d$ is singular, no unique solution for $\hat{\boldsymbol{\beta}}$ exists. In fact, there are an infinite number of vectors $\hat{\boldsymbol{\beta}}$ that satisfy the normal equations. All of these solutions are linear combinations of the vector $\mathbf{Y}$, but none of them is an unbiased estimator of $\boldsymbol{\beta}$. As Graybill (1961, p. 227) points out, no linear combination of the vector $\mathbf{Y}$ exists that produces an unbiased estimator of $\boldsymbol{\beta}$.

So how should we think of the term $\hat{\boldsymbol{\beta}}$? As Searle (1971) states, for a less than full rank model, $\hat{\boldsymbol{\beta}}$ provides "a solution" to the normal equations "and nothing more." Therefore, $\hat{\boldsymbol{\beta}}$ should be thought of as a nonunique solution to the system of p normal equations, rather than as an estimator of $\boldsymbol{\beta}$.

Although no linear combination of the vector $\mathbf{Y}$ produces an unbiased estimator of $\boldsymbol{\beta}$, unbiased estimators of $\mathbf{g}'\boldsymbol{\beta}$ do exist for certain $p \times 1$ nonzero vectors $\mathbf{g}$. Unfortunately, unbiased estimators of $\mathbf{g}'\boldsymbol{\beta}$ do not exist for all $\mathbf{g}$. For example, let the $p \times 1$ vector $\mathbf{g} = (1, 0, 0 \ldots, 0)'$. The parameter $\mathbf{g}'\boldsymbol{\beta}$ is not estimable in this case since $\boldsymbol{\beta}$ is not estimable and therefore no element of $\boldsymbol{\beta}$ is estimable. The term "estimable" has been introduced. Before continuing, we formally define estimability.

Definition 9.4.1 *Estimable*: A parameter (or function of parameters) is estimable if there exists an unbiased estimate of the parameter (or function of parameters).

Definition 9.4.2 *Linearly Estimable*: A parameter (or function of parameters) is linearly estimable if there exists a linear combination of the observations whose expected value is equal to the parameter (or function of parameters).

For the remainder of this chapter we confine our attention to linearly estimable functions. Therefore, the term *estimable* will subsequently imply *linearly estimable*.

The next example demonstrates that all linear combinations of the vector $\boldsymbol{\mu}$ from the mean model are estimable.

Example 9.4.1 For the mean model $\mathbf{Y} = \mathbf{R}\boldsymbol{\mu} + \mathbf{E}$

$$\mathrm{E}[\mathbf{t}'\hat{\boldsymbol{\mu}}] = \mathrm{E}[\mathbf{t}'\mathbf{D}^{-1}\mathbf{R}'\mathbf{Y}] = \mathbf{t}'\mathbf{D}^{-1}\mathbf{R}'\mathbf{R}\boldsymbol{\mu} = \mathbf{t}'\boldsymbol{\mu}$$

where $\mathbf{t}$ is any $k \times 1$ nonzero vector.

For the less than full rank model the question still remains: When is $\mathbf{g}'\boldsymbol{\beta}$ estimable? The following theorem addresses this question. The answer lies in the relationship that links $\boldsymbol{\mu}$ and $\boldsymbol{\beta}$, namely, $\boldsymbol{\mu} = \mathbf{X}_d\boldsymbol{\beta}$.

Theorem 9.4.1 *The linear combination* $\mathbf{g}'\boldsymbol{\beta}$ *is estimable if and only if there exists a* $k \times 1$ *vector* $\mathbf{t}$ *such that* $\mathbf{g} = \mathbf{X}_d'\mathbf{t}$.

Proof: By definition $\mathbf{g}'\boldsymbol{\beta}$ is estimable if and only if there exists an $n \times 1$ vector $\mathbf{b}$ such that $\mathrm{E}[\mathbf{b}'\mathbf{Y}] = \mathbf{g}'\boldsymbol{\beta}$. First assume that $\mathbf{g}'\boldsymbol{\beta}$ is estimable. Therefore, there exists an $n \times 1$ vector $\mathbf{b}$ such that $\mathbf{g}'\boldsymbol{\beta} = \mathrm{E}[\mathbf{b}'\mathbf{Y}] = \mathbf{b}'\mathbf{R}\mathbf{X}_d\boldsymbol{\beta}$ for all $\boldsymbol{\beta}$, which implies $\mathbf{g}' = \mathbf{b}'\mathbf{R}\mathbf{X}_d$ or $\mathbf{g} = \mathbf{X}_d'\mathbf{R}'\mathbf{b}$. Let $\mathbf{t} = \mathbf{R}'\mathbf{b}$ and there exists a $k \times 1$ vector $\mathbf{t}$ such that $\mathbf{g} = \mathbf{X}_d'\mathbf{t}$. Now assume there exists a $k \times 1$ vector $\mathbf{t}$ such that $\mathbf{g} = \mathbf{X}_d'\mathbf{t}$. Then $\mathrm{E}[\mathbf{t}'\hat{\boldsymbol{\mu}}] = \mathrm{E}[\mathbf{t}'\mathbf{D}^{-1}\mathbf{R}'\mathbf{Y}] = \mathbf{t}'\boldsymbol{\mu} = \mathbf{t}'\mathbf{X}_d\boldsymbol{\beta} = \mathbf{g}'\boldsymbol{\beta}$. Therefore, $\mathbf{g}'\boldsymbol{\beta}$ is estimable. ■

As mentioned earlier, the normal equations have an infinite number of solutions. If $\hat{\boldsymbol{\beta}}_0$ represents any one of the solutions then the next theorem shows that $\mathbf{g}'\hat{\boldsymbol{\beta}}_0$ is invariant to the choice of $\hat{\boldsymbol{\beta}}_0$ when $\mathbf{g}'\boldsymbol{\beta}$ is estimable.

Theorem 9.4.2 *If* $\mathbf{g}'\boldsymbol{\beta}$ *is estimable then* $\mathbf{g}'\hat{\boldsymbol{\beta}}_0 = \mathbf{t}'\hat{\boldsymbol{\mu}}$ *provides a unique, unbiased estimate of* $\mathbf{g}'\boldsymbol{\beta}$ *where* $\hat{\boldsymbol{\beta}}_0$ *is any solution to the normal equations and* $\mathbf{t}$ *is defined in Theorem 9.4.1.*

Proof: Solve the normal equations for $\mathbf{X}_d\hat{\boldsymbol{\beta}}_0$. Therefore,

$$\mathbf{X}_d'\mathbf{D}\mathbf{X}_d\hat{\boldsymbol{\beta}}_0 = \mathbf{X}_d'\mathbf{R}'\mathbf{Y},$$

or

$$\mathbf{X}_d\hat{\boldsymbol{\beta}}_0 = \mathbf{D}^{-1}\mathbf{R}'\mathbf{Y} = \hat{\boldsymbol{\mu}}.$$

Since $\mathbf{g}'\boldsymbol{\beta}$ is estimable, $\mathbf{g}'\hat{\boldsymbol{\beta}}_0 = \mathbf{t}'\mathbf{X}_d\hat{\boldsymbol{\beta}}_0 = \mathbf{t}'\hat{\boldsymbol{\mu}}$ where $\mathbf{t}'\hat{\boldsymbol{\mu}} = \mathbf{t}'\bar{\mathbf{Y}}$ is a unique estimate. Furthermore, $\mathbf{g}'\hat{\boldsymbol{\beta}}_0$ is an unbiased estimate of $\mathbf{g}'\boldsymbol{\beta}$ since $\mathrm{E}[\mathbf{g}'\hat{\boldsymbol{\beta}}_0] = \mathrm{E}[\mathbf{t}'\hat{\boldsymbol{\mu}}] = \mathbf{t}'\boldsymbol{\mu} = \mathbf{t}'\mathbf{X}_d\boldsymbol{\beta} = \mathbf{g}'\boldsymbol{\beta}$. ■

The Gauss–Markov theorem is applied to find the BLUE of $\mathbf{g}'\boldsymbol{\beta}$.

Theorem 9.4.3 *If* $\mathbf{g}'\boldsymbol{\beta}$ *is estimable then the BLUE of* $\mathbf{g}'\boldsymbol{\beta}$ *is* $\mathbf{g}'\hat{\boldsymbol{\beta}}_0 = \mathbf{t}'\hat{\boldsymbol{\mu}}$.

Proof: By Theorem 9.4.2, $\mathbf{t}'\hat{\boldsymbol{\mu}} = \mathbf{g}'\boldsymbol{\beta}_0$. By Theorem 9.4.1, $\mathbf{t}'\boldsymbol{\mu} = \mathbf{t}'\mathbf{X}_d\boldsymbol{\beta} = \mathbf{g}'\boldsymbol{\beta}$. By the Gauss–Markov theorem, $\mathbf{t}'\hat{\boldsymbol{\mu}}$ is the BLUE of $\mathbf{t}'\boldsymbol{\mu} = \mathbf{g}'\hat{\boldsymbol{\beta}}$. ■

The heart of the three previous proofs lies in the relationship $\boldsymbol{\mu} = \mathbf{X}_d\boldsymbol{\beta}$. Since $\mathbf{t}'\boldsymbol{\mu}$ is always estimable by $\mathbf{t}'\hat{\boldsymbol{\mu}}$, $\mathbf{t}'\mathbf{X}_d\boldsymbol{\beta} = \mathbf{t}'\boldsymbol{\mu}$ is also estimable by $\mathbf{t}'\hat{\boldsymbol{\mu}}$. Therefore, $\mathbf{g}'\boldsymbol{\beta}$ is estimable provided $\mathbf{g}'$ can be written as $\mathbf{t}'\mathbf{X}_d$ for some $k \times 1$ nonzero vector $\mathbf{t}$. One could argue that the whole topic of estimable functions is viable only in so far as $\boldsymbol{\beta}$ is related to $\boldsymbol{\mu}$ through the relationship $\boldsymbol{\mu} = \mathbf{X}_d\boldsymbol{\beta}$. Stated more strongly, estimable functions have little meaning without the mean model and, because of the mean model, estimable functions are at best redundant.

This section concludes with a SAS PROC GLM program that analyzes Federer's (1955) data from Example 9.1.1. Both the mean model and Searle's less than full rank model are run. The two models are then used to generate the same estimable functions.

Example 9.4.2 The data set is given in Figure 9.4.1. The SAS program and output are presented in Appendix 4. The SAS output provides the following parameter estimates for the model $Y_{ij} = \alpha + \alpha_i + E_{ij}$ (or equivalently for the model $\mathbf{Y} = \mathbf{R}\mathbf{X}_d\boldsymbol{\beta} + \mathbf{E}$).

$$\hat{\boldsymbol{\beta}}_0 = (\hat{\alpha}, \hat{\alpha}_1, \hat{\alpha}_2, \hat{\alpha}_3)' = (32, 68, 54, 0)'.$$

Note that although the notation $\hat{\alpha}, \hat{\alpha}_1, \hat{\alpha}_2, \hat{\alpha}_3$ is used, $\hat{\boldsymbol{\beta}}_0$ should not be viewed as an estimate of $\boldsymbol{\beta} = (\alpha, \alpha_1, \alpha_2, \alpha_3)'$. Rather $\hat{\boldsymbol{\beta}}_0$ is one of the normal equation solutions for $\boldsymbol{\beta}$. The SAS output also provides the following estimates for the mean model $Y_{ij} = \mu_i + E_{ij}$ (or equivalently for the model $\mathbf{Y} = \mathbf{R}\boldsymbol{\mu} + \mathbf{E}$).

$$\hat{\boldsymbol{\mu}} = (\hat{\mu}_1, \hat{\mu}_2, \hat{\mu}_3)' = (\bar{Y}_{1.}, \bar{Y}_{2.}, \bar{Y}_{3.})' = (100, 86, 32)'.$$

Suppose it is of interest to estimate $\alpha_1 - \alpha_2 = \mathbf{g}'\boldsymbol{\beta}$ for $\mathbf{g} = (0, 1, -1, 0)'$. By Theorem 9.4.1, $\mathbf{g}'\boldsymbol{\beta}$ is estimable since there exists a 3×1 vector $\mathbf{t} = (1, -1, 0)'$ such that

$$\mathbf{g}' = (0, 1, -1, 0) = (1, -1, 0) \begin{bmatrix} 1 & 1 & 0 & 0 \\ 1 & 0 & 1 & 0 \\ 1 & 0 & 0 & 1 \end{bmatrix} = \mathbf{t}'\mathbf{X}_d.$$

Therefore, by Theorems 9.4.2 and 9.4.3, the unique BLUE of $\mathbf{g}'\boldsymbol{\beta} = \alpha_1 - \alpha_2$ is provided by $\mathbf{g}'\hat{\boldsymbol{\beta}}_0 = \mathbf{t}'\hat{\boldsymbol{\mu}}$ where

$$\mathbf{g}'\hat{\boldsymbol{\beta}}_0 = (0, 1, -1, 0) \begin{bmatrix} 32 \\ 68 \\ 54 \\ 0 \end{bmatrix} = 68 - 54 = 14 = (1, -1, 0) \begin{bmatrix} 100 \\ 86 \\ 32 \end{bmatrix} = \mathbf{t}'\hat{\boldsymbol{\mu}}.$$

	Fixed Factor A		
	1	2	3
Replicates	101 105 94	84 88	32

Figure 9.4.1 Federer's (1955) Data Set.

9.5 MEAN MODEL ANALYSIS WHEN COV(E) = σ^2V

Previous sections of Chapter 9 have been limited to models where $\text{cov}(\mathbf{E}) = \sigma^2\mathbf{I}_n$. Such models occur when all the main factors in the design are fixed with random replicates nested in the combinations of the fixed factors. In these cases, the replication matrix $\mathbf{R}$ identifies the replicate observations in the k combinations of the fixed factors. Therefore, $\mathbf{R}$ is an $n \times k$ matrix, $\boldsymbol{\mu}$ is a $k \times 1$ vector and $\mathbf{Y} = \mathbf{R}\boldsymbol{\mu} + \mathbf{E}$ is a viable model.

We are now interested in extending our attention to models where $\text{cov}(\mathbf{E}) = \sigma^2\mathbf{V}$ and $\mathbf{V}$ is an $n \times n$ positive definite matrix. Such models can occur when some of the factors in the design are fixed and some are random, with random replicates nested in the combinations of the fixed and random factors. In these cases, the replication matrix identifies the replicate observations nested in the k combinations of the fixed and random factors. Therefore, $\mathbf{R}$ is an $n \times k$ matrix. However, $\boldsymbol{\mu}$ is a $v \times 1$ vector where $v < k$ is the number of fixed treatment combinations that contain data. In this case, $\mathbf{R}\boldsymbol{\mu}$ is not a viable structure since the number of columns of $\mathbf{R}(= k)$ is greater than the number of rows of $\boldsymbol{\mu}(= v)$. A solution to this problem is to use the more general mean model

$$\mathbf{Y} = \mathbf{R}\mathbf{C}\boldsymbol{\mu} + \mathbf{E}$$

where $\mathbf{C}$ is a $k \times v$ matrix of zeros and ones. The matrix $\mathbf{C}$ identifies which elements of the $v \times 1$ vector $\boldsymbol{\mu}$ correspond to the k combinations of the fixed and random factors. The next example illustrates how to construct the matrix $\mathbf{C}$.

Example 9.5.1 Consider the unbalanced experimental layout given in Figure 9.5.1 where the three blocks are random, the two treatments are fixed, and the nested replicates are random. In this experiment the number of observations is $n = 8$, the number of combinations of the fixed and random factors that contain data is $k = 5$, the number of fixed treatment combinations is $v = 2$, and the number of replicates in the ij^{th} combination of blocks and treatments is r_{ij} where

Fixed treatments (j)	Random blocks (i) 1	2	3
1	Y_{111}	Y_{211} Y_{212}	Y_{311}
2	Y_{121}		Y_{321} Y_{322} Y_{323}

Figure 9.5.1 Unbalanced Experimental Layout for Example 9.5.1.

$r_{11} = r_{12} = r_{31} = 1, r_{21} = 2$, and $r_{32} = 3$. The 8×5 replication matrix $\mathbf{R}$, the 5×2 matrix $\mathbf{C}$, and the 2×1 mean vector $\boldsymbol{\mu}$ are given by

$$\mathbf{R} = \begin{bmatrix} 1 & & & & \\ & 1 & \mathbf{0} & & \\ & & \mathbf{1}_2 & & \\ & \mathbf{0} & & 1 & \\ & & & & \mathbf{1}_3 \end{bmatrix}, \qquad \mathbf{C} = \begin{bmatrix} 1 & 0 \\ 0 & 1 \\ 1 & 0 \\ 1 & 0 \\ 0 & 1 \end{bmatrix}, \qquad \text{and } \boldsymbol{\mu} = \begin{bmatrix} \mu_1 \\ \mu_2 \end{bmatrix}.$$

Note that the $k \times v$ matrix $\mathbf{C}$ can also be constructed using the relationship $\mathbf{C} = \mathbf{M}\mathbf{X}^*$ where matrices $\mathbf{M}$ and $\mathbf{X}^*$ are defined according to the methods in Section 7.2. For Example 9.5.1, the 5×6 pattern matrix $\mathbf{M}$ and the 6×2 matrix $\mathbf{X}^*$ are given by

$$\mathbf{M} = \begin{bmatrix} 1 & 0 & 0 & 0 & 0 & 0 \\ 0 & 1 & 0 & 0 & 0 & 0 \\ 0 & 0 & 1 & 0 & 0 & 0 \\ 0 & 0 & 0 & 0 & 1 & 0 \\ 0 & 0 & 0 & 0 & 0 & 1 \end{bmatrix} \qquad \text{and} \qquad \mathbf{X}^* = \mathbf{1}_3 \otimes \mathbf{I}_2,$$

and the 5×2 matrix $\mathbf{C} = \mathbf{M}\mathbf{X}^*$.

The general mean model $\mathbf{Y} = \mathbf{R}\mathbf{C}\boldsymbol{\mu} + \mathbf{E}$ also applies to the examples in Sections 9.1 through 9.4 where all the main factors were fixed and the nested replicates were random. In such cases, $v = k$ and the $k \times v$ (i.e., the $k \times k$) matrix $\mathbf{C} = \mathbf{I}_k$.

The analysis approach presented in Section 9.3 is now summarized for the general mean model $\mathbf{Y} = \mathbf{R}\mathbf{C}\boldsymbol{\mu} + \mathbf{E}$ when $\mathbf{E} \sim \mathrm{N}_n(\mathbf{0}, \sigma^2\mathbf{V})$ and $\mathbf{V}$ is a known $n \times n$ positive definite matrix.

The weighted least squares estimator of $\boldsymbol{\mu}$ is given by

$$\hat{\boldsymbol{\mu}}_{\mathrm{w}} = (\mathbf{C}'\mathbf{R}'\mathbf{V}^{-1}\mathbf{R}\mathbf{C})^{-1}\mathbf{C}'\mathbf{R}'\mathbf{V}^{-1}\mathbf{Y}.$$

The weighted estimator $\hat{\boldsymbol{\mu}}_{\mathrm{w}}$ is unbiased for $\boldsymbol{\mu}$ since $\mathrm{E}(\hat{\boldsymbol{\mu}}_{\mathrm{w}}) = \mathrm{E}[(\mathbf{C}'\mathbf{R}'\mathbf{V}^{-1}\mathbf{R}\mathbf{C})^{-1}$ $\mathbf{C}'\mathbf{R}'\mathbf{V}^{-1}\mathbf{Y}] = (\mathbf{C}'\mathbf{R}'\mathbf{V}^{-1}\mathbf{R}\mathbf{C})^{-1}\mathbf{C}'\mathbf{R}'\mathbf{V}^{-1}\mathbf{R}\mathbf{C}\boldsymbol{\mu} = \boldsymbol{\mu}$. The covariance matrix of $\hat{\boldsymbol{\mu}}_{\mathrm{w}}$ is given by

$$\begin{aligned} \mathrm{cov}(\hat{\boldsymbol{\mu}}_{\mathrm{w}}) &= (\mathbf{C}'\mathbf{R}'\mathbf{V}^{-1}\mathbf{R}\mathbf{C})^{-1}\mathbf{C}'\mathbf{R}'\mathbf{V}^{-1}(\sigma^2\mathbf{V})\mathbf{V}^{-1}\mathbf{R}\mathbf{C}(\mathbf{C}'\mathbf{R}'\mathbf{V}^{-1}\mathbf{R}\mathbf{C})^{-1} \\ &= \sigma^2(\mathbf{C}'\mathbf{R}'\mathbf{V}^{-1}\mathbf{R}\mathbf{C})^{-1} \end{aligned}$$

and the weighted least squares estimator of σ^2 is

$$\hat{\sigma}^2_{\mathrm{w}} = \mathbf{Y}'[\mathbf{V}^{-1} - \mathbf{V}^{-1}\mathbf{R}\mathbf{C}(\mathbf{C}'\mathbf{R}'\mathbf{V}^{-1}\mathbf{R}\mathbf{C})^{-1}\mathbf{C}'\mathbf{R}'\mathbf{V}^{-1}]\mathbf{Y}/(n-k).$$

The ANOVA table for the weighted analysis is provided in Table 9.5.1.

In the weighted case, the $100(1-\gamma)\%$ confidence bands on $\mathbf{t}'\boldsymbol{\mu}$ are given by

$$\mathbf{t}'\hat{\boldsymbol{\mu}}_{\mathrm{w}} \pm t^{\gamma/2}_{n-k}\{[\mathbf{Y}'\mathbf{A}_{3\mathrm{w}}\mathbf{Y}/(n-k)][\mathbf{t}'(\mathbf{C}'\mathbf{R}'\mathbf{V}^{-1}\mathbf{R}\mathbf{C})^{-1}\mathbf{t}\}^{1/2}$$

Table 9.5.1
Mean Model Weighted ANOVA Table

Source	df	SS
Overall mean	1	$\mathbf{Y}'\mathbf{V}^{-1}\mathbf{1}_n(\mathbf{1}_n'\mathbf{V}^{-1}\mathbf{1}_n)^{-1}\mathbf{1}_n'\mathbf{V}^{-1}\mathbf{Y}$
Treatment combinations	$k-1$	$\mathbf{Y}'\mathbf{V}^{-1}[\mathbf{RC}(\mathbf{C}'\mathbf{R}'\mathbf{V}^{-1}\mathbf{RC})^{-1}\mathbf{C}'\mathbf{R}'$ $-\mathbf{1}_n(\mathbf{1}_n'\mathbf{V}^{-1}\mathbf{1}_n)^{-1}\mathbf{1}_n']\mathbf{V}^{-1}\mathbf{Y}$
Residual	$n-k$	$\mathbf{Y}'[\mathbf{V}^{-1}-\mathbf{V}^{-1}\mathbf{RC}(\mathbf{C}'\mathbf{R}'\mathbf{V}^{-1}\mathbf{RC})^{-1}\mathbf{C}'\mathbf{R}'\mathbf{V}^{-1}]\mathbf{Y}$
Total	n	$\mathbf{Y}'\mathbf{V}^{-1}\mathbf{Y}$

where $\mathbf{A}_{3\mathrm{w}}$ is the residual sum of squares matrix from Table 9.5.1. A γ level test of $\mathrm{H}_0 : \boldsymbol{\mu} = c\mathbf{1}_k$ versus $\mathrm{H}_1 : \boldsymbol{\mu} \neq c\mathbf{1}_k$ is to reject H_0 if $F_{\mathrm{w}}^* > F_{k-1,n-k}^{\gamma}$ where

$$F_{\mathrm{w}}^* = \frac{\mathbf{Y}'\mathbf{A}_{2\mathrm{w}}\mathbf{Y}/(k-1)}{\mathbf{Y}'\mathbf{A}_{3\mathrm{w}}\mathbf{Y}/(n-k)}.$$

and $\mathbf{A}_{2\mathrm{w}}$ is the treatment combination sum of squares matrix from Table 9.5.1. The derivations of the confidence band and the test statistic are left to the reader.

Models of the form $\mathbf{Y} = \mathbf{RC}\boldsymbol{\mu} + \mathbf{E}$ with $\mathbf{E} \sim \mathrm{N}_n(\mathbf{0}, \sigma^2\mathbf{V})$ are also encountered when $\mathbf{V}$ is an $n \times n$ positive definite matrix whose elements are functions of m unknown variance parameters. The analysis for this type of model is discussed in Chapter 10.

EXERCISES

Use the data in Table 5.1.1 to solve Exercises 1–5.

1. A researcher wants to fit the model

$$Y_i = \beta_0 + \beta_1 x_{i1} + \beta_2 x_{i2} + \beta_3 x_{i3} + \beta_4 x_{i4} + E_i$$

where $i = 1, \ldots, 10$, x_{i1} is the speed of the ith vehicle, x_{i2} is the speed $\times$ grade of the ith vehicle, x_{i3} is the speed $\times$ speed of the ith vehicle, and x_{i4} is the speed $\times$ speed $\times$ grade of the ith vehicle.

(a) Write the above model in matrix form $\mathbf{Y} = \mathbf{RX}_{\mathrm{d}}\boldsymbol{\beta} + \mathbf{E}$ where the 5×1 vector $\boldsymbol{\beta} = (\beta_0, \beta_1, \beta_2, \beta_3, \beta_4)'$. Define all terms and distributions explicitly.

(b) Is $\boldsymbol{\beta}$ estimable? Explain.

(c) Write the mean model $\mathbf{Y} = \mathbf{R}\boldsymbol{\mu} + \mathbf{E}$ for this problem. Define all terms and distributions explicitly.

(d) What is the rank and dimension of $\mathbf{X}_d$?

2. Find the ordinary least-squares estimates of $\boldsymbol{\mu}$, σ^2, and $\text{cov}(\hat{\boldsymbol{\mu}})$ for the mean model $\mathbf{Y} = \mathbf{R}\boldsymbol{\mu} + \mathbf{E}$.

3. Use the mean model to do the following:

   (a) Construct the ANOVA table.

   (b) At the $\gamma = 0.05$ level, test the hypothesis $\text{H}_0 : \boldsymbol{\mu} = \alpha\mathbf{1}_4$ versus $\text{H}_1 : \boldsymbol{\mu} \neq \alpha\mathbf{1}_4$ where $\boldsymbol{\mu} = (\mu_1, \mu_2, \mu_3, \mu_4)'$.

   (c) Place 99% confidence bands on $\sum_{k=1}^{4} \mu_k$.

4. Show that $\beta_2 + 50\beta_4$ is estimable by defining vectors $\mathbf{g}$ and $\mathbf{t}$ such that $\mathbf{g} = \mathbf{X}_d'\mathbf{t}$. The parameters β_0, β_1, and β_2 are defined in Exercise 1.

5. For the mean model, assume $\text{cov}(\mathbf{E}) = \sigma^2\mathbf{V}$ where

$$\mathbf{V} = \begin{bmatrix} 1 & 0 \\ 0 & 2 \end{bmatrix} \otimes \mathbf{I}_5.$$

   (a) Find the weighted least-squares estimates of $\boldsymbol{\mu}$, σ^2, and $\text{cov}(\hat{\boldsymbol{\mu}})$.

   (b) Construct the weighted ANOVA table.

   (c) At the $\gamma = 0.05$ level, test the hypothesis $\text{H}_0 : \mu_1 = \mu_2 = \mu_3 = \mu_4$ versus H_1 : at least one of the μ_k's is not equal to the others where $\boldsymbol{\mu} = (\mu_1, \mu_2, \mu_3, \mu_4)'$.

   (d) Place 99% confidence bands on $\sum_{k=1}^{4} \mu_k$.

6. Consider the experiment described in Figure 7.3.1.

   (a) Write the general mean model $\mathbf{Y} = \mathbf{R}\mathbf{C}\boldsymbol{\mu} + \mathbf{E}$ defining all terms and distributions explicitly.

   (b) In Section 7.3, the model for the experiment was written

$$\mathbf{Y} = \mathbf{R}\mathbf{M}\mathbf{X}^*\boldsymbol{\beta} + \mathbf{E}$$

   where $\mathbf{E} \sim \text{N}_9(\mathbf{0}, \Sigma)$. Define relationships between the terms $\mathbf{R}$, $\mathbf{C}$, $\boldsymbol{\mu}$, and $\mathbf{E}$ from the mean model in part a and the terms $\mathbf{R}$, $\mathbf{M}$, $\mathbf{X}^*$, $\boldsymbol{\beta}$, and $\mathbf{E}$ from the model in Section 7.3. In particular, define the matrix $\mathbf{X}_d$ such that $\boldsymbol{\mu} = \mathbf{X}_d\boldsymbol{\beta}$. What is the rank and dimension of $\mathbf{X}_d$?

   (c) Is the model $\mathbf{Y} = \mathbf{R}\mathbf{M}\mathbf{X}^*\boldsymbol{\beta} + \mathbf{E}$ from Section 7.3 a less than full rank model?

   (d) If $\hat{\boldsymbol{\mu}}$ and $\hat{\boldsymbol{\beta}}$ are the ordinary least-squares estimators of $\boldsymbol{\mu}$ and $\boldsymbol{\beta}$, respectively, find the relationship between $\hat{\boldsymbol{\mu}}$ and $\hat{\boldsymbol{\beta}}$.

7. Consider the balanced incomplete block experiment described in Figure 8.1.1.

   (a) Write the general mean model $\mathbf{Y} = \mathbf{RC}\boldsymbol{\mu} + \mathbf{E}$ defining all terms and distributions explicitly.

   (b) In Section 8.1, the model for the design was $\mathbf{Y} = \mathbf{MX}^*\boldsymbol{\beta} + \mathbf{E}$ where $\mathbf{E} \sim \mathrm{N}_{bk}(\mathbf{0}, \Sigma)$. Relate the terms $\mathbf{R}$, $\mathbf{C}$, $\boldsymbol{\mu}$, and $\mathbf{E}$ from the general mean model in part a to the terms $\mathbf{M}$, $\mathbf{X}^*$, $\boldsymbol{\beta}$, and $\mathbf{E}$ from the model in Section 8.1.

   (c) Is the model $\mathbf{Y} = \mathbf{MX}^*\boldsymbol{\beta} + \mathbf{E}$ from Section 8.1 a less than full rank model?

10 The General Mixed Model

Any model that includes some fixed factors and some random factors is called a mixed model. Numerous examples of mixed models have been presented throughout the text. For example, the model for the experiment described in Figure 4.1.1 is an example of a balanced mixed model. The general class of mixed models applies to both balanced and unbalanced data structures. Furthermore, the general mixed model covariance structure contains a broad class of matrix patterns including those discussed in Chapter 4. In this chapter the analysis of the general mixed model is presented. Balanced mixed model examples from previous chapters are reviewed and new unbalanced examples are presented to illustrate the general mixed model approach.

10.1 THE MIXED MODEL STRUCTURE AND ASSUMPTIONS

The mixed model is applicable whenever an experiment contains fixed and random factors. Consider the experiment presented in Table 4.1.1. The experiment has three factors where B and R are random factors and T is a fixed factor. The model is

$$Y_{ijs} = \mu_j + B_i + BT_{ij} + R(BT)_{(ij)s}$$

for $i = 1, \ldots, b$, $j = 1, \ldots, t$, and $s = 1, \ldots, r$ where μ_j represents the fixed portion of the model and $B_i + BT_{ij} + R(BT)_{(ij)s}$ represents the random portion. Assume a finite model error structure where $B_i \sim iid\, \mathrm{N}_1(0, \sigma_B^2)$,$(\mathbf{I}_b \otimes \mathbf{P}_t')$ $(BT_{11}, \ldots, BT_{bt}) \sim \mathrm{N}_{b(t-1)}(\mathbf{0}, \sigma_{BT}^2 \mathbf{I}_b \otimes \mathbf{I}_{t-1})$ and $R(BT)_{(ij)_s} \sim iid\, \mathrm{N}_1(0, \sigma_{R(BT)}^2)$. Furthermore, assume the three sets of random variables are mutually independent.

In matrix form the mixed model is

$$\mathbf{Y} = \mathbf{RC}\boldsymbol{\mu} + \mathbf{U}_1\mathbf{a}_1 + \mathbf{U}_2\mathbf{a}_2 + \mathbf{U}_3\mathbf{a}_3$$

where the $btr \times 1$ vector $\mathbf{Y} = (Y_{111}, \ldots, Y_{11r}, \ldots, Y_{bt1}, \ldots, Y_{btr})'$. The fixed portion of the model is given by $\mathbf{RC}\boldsymbol{\mu}$ where the $btr \times bt$ replication matrix $\mathbf{R} = \mathbf{I}_b \otimes \mathbf{I}_t \otimes \mathbf{1}_r$, the $bt \times t$ matrix $\mathbf{C} = \mathbf{1}_b \otimes \mathbf{I}_t$, and the $t \times 1$ mean vector $\boldsymbol{\mu} = (\mu_1, \ldots, \mu_t)'$. The random portion of the model is $\mathbf{U}_1\mathbf{a}_1 + \mathbf{U}_2\mathbf{a}_2 + \mathbf{U}_3\mathbf{a}_3$ where the $b \times 1$ random vector $\mathbf{a}_1 \sim \mathrm{N}_b(\mathbf{0}, \sigma_B^2 \mathbf{I}_b)$, the $btr \times b$ matrix $\mathbf{U}_1 = \mathbf{I}_b \otimes \mathbf{1}_t \otimes \mathbf{1}_r$, the $b(t-1) \times 1$ random vector $\mathbf{a}_2 \sim \mathrm{N}_{b(t-1)}(\mathbf{0}, \sigma_{BT}^2 \mathbf{I}_b \otimes \mathbf{I}_{t-1})$, the $btr \times b(t-1)$ matrix $\mathbf{U}_2 = \mathbf{I}_b \otimes \mathbf{P}_t \otimes \mathbf{1}_r$, the $(t-1) \times t$ matrix $\mathbf{P}_t'$ is the lower portion of a t-dimensional Helmert matrix, the $btr \times 1$ random vector $\mathbf{a}_3 \sim \mathrm{N}_{btr}(\mathbf{0}, \sigma_{R(BT)}^2 \mathbf{I}_b \otimes \mathbf{I}_t \otimes \mathbf{I}_r)$, and the $btr \times btr$ matrix $\mathbf{U}_3 = \mathbf{I}_b \otimes \mathbf{I}_t \otimes \mathbf{I}_r$. Furthermore, vectors $\mathbf{a}_1$, $\mathbf{a}_2$, and $\mathbf{a}_3$ are mutually independent. Therefore, the $btr \times 1$ random vector $\mathbf{Y} \sim \mathrm{N}_{btr}(\mathbf{RC}\boldsymbol{\mu}, \Sigma)$ where

$$\begin{aligned}\Sigma &= \sigma_B^2 \mathbf{U}_1\mathbf{U}_1' + \sigma_{BT}^2 \mathbf{U}_2\mathbf{U}_2' + \sigma_{R(BT)}^2 \mathbf{U}_3\mathbf{U}_3' \\ &= \sigma_B^2[\mathbf{I}_b \otimes \mathbf{J}_t \otimes \mathbf{J}_r] + \sigma_{BT}^2\left[\mathbf{I}_b \otimes \left(\mathbf{I}_t - \frac{1}{t}\mathbf{J}_t\right) \otimes \mathbf{J}_r\right] \\ &\quad + \sigma_{R(BT)}^2[\mathbf{I}_b \otimes \mathbf{I}_t \otimes \mathbf{I}_r].\end{aligned}$$

Note that Σ matches the finite model covariance structure given in Section 4.2. If an infinite model is assumed, the same mixed model can be used with $\mathbf{U}_2$ replaced by $\mathbf{I}_b \otimes \mathbf{I}_t \otimes \mathbf{1}_r$.

The mixed model can be generalized as

$$\mathbf{Y} = \mathbf{RC}\boldsymbol{\mu} + \sum_{f=1}^{m} \mathbf{U}_f\mathbf{a}_f$$

where $\mathbf{Y}$ is an $n \times 1$ random vector of observations. The fixed portion of the model is given by $\mathbf{RC}\boldsymbol{\mu}$ where $\mathbf{R}$ is an $n \times k$ replication matrix of rank k, $\mathbf{C}$ is a $k \times v$ matrix of rank v, and $\boldsymbol{\mu}$ is a $v \times 1$ mean vector. The random portion of the model is $\sum_{f=1}^{m} \mathbf{U}_f\mathbf{a}_f$ where $\mathbf{U}_f$ is an $n \times q_f$ matrix of rank q_f, $\mathbf{a}_f$ is a $q_f \times 1$ random vector distributed $\mathrm{N}_{q_f}(\mathbf{0}, \sigma_f^2 \mathbf{I}_{q_f})$, and $\mathbf{a}_1, \ldots, \mathbf{a}_m$ are mutually independent. The general mixed model can also be written in the form

$$\mathbf{Y} = \mathbf{RC}\boldsymbol{\mu} + \mathbf{E}$$

where the $n \times 1$ random vector $\mathbf{E} \sim \mathrm{N}_n(\mathbf{0}, \Sigma)$ and $\Sigma = \sum_{f=1}^{m} \sigma_f^2 \mathbf{U}_f \mathbf{U}_f'$.

The $n \times v$ matrix $\mathbf{RC}$ has full column rank v. Therefore, the elements of $\boldsymbol{\mu}$ are the expected values of the random, variables in the v distinct fixed factor combinations that contain data. Consequently, this form of the mixed model is a full rank model and could logically be called the mixed mean model. Similarly, the random portion of the model $\sum_{f=1}^{m} \mathbf{U}_f \mathbf{a}_f$ is sometimes called the variance components portion of the model since the distributions of the m random vectors $\mathbf{a}_1, \ldots, \mathbf{a}_m$ are functions of the m variance components of $\sigma_1^2, \ldots, \sigma_m^2$.

Although the general mixed model was motivated using a complete, balanced design with three factors, the model applies to a broad class of balanced and unbalanced designs with a wide variety of covariance structures. In fact, every experimental layout and covariance structure covered in this text can be modeled in the general mixed model format. However, because the general mixed model format applies to such a broad class of experiments, no one analysis approach is "best" for all cases. A number of different analysis approaches are available. Two approaches for analyzing the random portion of the model are presented in Sections 10.2 and 10.3. A numerical example of the random portion analysis is provided in Section 10.4. An analysis of the fixed portion of the model is presented in Section 10.5.

10.2 RANDOM PORTION ANALYSIS: TYPE I SUM OF SQUARES METHOD

In some mixed models, estimates of the variance components can be derived by calculating the Type I sums of squares for the fixed portion first and then calculating m Type I sums of squares for the subsequent random factors. The expectations of the Type I sums of squares for the subsequent m random factors do not involve $\boldsymbol{\mu}$ since they are calculated after the fixed portion has been removed. Setting these m expectations equal to the corresponding Type I sums of squares produces m equations and the m unknowns $\sigma_1^2, \ldots, \sigma_m^2$. Estimates of the m unknown variance parameters can then be derived provided the m equations are linearly independent. Balanced and unbalanced examples of the method are given next.

Example 10.2.1 Consider the complete, balanced design described in Section 10.1 where B and R are random factors and T is a fixed factor. Since the design is complete and balanced, the Type I sums of squares equal the sums of squares defined by the algorithm in Section 4.3. Therefore, the Type I sums of squares due to the three random effects are given by

$$\begin{aligned}
\text{SS }(B|\text{ overall mean}, T) &= \mathbf{Y}'\left[\left(\mathbf{I}_b - \frac{1}{b}\mathbf{J}_b\right) \otimes \frac{1}{t}\mathbf{J}_t \otimes \frac{1}{r}\mathbf{J}_r\right]\mathbf{Y} \\
&= \mathbf{Y}'\mathbf{A}_3\mathbf{Y}, \\
\text{SS }(BT|\text{ overall mean}, T, B) &= \mathbf{Y}'\left[\left(\mathbf{I}_b - \frac{1}{b}\mathbf{J}_b\right) \otimes \left(\mathbf{I}_t - \frac{1}{t}\mathbf{J}_t\right)\right. \\
&\qquad \left.\otimes \frac{1}{r}\mathbf{J}_r\right]\mathbf{Y} = \mathbf{Y}'\mathbf{A}_4\mathbf{Y} \\
\text{SS }(R(BT)|\text{ overall mean}, T, B, BT) &= \mathbf{Y}'\left[\mathbf{I}_b \otimes \mathbf{I}_t \otimes \left(\mathbf{I}_r - \frac{1}{r}\mathbf{J}_r\right)\right]\mathbf{Y} \\
&= \mathbf{Y}'\mathbf{A}_5\mathbf{Y}.
\end{aligned}$$

The expectations of these three Type I sums of squares are

$$\begin{aligned}
\mathrm{E}[\mathbf{Y}'\mathbf{A}_3\mathbf{Y}] &= (b-1)[rt\sigma_B^2 + \sigma_{R(BT)}^2], \\
\mathrm{E}[\mathbf{Y}'\mathbf{A}_4\mathbf{Y}] &= (b-1)(t-1)[r\sigma_{BT}^2 + \sigma_{R(BT)}^2] \\
\mathrm{E}[\mathbf{Y}'\mathbf{A}_5\mathbf{Y}] &= bt(r-1)\sigma_{R(BT)}^2.
\end{aligned}$$

Setting the three Type I sums of squares equal to their expectations and solving for $\sigma_B^2, \sigma_{BT}^2$, and $\sigma_{R(BT)}^2$ produces the estimators

$$\begin{aligned}
\hat{\sigma}_B^2 &= \{\mathbf{Y}'\mathbf{A}_3\mathbf{Y}/(b-1) - \mathbf{Y}'\mathbf{A}_5\mathbf{Y}/[bt(r-1)]\}/(rt) \\
\hat{\sigma}_{BT}^2 &= \{\mathbf{Y}'\mathbf{A}_4\mathbf{Y}/[(b-1)(t-1)] - \mathbf{Y}'\mathbf{A}_5\mathbf{Y}/[bt(r-1)]\}/r \\
\hat{\sigma}_{R(BT)}^2 &= \mathbf{Y}'\mathbf{A}_5\mathbf{Y}/[bt(r-1)].
\end{aligned}$$

In general, if any of the variance estimates is less than zero, set it equal to zero.

Example 10.2.2 Consider the unbalanced experimental layout given in Figure 9.5.1. Let the 8×1 random vector $\mathbf{Y} = (Y_{111}, Y_{121}, Y_{211}, Y_{212}, Y_{311}, Y_{321}, Y_{322}, Y_{323})'$. The mean model for this experiment is

$$\mathbf{Y} = \mathbf{RC}\boldsymbol{\mu} + \mathbf{E}$$

where the 2×1 mean vector $\boldsymbol{\mu} = (\mu_1, \mu_2)'$ and the 8×1 random error vector $\mathbf{E} = (E_{111}, E_{121}, E_{211}, E_{212}, E_{311}, E_{321}, E_{322}, E_{323})' \sim \mathrm{N}_8(\mathbf{0}, \Sigma)$. The 8×5 replication matrix $\mathbf{R}$ and the 5×2 matrix $\mathbf{C}$ are given in Example 9.5.1. Let $\mathbf{Y}'\mathbf{A}_3\mathbf{Y}, \mathbf{Y}'\mathbf{A}_4\mathbf{Y}$, and $\mathbf{Y}'\mathbf{A}_5\mathbf{Y}$ equal the SS ($B|$ overall mean, T), SS ($BT|$ overall mean, T, B) and SS ($R(BT)|$ overall mean, T, B, BT), respectively. The matrices $\mathbf{A}_3, \mathbf{A}_4, \mathbf{A}_5$, and Σ were generated using the SAS PROC IML program listed in Section A5.1 of Appendix 5. The program output and derivations of the various results are also given in Section A5.1. From Appendix 5 the expected values of the three Type I sums of squares are

$$\mathrm{E}[\mathbf{Y}'\mathbf{A}_3\mathbf{Y}] = 4\sigma_B^2 + 0.8\sigma_{BT}^2 + 2\sigma_{R(BT)}^2$$
$$\mathrm{E}[\mathbf{Y}'\mathbf{A}_4\mathbf{Y}] = 1.2\sigma_{BT}^2 + \sigma_{R(BT)}^2$$
$$\mathrm{E}[\mathbf{Y}'\mathbf{A}_5\mathbf{Y}] = 3\sigma_{R(BT)}^2.$$

Setting these three Type I sums of squares equal to their expectations and solving for σ_B^2, σ_{BT}^2, and $\sigma_{R(BT)}^2$ produces the estimators

$$\hat{\sigma}_B^2 = \{\mathbf{Y}'\mathbf{A}_3\mathbf{Y} - (2\mathbf{Y}'\mathbf{A}_4\mathbf{Y}/3) - (4\mathbf{Y}'\mathbf{A}_5\mathbf{Y}/9)\}/4$$
$$\hat{\sigma}_{BT}^2 = \{\mathbf{Y}'\mathbf{A}_4\mathbf{Y} - (\mathbf{Y}'\mathbf{A}_5\mathbf{Y}/3)\}/1.2$$
$$\hat{\sigma}_{R(BT)}^2 = \mathbf{Y}'\mathbf{A}_5\mathbf{Y}/3.$$

We now demonstrate that the covariance structure for the unbalanced experiment in Example 10.2.2 follows the mixed model covariance format. Following the methods described in Section 7.3, the covariance matrix of $\mathbf{E}$ from Example 10.2.2 is given by

$$\Sigma = \mathbf{R}\mathbf{M}\Sigma^*\mathbf{M}'\mathbf{R}' + \sigma_{R(BT)}^2\mathbf{I}_8$$

where the 8×5 replication matrix $\mathbf{R}$ is provided in Example 9.5.1. The 6×6 matrix Σ^* and the 5×6 pattern matrix $\mathbf{M}$ are given by

$$\Sigma^* = \sigma_B^2\mathbf{I}_3 \otimes \mathbf{J}_2 + \sigma_{BT}^2\mathbf{I}_3 \otimes \left(\mathbf{I}_2 - \frac{1}{2}\mathbf{J}_2\right)$$

and

$$\mathbf{M} = \begin{bmatrix} 1 & 0 & 0 & 0 & 0 & 0 \\ 0 & 1 & 0 & 0 & 0 & 0 \\ 0 & 0 & 1 & 0 & 0 & 0 \\ 0 & 0 & 0 & 0 & 1 & 0 \\ 0 & 0 & 0 & 0 & 0 & 1 \end{bmatrix}.$$

Therefore,

$$\begin{aligned}\Sigma &= \sigma_B^2\mathbf{R}\mathbf{M}(\mathbf{I}_3 \otimes \mathbf{J}_2)\mathbf{M}'\mathbf{R}' + \sigma_{BT}^2\mathbf{R}\mathbf{M}\left[\mathbf{I}_3 \otimes \left(\mathbf{I}_2 - \frac{1}{2}\mathbf{J}_2\right)\right]\mathbf{M}'\mathbf{R}' + \sigma_{R(BT)}^2\mathbf{I}_8 \\ &= \sigma_B^2\mathbf{R}\mathbf{M}(\mathbf{I}_3 \otimes \mathbf{1}_2)(\mathbf{I}_3 \otimes \mathbf{1}_2')\mathbf{M}'\mathbf{R}' + \sigma_{BT}^2\mathbf{R}\mathbf{M}[\mathbf{I}_3 \otimes \mathbf{P}_2][\mathbf{I}_3 \otimes \mathbf{P}_2']\mathbf{M}'\mathbf{R}' \\ &\quad + \sigma_{R(BT)}^2\mathbf{I}_8 \\ &= \sigma_1^2\mathbf{U}_1\mathbf{U}_1' + \sigma_2^2\mathbf{U}_2\mathbf{U}_2' + \sigma_3^2\mathbf{U}_3\mathbf{U}_3'\end{aligned}$$

where $\sigma_1^2 = \sigma_B^2, \sigma_2^2 = \sigma_{BT}^2, \sigma_3^2 = \sigma_{R(BT)}^2$, the 8×3 matrix $\mathbf{U}_1 = \mathbf{RM}(\mathbf{I}_3 \otimes \mathbf{1}_2)$, the 8×3 matrix $\mathbf{U}_2 = \mathbf{RM}(\mathbf{I}_3 \otimes \mathbf{P}_2)$, the 8×8 matrix $\mathbf{U}_3 = \mathbf{I}_8$, and the 1×2 matrix $\mathbf{P}_2'$ is the lower portion of a two-dimensional Helmert matrix with $\mathbf{P}_2\mathbf{P}_2' = \mathbf{I}_2 - \frac{1}{2}\mathbf{J}_2$.

10.3 RANDOM PORTION ANALYSIS: RESTRICTED MAXIMUM LIKELIHOOD METHOD

The maximum likelihood estimators (MLEs) of the variance components are the values of $\sigma_1^2, \ldots, \sigma_m^2$ that maximize the function

$$\ell\,(\boldsymbol{\mu}, \sigma_1^2, \ldots, \sigma_m^2, \mathbf{Y}) = (2\pi)^{-n/2}|\boldsymbol{\Sigma}|^{-1/2}e^{-\{(\mathbf{Y}-\mathbf{RC}\boldsymbol{\mu})'\boldsymbol{\Sigma}^{-1}(\mathbf{Y}-\mathbf{RC}\boldsymbol{\mu})\}/2}$$

where the maximization is performed simultaneously with respect to the $k + m$ terms $\boldsymbol{\mu}, \sigma_1^2, \ldots, \sigma_m^2$. By Theorem 6.3.1, for complete, balanced designs, the MLE of $\boldsymbol{\mu}$ equals the ordinary least-squares estimator $(\mathbf{C}'\mathbf{DC})^{-1}\mathbf{C}'\mathbf{R}'\mathbf{Y}$ and the MLE of σ_f^2 for $f = 1, \ldots, m$ equals a linear combination of the sums of squares for the m random effects and interactions. However, in general, for unbalanced designs, derivation of the MLEs of $\boldsymbol{\mu}, \sigma_1^2, \ldots, \sigma_m^2$ can be tedious and usually involves numerical techniques such as the Newton–Raphson method.

Russell and Bradley (1958), Anderson and Bancroft (1952), W. A. Thompson (1962), and later Patterson and R. Thompson (1971, 1974) suggested what is called a restricted maximum likelihood (REML) approach. The REML estimators of $\sigma_1^2, \ldots, \sigma_m^2$ are derived by first expressing the likelihood function in two parts, one involving the fixed parameters, $\boldsymbol{\mu}$, and the second free of these fixed parameters. To construct the REMLs, let $\mathbf{G}$ be any $n \times (n-k)$ matrix of rank $n-k$ such that $\mathbf{G}'\mathbf{RC} = \mathbf{0}_{(n-k)\times k}$. For example, $\mathbf{G}$ can be defined such that $\mathbf{GG}' = \mathbf{I}_n - \mathbf{RC}(\mathbf{C}'\mathbf{DC})^{-1}\mathbf{C}'\mathbf{R}'$ and $\mathbf{G}'\mathbf{G} = \mathbf{I}_{n-k}$ where $\mathbf{D} = \mathbf{R}'\mathbf{R}$. Next, transform the $n \times 1$ random vector $\mathbf{Y}$ by the $n \times 1$ nonsingular matrix $[\mathbf{RC}, \mathbf{G}]'$ where

$$\begin{bmatrix} \mathbf{C}'\mathbf{R}' \\ \mathbf{G}' \end{bmatrix} \mathbf{Y} \sim \mathrm{N}_n\left(\begin{bmatrix} \mathbf{C}'\mathbf{DC}\boldsymbol{\mu} \\ \mathbf{0} \end{bmatrix}, \begin{bmatrix} \mathbf{C}'\mathbf{R}'\boldsymbol{\Sigma}\mathbf{RC} & \mathbf{C}'\mathbf{R}'\boldsymbol{\Sigma}\mathbf{G} \\ \mathbf{G}'\boldsymbol{\Sigma}\mathbf{RC} & \mathbf{G}'\boldsymbol{\Sigma}\mathbf{G} \end{bmatrix} \right).$$

The distribution of $\mathbf{G}'\mathbf{Y}$ is free of the fixed parameters $\boldsymbol{\mu}$. The REML estimators of the variance components are the values of $\sigma_1^2, \ldots, \sigma_m^2$ that maximize the marginal likelihood of $\mathbf{G}'\mathbf{Y}$:

$$\ell\,(\sigma_1^2, \ldots, \sigma_m^2, \mathbf{Y}) = (2\pi)^{-(n-k)/2}|\mathbf{G}'\boldsymbol{\Sigma}\mathbf{G}|^{-1/2}e^{-\mathbf{Y}'(\mathbf{G}'\boldsymbol{\Sigma}\mathbf{G})^{-1}\mathbf{Y}/2}.$$

As with the maximum likelihood approach, numerical techniques are often necessary to determine the REML estimates of $\sigma_1^2, \ldots, \sigma_m^2$. Furthermore, the maximization is performed under the restriction that the estimators of $\sigma_1^2, \ldots, \sigma_m^2$ are all positive.

The SAS PROC VARCOMP routine has a Type I sum of squares and a restricted maximum likelihood option. In the next section a numerical example is analyzed using PROC VARCOMP.

10.4 RANDOM PORTION ANALYSIS: A NUMERICAL EXAMPLE

Suppose that the observations in Table 9.5.1 take the values $\mathbf{Y} = (Y_{111}, Y_{121}, Y_{211}, Y_{212}, Y_{311}, Y_{321}, Y_{322}, Y_{323})' = (237, 178, 249, 268, 186, 183, 182, 165)'$. All of the numerical calculations are performed using the PROC VARCOMP output listed in Section A5.2 of Appendix 5. The Type I sums of squares option of the PROC VARCOMP program provides the SS ($B|$ overall mean, T) $= \mathbf{Y}'\mathbf{A}_3\mathbf{Y} = 2770.8$, SS ($BT|$ overall mean, T, B) $= \mathbf{Y}'\mathbf{A}_4\mathbf{Y} = 740.03$ and SS ($R(BT)|$ overall mean, T, B, BT) $= \mathbf{Y}'\mathbf{A}_5\mathbf{Y} = 385.17$. These values are now substituted into the variance parameter relationships derived in Example 10.2.2. Therefore, the finite model estimates of $\sigma_B^2, \sigma_{BT}^2$, and $\sigma_{R(BT)}^2$ are

$$\hat{\sigma}_B^2 = [2770.8 - 2(740.03/3) - 4(385.17/9)]/4 = 526.56$$

$$\hat{\sigma}_{BT}^2 = [740.03 - (385.17/3)]/1.2 = 509.70$$

$$\hat{\sigma}_{R(BT)}^2 = 385.17/3 = 128.39.$$

The PROC VARCOMP program will actually provide variance parameter estimates automatically for both the Type I sum of squares and the REML options. However, when calculating the variance parameter estimates, PROC VARCOMP assumes an infinite model for both options. Conversion between the infinite and finite model estimates can be accomplished quite easily, however, since one model is simply a reparameterization of the other. For example, let the variance parameters for the finite model remain $\sigma_B^2, \sigma_{BT}^2$, and $\sigma_{R(BT)}^2$. Let $\sigma_{B^*}^2, \sigma_{BT^*}^2$, and $\sigma_{R(BT)^*}^2$ be the corresponding variance parameters for the infinite model. Then $\sigma_B^2 = \sigma_{B^*}^2 + \frac{1}{t}\sigma_{BT^*}^2, \sigma_{BT}^2 = \sigma_{BT^*}^2$, and $\sigma_{R(BT)}^2 = \sigma_{R(BT)^*}^2$ where t is the number of fixed treatment levels. Therefore, using the preceding relationships, infinite model variance parameter estimates provided by PROC VARCOMP can be converted directly to finite model estimates.

From the PROC VARCOMP output in Section A5.2, the Type I sum of squares estimates of the infinite model variance parameters are $\hat{\sigma}_{B^*}^2 = 271.71, \hat{\sigma}_{BT^*}^2 = 509.70$, and $\hat{\sigma}_{R(BT)^*}^2 = 128.39$. Therefore, the corresponding finite model variance parameter estimates are

$$\hat{\sigma}_B^2 = \hat{\sigma}_{B^*}^2 + \frac{1}{2}\hat{\sigma}_{BT^*}^2$$

$$= 271.71 + \frac{1}{2}(509.7) = 526.56$$

$$\hat{\sigma}^2_{BT} = \hat{\sigma}^2_{BT*} = 509.7$$
$$\hat{\sigma}^2_{R(BT)} = \hat{\sigma}^2_{R(BT)*} = 128.39.$$

These finite model estimates agree exactly with the estimates derived earlier using the Example 10.2.2 variance parameter relationships.

The expected mean squares (EMSs) for the infinite model are also automatically given by the Type I sum of squares option of the PROC VARCOMP program. The three EMSs are

$$\mathrm{E}[\mathbf{Y}'\mathbf{A}_3\mathbf{Y}/2] = 2\sigma^2_{B*} + 1.4\sigma^2_{BT*} + \sigma^2_{R(BT)*}$$
$$\mathrm{E}[\mathbf{Y}'\mathbf{A}_4\mathbf{Y}/1] = 1.2\sigma^2_{BT*} + \sigma^2_{R(BT)*}$$
$$\mathrm{E}[\mathbf{Y}'\mathbf{A}_5\mathbf{Y}/3] = \sigma^2_{R(BT)*}.$$

Using the relationships that link the finite and infinite model parameters, the expected mean squares for the finite model are

$$\begin{aligned}\mathrm{E}[\mathbf{Y}'\mathbf{A}_3\mathbf{Y}/2] &= 2\left(\sigma^2_B - \frac{1}{2}\sigma^2_{BT}\right) + 1.4\sigma^2_{BT} + \sigma^2_{R(BT)} \\ &= 2\sigma^2_B + 0.4\sigma^2_{BT} + \sigma^2_{R(BT)} \\ \mathrm{E}[\mathbf{Y}'\mathbf{A}_4\mathbf{Y}/1] &= 1.2\sigma^2_{BT} + \sigma^2_{R(BT)} \\ \mathrm{E}[\mathbf{Y}'\mathbf{A}_5\mathbf{Y}/3] &= \sigma^2_{R(BT)}.\end{aligned}$$

These finite model EMSs are equivalent to the expectations derived in Example 10.2.2.

The REML option of PROC VARCOMP was also run on the data set. Assuming the infinite model, the program provides the REML variance parameter estimates $\hat{\sigma}^2_{B*} = 88.60$, $\hat{\sigma}^2_{BT*} = 726.62$, and $\hat{\sigma}^2_{R(BT)*} = 129.15$. The corresponding REML variance parameter estimates for the finite model are therefore

$$\begin{aligned}\hat{\sigma}^2_B &= \hat{\sigma}^2_{B*} + \frac{1}{2}\hat{\sigma}^2_{BT*} = 88.6 + \frac{1}{2}(726.62) = 451.91 \\ \hat{\sigma}^2_{BT} &= \hat{\sigma}^2_{BT*} = 726.62 \\ \hat{\sigma}^2_{R(BT)} &= \hat{\sigma}^2_{R(BT)*} = 129.15.\end{aligned}$$

10.5 FIXED PORTION ANALYSIS

In Section 10.4 REML and Type I sum of squares estimates of the variance parameters $\sigma^2_1, \ldots, \sigma^2_m$ were derived for both the finite and infinite models. Therefore, the covariance matrix Σ can be estimated by

$$\hat{\Sigma} = \hat{\sigma}_1^2 \mathbf{U}_1\mathbf{U}_1' + \hat{\sigma}_2^2 \mathbf{U}_2\mathbf{U}_2' + \cdots + \hat{\sigma}_m^2 \mathbf{U}_m\mathbf{U}_m'$$

where $\hat{\sigma}_1^2, \ldots, \hat{\sigma}_m^2$ are a set of variance parameter estimates. The weighted least-squares estimator of $\boldsymbol{\mu}$ can therefore be estimated by

$$\hat{\boldsymbol{\mu}}_{\mathrm{w}} = (\mathbf{C}'\mathbf{R}'\hat{\Sigma}^{-1}\mathbf{R}\mathbf{C})^{-1}\mathbf{C}'\mathbf{R}\hat{\Sigma}^{-1}\mathbf{Y}.$$

The BLUE of $\mathbf{t}'\boldsymbol{\mu}$ can be estimated by $\mathbf{t}'\hat{\hat{\boldsymbol{\mu}}}_{\mathrm{w}}$ where $\mathbf{t}$ is any nonzero $v \times 1$ vector of constants. The covariance matrix of the weighted least- squares estimator of $\boldsymbol{\mu}$ can be estimated by

$$\widehat{\mathrm{cov}}(\hat{\boldsymbol{\mu}}_{\mathrm{w}}) = (\mathbf{C}'\mathbf{R}'\hat{\Sigma}^{-1}\mathbf{R}\mathbf{C})^{-1}$$

and the variance of the BLUE of $\mathbf{t}'\boldsymbol{\mu}$ is estimated by

$$\widehat{\mathrm{var}}(\mathbf{t}'\hat{\boldsymbol{\mu}}_{\mathrm{w}}) = \mathbf{t}'(\mathbf{C}'\mathbf{R}'\hat{\Sigma}^{-1}\mathbf{R}\mathbf{C})^{-1}\mathbf{t}.$$

For large samples, $(\mathbf{t}'\hat{\hat{\boldsymbol{\mu}}}_{\mathrm{w}} - \mathbf{t}'\boldsymbol{\mu})/\sqrt{\widehat{\mathrm{var}}(\mathbf{t}'\hat{\boldsymbol{\mu}}_{\mathrm{w}})} \sim \mathrm{N}_1(0, 1)$. Therefore, a $100(1-\gamma)\%$ confidence band on $\mathbf{t}'\boldsymbol{\mu}$ is

$$\mathbf{t}'\hat{\hat{\boldsymbol{\mu}}}_{\mathrm{w}} \pm Z_{\gamma/2}\sqrt{\widehat{\mathrm{var}}(\mathbf{t}'\hat{\boldsymbol{\mu}}_{\mathrm{w}})}.$$

Hypothesis tests on various fixed effects can be constructed using a Satterthwaite approximation. Suppose the expected mean square of a certain fixed effect is

$$k_1\sigma_1^2 + k_2\sigma_2^2 + \cdots + k_m\sigma_m^2 + \Phi(F)$$

where $\Phi(F)$ is the fixed portion of the expected mean square and $k_1, \ldots, k_m$ are constants. An unbiased estimator of the first m terms of the expected mean square can be constructed using a linear combination of random effect mean squares. That is,

$$W = c_1\mathrm{MS}_1 + c_2\mathrm{MS}_2 + \cdots + c_m\mathrm{MS}_m$$

where $\mathrm{E}(W) = k_1\sigma_1^2 + k_2\sigma_2^2 + \cdots + k_m\sigma_m^2$ and where MS_f is the mean square of a particular random effect for $f = 1, \ldots, m$. If MSF is the mean square for a specific fixed effect with v_F degrees of freedom, then the statistic MSF/W is approximately distributed as an F random variable with v_F and $\mathfrak{f}$ degrees of freedom where

$$\mathfrak{f} = \frac{W^2}{\left\{\frac{(c_1\mathrm{MS}_1)^2}{v_1} + \cdots + \frac{(c_m\mathrm{MS}_m)^2}{v_m}\right\}}$$

and $v_1, \ldots, v_m$ are the degrees of freedom associated with $\mathrm{MS}_1, \ldots, \mathrm{MS}_m$. Therefore, a γ level test on the equality of the fixed treatment means is to reject if $\mathrm{MSF}/W > F^{\gamma}_{v_F, \mathfrak{f}}$.

All of these fixed portion calculations are performed for an example problem in the next section.

10.6 FIXED PORTION ANALYSIS: A NUMERICAL EXAMPLE

An estimate of $\boldsymbol{\mu}$, a confidence band on $\mathbf{t}'\boldsymbol{\mu}$, and a hypothesis test on the fixed treatment effect are now calculated using the data in Section 10.4. The numerical calculations are performed using the SAS PROC IML procedure detailed in Section A5.3 of Appendix 5. All calculations are made assuming a finite model with Type I sums of squares estimates for σ_B^2, σ_{BT}^2, and $\sigma_{R(BT)}^2$.

From Section A5.3 the estimate of $\boldsymbol{\mu} = (\mu_1, \mu_2)'$ is given by

$$\hat{\boldsymbol{\mu}}_{\mathrm{w}} = (\hat{\mu}_1, \hat{\mu}_2)' = (\mathbf{C}'\mathbf{R}'\hat{\Sigma}^{-1}\mathbf{R}\mathbf{C})^{-1}\mathbf{C}'\mathbf{R}'\hat{\Sigma}^{-1}\mathbf{Y} = \begin{bmatrix} 227.94 \\ 182.62 \end{bmatrix}.$$

The estimated covariance matrix of the weighted least squares estimator of $\boldsymbol{\mu}$ is

$$\widehat{\mathrm{cov}}(\hat{\boldsymbol{\mu}}_{\mathrm{w}}) = (\mathbf{C}'\mathbf{R}'\hat{\Sigma}^{-1}\mathbf{R}\mathbf{C})^{-1} = \begin{bmatrix} 295.78 & 88.33 \\ 88.33 & 418.14 \end{bmatrix}.$$

An estimate of the BLUE of $\mu_1 - \mu_2 = \mathbf{t}'\boldsymbol{\mu}$ for $\mathbf{t}' = (1, -1)$ is given by

$$\mathbf{t}'\hat{\boldsymbol{\mu}}_{\mathrm{w}} = (1, -1) \begin{bmatrix} 227.94 \\ 182.62 \end{bmatrix} = 45.32.$$

The variance of the BLUE of $\mu_1 - \mu_2$ is estimated by

$$\begin{aligned} \widehat{\mathrm{var}}(\mathbf{t}'\hat{\boldsymbol{\mu}}_{\mathrm{w}}) &= \mathbf{t}'(\mathbf{C}'\mathbf{R}'\hat{\Sigma}^{-1}\mathbf{R}\mathbf{C})^{-1}\mathbf{t} \\ &= (1, -1) \begin{bmatrix} 295.78 & 88.33 \\ 88.33 & 418.14 \end{bmatrix} \begin{bmatrix} 1 \\ -1 \end{bmatrix} = 537.26. \end{aligned}$$

Therefore, 95% confidence bands on $\mu_1 - \mu_2$ are

$$\mathbf{t}'\hat{\boldsymbol{\mu}}_{\mathrm{w}} \pm Z_{0.025}\sqrt{\widehat{\mathrm{var}}(\mathbf{t}'\hat{\boldsymbol{\mu}}_{\mathrm{w}})} = 45.32 \pm 1.96\sqrt{537.26} = (-0.11, 90.75).$$

A Satterthwaite test on the treatment means can be constructed. Note that

$$\mathrm{E}[\mathbf{Y}'\mathbf{A}_2\mathbf{Y}/1] = \sigma_B^2 + 1.5\sigma_{BT}^2 + \sigma_{R(BT)}^2 + \Phi(T)$$

where $\mathbf{Y}'\mathbf{A}_2\mathbf{Y}$ is the Type I sum of squares due to treatments given the overall mean and $\Phi(T)$ is the fixed portion of the treatment EMS. Furthermore,

$$\sigma_B^2 + 1.5\sigma_{BT}^2 + \sigma_{R(BT)}^2 = c_1\mathrm{E}[\mathbf{Y}'\mathbf{A}_3\mathbf{Y}/2] + c_2\mathrm{E}[\mathbf{Y}'\mathbf{A}_4\mathbf{Y}] + c_3\mathrm{E}[\mathbf{Y}'\mathbf{A}_5\mathbf{Y}/3]$$

where $c_1 = 1/2$, $c_2 = 1.3/1.2$, and $c_3 = -0.7/1.2$. Therefore, the Satterthwaite statistic to test a significant treatment effect is

$$\begin{aligned} \text{MSF}/W &= \frac{\mathbf{Y}'\mathbf{A}_2\mathbf{Y}/1}{c_1(\mathbf{Y}'\mathbf{A}_3\mathbf{Y}/2) + c_2(\mathbf{Y}'\mathbf{A}_4\mathbf{Y}/1) + c_3(\mathbf{Y}'\mathbf{A}_5\mathbf{Y}/3)} \\ &= \frac{6728}{1419.51} = 4.74. \end{aligned}$$

The degree of freedom for the denominator of the Satterthwaite statistic is

$$\begin{aligned} f &= \frac{W^2}{\left\{\frac{(c_1\mathbf{Y}'\mathbf{A}_3\mathbf{Y}/2)^2}{2} + \frac{(c_2\mathbf{Y}'\mathbf{A}_4\mathbf{Y}/1)^2}{1} + \frac{(c_3\mathbf{Y}'\mathbf{A}_5\mathbf{Y}/3)^2}{3}\right\}} \\ &= \frac{(1419.51)^2}{\left\{\frac{(2770.8/4)^2}{2} + \frac{[(1.3/1.2)(740.03)]^2}{1} + \frac{[(-0.7/1.2)(128.4)]^2}{3}\right\}} = 2.28. \end{aligned}$$

Since $\text{MSF}/W = 4.74 < 18.5 = F_{1,2}^{0.05}$, do not reject the hypothesis that there is no treatment effect.

EXERCISES

For Exercises 1–6, assume that $\mathbf{E}$ has a multivariate normal distribution with mean vector $\mathbf{0}$ and covariance matrix $\Sigma = \sum_{f=1}^{m} \sigma_f^2 \mathbf{U}_f \mathbf{U}_f'$.

1. Write the general mixed model $\mathbf{Y} = \mathbf{RC}\boldsymbol{\mu} + \mathbf{E}$ for the two-way cross classification described in Example 4.5.2. Define all terms and distributions explicitly.
2. Write the general mixed model $\mathbf{Y} = \mathbf{RC}\boldsymbol{\mu} + \mathbf{E}$ for the split plot design described in Example 4.5.3. Define all terms and distributions explicitly.
3. Write the general mixed model $\mathbf{Y} = \mathbf{RC}\boldsymbol{\mu} + \mathbf{E}$ for the experiment described in Example 4.5.4. Define all terms and distributions explicitly.
4. Write the general mixed model $\mathbf{Y} = \mathbf{RC}\boldsymbol{\mu} + \mathbf{E}$ for the experiment described in Figure 7.3.1. Define all terms and distributions explicitly.
5. Write the general mixed model $\mathbf{Y} = \mathbf{RC}\boldsymbol{\mu} + \mathbf{E}$ for the experiment described in Exercise 12 in Chapter 5. Define all terms and distributions explicitly.
6. Write the general mixed model $\mathbf{Y} = \mathbf{RC}\boldsymbol{\mu} + \mathbf{E}$ for the experiment described in Exercise 7 in Chapter 4. Define all terms and distributions explicitly.
7. Let the 8×1 vector of observations $\mathbf{Y} = (Y_{111}, Y_{112}, Y_{121}, Y_{122}, Y_{211}, Y_{212}, Y_{221}, Y_{222})' = (2, 5, 10, 15, 17, 14, 39, 41)'$ represent the data for the experiment described in Example 10.2.1 with $b = t = r = 2$.

		Replicates (i)					
		1		2		3	
		W. Plot (j)		W. Plot (j)		W. Plot (j)	
		1	2	1	2	1	2
Split plot (k)	1	23.7	37.2	21.4		21.9	32.6
	2	28.9		27.9	40.5	21.2	
	3	17.8	60.6		45.1		41.6

Figure 10.6.1 Split Plot Data Set for Exercise 9.

(a) Find the Type I sum of squares estimates of σ_B^2, σ_{BT}^2, and $\sigma_{R(BT)}^2$ assuming a finite model.

(b) Find the REML estimates of σ_B^2, σ_{BT}^2, and $\sigma_{R(BT)}^2$ assuming a finite model. Are the Type I sum of squares and REML estimates of the three variance parameters equal?

8. The observations in Table E10.1 represent the data from the split plot experiment described in Example 4.5.3 with $r = s = 3, t = 2$, and with some of the observations missing.

   (a) Write the general mixed model $\mathbf{Y} = \mathbf{RC}\boldsymbol{\mu} + \mathbf{E}$ for this experiment where $\mathbf{E} \sim \mathrm{N}_{13}(\mathbf{0}, \Sigma)$ and $\Sigma = \sum_{f=1}^{3} \sigma_f^2 \mathbf{U}_f \mathbf{U}_f'$. Define all terms and distributions explicitly.

   (b) Find the Type I sum of squares and REML estimates of the variance parameters defined in part a assuming a finite model.

   (c) Find the Type I sum of squares and REML estimates of the variance parameters defined in part a assuming an infinite model.

   (d) Calculate the estimates $\hat{\hat{\boldsymbol{\mu}}}_{\mathrm{w}}$ and $\widehat{\mathrm{cov}}(\hat{\boldsymbol{\mu}}_{\mathrm{w}})$.

   (e) Construct a 95% confidence band on the difference between the two whole plot treatment means.

   (f) Construct a Satterthwaite test for the hypothesis that there is no significant whole plot treatment effect.

   (g) Construct a Satterthwaite test for the hypothesis that there is no significant split plot treatment effect.

APPENDIX 1: COMPUTER OUTPUT FOR CHAPTER 5

Table A1.1 provides a summary of the notation used in the Chapter 5 text and the coresponding names used in the SAS PROC IML computer program.

Table A1.1

Chapter 5 Notation and Corresponding PROC IML Program Names

Chapter 5 Notation	PROC IML Program Name
$\mathbf{Y}, \mathbf{X}_1, \mathbf{X}_c, \mathbf{X}$	Y, X1, XC, X
$\hat{\beta}, \hat{\sigma}^2$	BHAT, SIG2HAT
$\mathbf{Y}'\frac{1}{n}\mathbf{J}_n\mathbf{Y}$	SSMEAN
$\mathbf{Y}'\mathbf{X}_c(\mathbf{X}_c'\mathbf{X}_c)^{-1}\mathbf{X}_c'\mathbf{Y}$	SSREG
$\mathbf{Y}'[\mathbf{I}_n - \frac{1}{n}\mathbf{J}_n - \mathbf{X}_c(\mathbf{X}_c'\mathbf{X}_c)^{-1}\mathbf{X}_c']\mathbf{Y}$	SSRES
$\mathbf{Y}'\mathbf{Y}$	SSTOT
$\hat{\beta}_w, \hat{\sigma}_w^2$	WBETAHAT, WSIG2HAT
$\mathbf{Y}'\mathbf{V}^{-1}[\mathbf{1}_n(\mathbf{1}_n'\mathbf{V}^{-1}\mathbf{1}_n)^{-1}\mathbf{1}_n']\mathbf{V}^{-1}\mathbf{Y}$	SSWMEAN
$\mathbf{Y}'\mathbf{V}^{-1}[\mathbf{X}(\mathbf{X}'\mathbf{V}^{-1}\mathbf{X})^{-1}\mathbf{X}' - \mathbf{1}_n(\mathbf{1}_n'\mathbf{V}^{-1}\mathbf{1}_n)^{-1}\mathbf{1}_n']\mathbf{V}^{-1}\mathbf{Y}$	SSWREG
$\mathbf{Y}'[\mathbf{V}^{-1} - \mathbf{V}^{-1}\mathbf{X}(\mathbf{X}'\mathbf{V}^{-1}\mathbf{X})^{-1}\mathbf{X}'\mathbf{V}^{-1}]\mathbf{Y}$	SSWRES
$\mathbf{Y}'\mathbf{V}^{-1}\mathbf{Y}$	SSWTOT
$\mathbf{Y}'\mathbf{A}_{pe}\mathbf{Y}$	SSPE
$\mathbf{Y}'[\mathbf{I}_n - \mathbf{X}(\mathbf{X}'\mathbf{X})^{-1}\mathbf{X}' - \mathbf{A}_{pe}]\mathbf{Y}$	SSLOF
SS(x_1| overall mean)	SSX1
SS(x_2| overall mean, x_1)	SSX2

The SAS PROC IML program and output follow:

```
PROC IML;
Y={1.7, 2.0, 1.9, 1.6, 3.2, 2.0, 2.5, 5.4, 5.7, 5.1};
X1 = J(10, 1, 1);
 XC = { −15 −102,
         −15 −102,
         −15 −102,
         −15 −102,
         −15    18,
          15   102,
          15  −102,
          15   198,
          15   198,
          15   198 };
X = X1| |XC;
BHAT = INV(T(X)*X)*T(X)*Y;
```

```
SIG2HAT = (1/7)#T(Y)*(I(10) - X*INV(T(X)*X)*T(X))*Y;
A1 = X1*INV(T(X1)*X1)*T(X1);
A2 = XC*INV(T(XC)*XC)*T(XC);
A3 = I(10) – A1 –A2;
SSMEAN = T(Y)*A1*Y;
SSREG = T(Y)*A2*Y;
SSRES = T(Y)*A3*Y;
SSTOT = T(Y)*Y;
V = BLOCK(1, 2) @ I(5);
WBETAHAT = INV(T(X)*INV(V)*X)*T(X)*INV(V)*Y;
WSIG2HAT=(1/7)#T(Y)*(INV(V)–INV(V)*X*INV(T(X)*INV(V)*X)
     *T(X)*INV(V))*Y;
AV1 = INV(V)*X1*INV(T(X1)*INV(V)*X1)*T(X1)*INV(V);
AV2 = INV(V)*X*INV(T(X)*INV(V)*X)*T(X)*INV(V) – AV1;
AV3 = INV(V) – AV1 –AV2;
SSWMEAN = T(Y)*AV1*Y;
SSWREG = T(Y)*AV2*Y;
SSWRES = T(Y)*AV3*Y;
SSWTOT = T(Y)*INV(V)*Y;
B1 = I(4) – J(4,4,1/4);
B2 = 0;
B3 = I(2) – J(2,2,1/2);
B4 = I(3)– J(3,3,1/3);
APE = BLOCK(B1,B2,B3,B4);
ALOF = A3–APE;
SSLOF = T(Y)*ALOF*Y;
SSPE = T(Y)*APE*Y;
R1 = X(|1:10,1|);
R2 = X(|1:10,1:2|);
T1 = R1*INV(T(R1)*R1)*T(R1);
T2 = R2*INV(T(R2)*R2)*T(R2)–T1;
T3 = X*INV(T(X)*X)*T(X)–R2*INV(T(R2)*R2)*T(R2);
SSX1 = T(Y)*T2*Y;
SSX2 = T(Y)*T3*Y;
PRINT  BHAT SIG2HAT SSMEAN SSREG SSRES SSTOT
       WBETAHAT WSIG2HAT SSWMEAN SSWREG SSWRES
       SSWTOT SSLOF SSPE SSX1 SSX2;
```

```
      BHAT   SIG2HAT    SSMEAN     SSREG      SSRES     SSTOT
      3.11 0.0598812    96.721 24.069831  0.4191687    121.21
 0.0134819
 0.0106124

  WBETAHAT  WSIG2HAT   SSWMEAN    SSWREG     SSWRES   SSWTOT
      3.11 0.0379176 57.408333 14.581244  0.2654231    72.255
     0.013
 0.0107051

               SSLOF      SSPE      SSX1       SSX2
           0.0141687     0.405    10.609  13.460831

QUIT;
RUN;
```

APPENDIX 2: COMPUTER OUTPUT FOR CHAPTER 7

A2.1 COMPUTER OUTPUT FOR SECTION 7.2

Table A2.1 provides a summary of the notation used in Section 7.2 of the text and the corresponding names used in the SAS PROC IML computer program.

Table A2.1

Section 7.2 Notation and Corresponding PROC IML Program Names

Section 7.2 Notation	PROC IML Program Name
$\mathbf{X}^*$	XSTAR
$\mathbf{I}_3 \otimes \mathbf{J}_3$	SIG1
$\mathbf{I}_3 \otimes (\mathbf{I}_3 - \frac{1}{3}\mathbf{J}_3)$	SIG2
$\mathbf{X}_1, \mathbf{Q}_3, \mathbf{Z}_1, \mathbf{X}_2, \mathbf{Z}_2$	X1, Q3, Z1, X2, Z2
$\mathbf{M}, \mathbf{X} = \mathbf{M}\mathbf{X}^*$	M, X
$\mathbf{M}[\mathbf{I}_3 \otimes \mathbf{J}_3]\mathbf{M}'$	SIGM1
$\mathbf{M}[\mathbf{I}_3 \otimes (\mathbf{I}_3 - \frac{1}{3}\mathbf{J}_3)]\mathbf{M}'$	SIGM2
$\mathbf{S}_1, \mathbf{T}_1$	S1, T1
$\mathbf{A}_1, \ldots, \mathbf{A}_4$	A1, . . . , A4
$\mathbf{A}_1\mathbf{M}[\mathbf{I}_3 \otimes \mathbf{J}_3]\mathbf{M}', \ldots, \mathbf{A}_4\mathbf{M}[\mathbf{I}_3 \otimes \mathbf{J}_3]\mathbf{M}'$	A1SIGM1, . . . , A4SIGM1
$\mathbf{A}_1\mathbf{M}[\mathbf{I}_3 \otimes (\mathbf{I}_3 - \frac{1}{3}\mathbf{J}_3)]\mathbf{M}', \ldots, \mathbf{A}_4\mathbf{M}[\mathbf{I}_3 \otimes (\mathbf{I}_3 - \frac{1}{3}\mathbf{J}_3)]\mathbf{M}'$	A1SIGM2, . . . , A4SIGM2
$\mathbf{X}'\mathbf{A}_1\mathbf{X}, \ldots, \mathbf{X}'\mathbf{A}_4\mathbf{X}$	XA1X, . . . , XA4X
$\mathbf{S}_2^*, \mathbf{T}_1^*$	S2STAR, T1STAR
$\mathbf{A}_2^*, \mathbf{A}_3^*$	A2STAR, A3STAR
$\mathbf{X}'\mathbf{A}_3^*\mathbf{X}$	XA3STARX
$\mathrm{tr}[\mathbf{A}_3^*\mathbf{M}(\mathbf{I}_3 \otimes \mathbf{J}_3)\mathbf{M}']$	TRA3ST1
$\mathrm{tr}\{\mathbf{A}_3^*\mathbf{M}[\mathbf{I}_3 \otimes (\mathbf{I}_3 - \frac{1}{3}\mathbf{J}_3)]\mathbf{M}'\}$	TRA3ST2

The SAS PROC IML program for Section 7.2 and the output follow:

```
PROC IML;
XSTAR=J(3, 1, 1)@I(3)
SIG1=I(3)@J(3, 3, 1);
SIG2=I(3)@(I(3)–(1/3)#J(3, 3, 1));
X1=J(9, 1, 1);
Q3 = {  1   1 ,
       −1   1 ,
        0  −2 };
Z1=Q3@J(3, 1, 1);
X2=J(3, 1, 1)@Q3;
Z2=Q3@Q3;
```

```
M={0  1  0  0  0  0  0  0  0,
   0  0  1  0  0  0  0  0  0,
   0  0  0  1  0  0  0  0  0,
   0  0  0  0  0  1  0  0  0,
   0  0  0  0  0  0  1  0  0,
   0  0  0  0  0  0  0  1  0};
X=M*XSTAR;
SIGM1=M*SIG1*T(M);
SIGM2=M*SIG2*T(M);
S1=M*X1;
T1=M*(X1||Z1);
S2=M*(X1||Z1||X2);
A1=S1*INV(T(S1)*S1)*T(S1);
A2=T1*INV(T(T1)*T1)*T(T1)-A1;
A3=S2*INV(T(S2)*S2)*T(S2)-A1-A2;
A4=I(3)@I(2)-A1-A2-A3;
A1SIGM1=A1*SIGM1;
A1SIGM2=A1*SIGM2;
A2SIGM1=A2*SIGM1;
A2SIGM2=A2*SIGM2;
A3SIGM1=A3*SIGM1;
A3SIGM2=A3*SIGM2;
A4SIGM1=A4*SIGM1;
A4SIGM2=A4*SIGM2;
XA1X=T(X)*A1*X;
XA2X=T(X)*A2*X;
XA3X=T(X)*A3*X;
XA4X=T(X)*A4*X;
S2STAR=M*(X1||X2);
T1STAR=M*(X1||X2||Z1);
A2STAR=S2STAR*INV(T(S2STAR)*S2STAR)*T(S2STAR)-A1;
A3STAR=T1STAR*INV(T(T1STAR)*T1STAR)*T(T1STAR)-A2STAR-A1;
XA3STARX=T(X)*A3STAR*X;
TRA3ST1=TRACE(A3STAR*SIGM1);
TRA3ST2=TRACE(A3STAR*SIGM2);
PRINT  SIG1 SIG2 SIGM1 SIGM2 A1 A2 A3 A4 A1SIGM1 A1SIGM2
       A2SIGM1 A2SIGM2 A3SIGM1 A3SIGM2 A4SIGM1 A4SIGM2
       XA1X XA2X XA3X XA4X A3STAR XA3STARX TRA3ST1
       TRA3ST2
```

SIG1

1	1	1	0	0	0	0	0	0
1	1	1	0	0	0	0	0	0
1	1	1	0	0	0	0	0	0
0	0	0	1	1	1	0	0	0
0	0	0	1	1	1	0	0	0
0	0	0	1	1	1	0	0	0
0	0	0	0	0	0	1	1	1
0	0	0	0	0	0	1	1	1
0	0	0	0	0	0	1	1	1

SIG2

0.6667	−0.333	−0.333	0	0	0	0	0	0
−0.333	0.6667	−0.333	0	0	0	0	0	0
−0.333	−0.333	0.6667	0	0	0	0	0	0
0	0	0	0.6667	−0.333	−0.333	0	0	0
0	0	0	−0.333	0.6667	−0.333	0	0	0
0	0	0	−0.333	−0.333	0.6667	0	0	0
0	0	0	0	0	0	0.6667	−0.333	−0.333
0	0	0	0	0	0	−0.333	0.6667	−0.333
0	0	0	0	0	0	−0.333	−0.333	0.6667

SIGM1

1	1	0	0	0	0
1	1	0	0	0	0
0	0	1	1	0	0
0	0	1	1	0	0
0	0	0	0	1	1
0	0	0	0	1	1

SIGM2

0.6667	−0.333	0	0	0	0
−0.333	0.6667	0	0	0	0
0	0	0.6667	−0.333	0	0
0	0	−0.333	0.6667	0	0
0	0	0	0	0.6667	−0.333
0	0	0	0	−0.333	0.6667

A1

0.166667	0.166667	0.166667	0.166667	0.166667	0.166667
0.166667	0.166667	0.166667	0.166667	0.166667	0.166667
0.166667	0.166667	0.166667	0.166667	0.166667	0.166667
0.166667	0.166667	0.166667	0.166667	0.166667	0.166667
0.166667	0.166667	0.166667	0.166667	0.166667	0.166667
0.166667	0.166667	0.166667	0.166667	0.166667	0.166667

A2

0.333333	0.333333	−0.16667	−0.16667	−0.16667	−0.16667
0.333333	0.333333	−0.16667	−0.16667	−0.16667	−0.16667
−0.16667	−0.16667	0.333333	0.333333	−0.16667	−0.16667
−0.16667	−0.16667	0.333333	0.333333	−0.16667	−0.16667
−0.16667	−0.16667	−0.16667	−0.16667	0.333333	0.333333
−0.16667	−0.16667	−0.16667	−0.16667	0.333333	0.333333

A3

0.333333	−0.33333	0.166667	−0.16667	−0.16667	0.166667
−0.33333	0.333333	−0.16667	0.166667	0.166667	−0.16667
0.166667	−0.16667	0.333333	−0.33333	0.166667	−0.16667
−0.16667	0.166667	−0.33333	0.333333	−0.16667	0.166667
−0.16667	0.166667	0.166667	−0.16667	0.333333	−0.33333
0.166667	−0.16667	−0.16667	0.166667	−0.33333	0.333333

A4

0.166667	−0.16667	−0.16667	0.166667	0.166667	−0.16667
−0.16667	0.166667	0.166667	−0.16667	−0.16667	0.166667
−0.16667	0.166667	0.166667	−0.16667	−0.16667	0.166667
0.166667	−0.16667	−0.16667	0.166667	0.166667	−0.16667
0.166667	−0.16667	−0.16667	0.166667	0.166667	−0.16667
−0.16667	0.166667	0.166667	−0.16667	−0.16667	0.166667

A1SIGM1

0.333333	0.333333	0.333333	0.333333	0.333333	0.333333
0.333333	0.333333	0.333333	0.333333	0.333333	0.333333
0.333333	0.333333	0.333333	0.333333	0.333333	0.333333
0.333333	0.333333	0.333333	0.333333	0.333333	0.333333
0.333333	0.333333	0.333333	0.333333	0.333333	0.333333
0.333333	0.333333	0.333333	0.333333	0.333333	0.333333

A1SIGM2

0.055556	0.055556	0.055556	0.055556	0.055556	0.055556
0.055556	0.055556	0.055556	0.055556	0.055556	0.055556
0.055556	0.055556	0.055556	0.055556	0.055556	0.055556
0.055556	0.055556	0.055556	0.055556	0.055556	0.055556
0.055556	0.055556	0.055556	0.055556	0.055556	0.055556
0.055556	0.055556	0.055556	0.055556	0.055556	0.055556

A2SIGM1

0.666667	0.666667	−0.33333	−0.33333	−0.33333	−0.33333
0.666667	0.666667	−0.33333	−0.33333	−0.33333	−0.33333
−0.33333	−0.33333	0.666667	0.666667	−0.33333	−0.33333
−0.33333	−0.33333	0.666667	0.666667	−0.33333	−0.33333
−0.33333	−0.33333	−0.33333	−0.33333	0.666667	0.666667
−0.33333	−0.33333	−0.33333	−0.33333	0.666667	0.666667

A2SIGM2

0.111111	0.111111	−0.05556	−0.05556	−0.05556	−0.05556
0.111111	0.111111	−0.05556	−0.05556	−0.05556	−0.05556
−0.05556	−0.05556	0.111111	0.111111	−0.05556	−0.05556
−0.05556	−0.05556	0.111111	0.111111	−0.05556	−0.05556
−0.05556	−0.05556	−0.05556	−0.05556	0.111111	0.111111
−0.05556	−0.05556	−0.05556	−0.05556	0.111111	0.111111

A3SIGM1

1.1E − 16	1.1E − 16	5.51E − 17	5.51E − 17	0	0
1.65E − 16	1.65E − 16	5.51E − 17	5.51E − 17	0	0
5.51E − 17	5.51E − 17	1.1E − 16	1.1E − 16	0	0
5.51E − 17	5.51E − 17	1.65E − 16	1.65E − 16	0	0
0	0	0	0	−1.1E − 16	−1.1E − 16
0	0	0	0	−1.1E − 16	−1.1E − 16

A3SIGM2

0.333333	−0.33333	0.166667	−0.16667	−0.16667	0.166667
−0.33333	0.333333	−0.16667	0.166667	0.166667	−0.16667
0.166667	−0.16667	0.333333	−0.33333	0.166667	−0.16667
−0.16667	0.166667	−0.33333	0.333333	−0.16667	0.166667
−0.16667	0.166667	0.166667	−0.16667	0.333333	−0.33333
0.166667	−0.16667	−0.16667	0.166667	−0.33333	0.333333

A4SIGM1

1.1E − 16	1.1E − 16	0	0	0	0
5.51E − 17	5.51E − 17	0	0	0	0
0	0	1.1E − 16	1.1E − 16	0	0
0	0	5.51E − 17	5.51E − 17	0	0
0	0	0	0	1.1E − 16	1.1E − 16
0	0	0	0	1.1E − 16	1.1E − 16

A4SIGM2

0.166667	−0.16667	−0.16667	0.166667	0.166667	−0.16667
−0.16667	0.166667	0.166667	−0.16667	−0.16667	0.166667
−0.16667	0.166667	0.166667	−0.16667	−0.16667	0.166667
0.166667	−0.16667	−0.16667	0.166667	0.166667	−0.16667
0.166667	−0.16667	−0.16667	0.166667	0.166667	−0.16667
−0.16667	0.166667	0.166667	−0.16667	−0.16667	0.166667

XA1X

0.666667	0.666667	0.666667
0.666667	0.666667	0.666667
0.666667	0.666667	0.666667

XA2X

0.333333	−0.16667	−0.16667
−0.16667	0.333333	−0.16667
−0.16667	−0.16667	0.333333

XA3X

1	−0.5	−0.5
−0.5	1	−0.5
−0.5	−0.5	1

XA4X

1.65E − 16	2.76E − 17	2.76E − 17
2.76E − 17	1.65E − 16	2.76E − 17
−2.8E − 17	−2.7E − 17	1.65E − 16

A3STAR

0.333333	0.166667	0.166667	−0.16667	−0.16667	−0.33333
0.166667	0.333333	−0.16667	−0.33333	0.166667	−0.16667
0.166667	−0.16667	0.333333	0.166667	−0.33333	−0.16667
−0.16667	−0.33333	0.166667	0.333333	−0.16667	0.166667
−0.16667	0.166667	−0.33333	−0.16667	0.333333	0.166667
−0.33333	−0.16667	−0.16667	0.166667	0.166667	0.333333

XA3STARX

2.76E − 16	1.65E − 16	0
1.65E − 16	2.76E − 16	0
0	0	−2.2E − 16

TRA3ST1	TRA3ST2
3	1

```
QUIT;
RUN;
```

In the following discussion, the computer output is translated into the results that appear in Section 7.2.

$$\mathbf{I}_3 \otimes \mathbf{J}_3 = \text{SIG1}$$

$$\mathbf{I}_3 \otimes \left(\mathbf{I}_3 - \frac{1}{3}\mathbf{J}_3\right) = \text{SIG2}$$

so

$$\sigma_B^2[\mathbf{I}_3 \otimes \mathbf{J}_3] + \sigma_{BT}^2\left[\mathbf{I}_3 \otimes \left(\mathbf{I}_3 - \frac{1}{3}\mathbf{J}_3\right)\right] = \sigma_B^2 \text{ SIG1} + \sigma_{BT}^2 \text{ SIG2}$$

$$\mathbf{M}[\mathbf{I}_3 \otimes \mathbf{J}_3]\mathbf{M}' = \mathbf{I}_3 \otimes \mathbf{J}_2 = \text{SIGM1}$$

$$\mathbf{M}\left[\mathbf{I}_3 \otimes \left(\mathbf{I}_3 - \frac{1}{3}\mathbf{J}_3\right)\right]\mathbf{M}' = \left[\mathbf{I}_3 \otimes \left(\mathbf{I}_2 - \frac{1}{3}\mathbf{J}_2\right)\right] = \text{SIGM2}$$

so

$$\mathbf{M}\left\{\sigma_B^2[\mathbf{I}_3 \otimes \mathbf{J}_3] + \sigma_{BT}^2\left[\mathbf{I}_3 \otimes \left(\mathbf{I}_3 - \frac{1}{3}\mathbf{J}_3\right)\right]\right\}\mathbf{M}' = \sigma_B^2 \text{ SIGM1} + \sigma_{BT}^2 \text{ SIGM2}$$

$$\frac{1}{3}\mathbf{J}_3 \otimes \frac{1}{2}\mathbf{J}_2 = \text{A1}$$

$$\left(\mathbf{I}_3 - \frac{1}{3}\mathbf{J}_3\right) \otimes \frac{1}{2}\mathbf{J}_2 = \text{A2}$$

$$\frac{1}{3}\begin{bmatrix} 2 & 1 & -1 \\ 1 & 2 & 1 \\ -1 & 1 & 2 \end{bmatrix} \otimes \frac{1}{2}\mathbf{J}_2 = \text{A3}$$

$$\frac{1}{3}\begin{bmatrix} 1 & -1 & 1 \\ 1 & 1 & -1 \\ 1 & -1 & 1 \end{bmatrix} \otimes \frac{1}{2}\mathbf{J}_2 = \text{A4}$$

$$\mathbf{A}_1\mathbf{M}[\mathbf{I}_3 \otimes \mathbf{J}_3]\mathbf{M}' = \text{A1SIGM1} = 2\text{A1}$$

$$\mathbf{A}_1\mathbf{M}\left[\mathbf{I}_3 \otimes \left(\mathbf{I}_3 - \frac{1}{3}\mathbf{J}_3\right)\right]\mathbf{M}' = \text{A1SIGM2} = \frac{1}{3}\text{A1}$$

so

$$\begin{aligned} \mathbf{A}_1\mathbf{M}\left\{\sigma_B^2[\mathbf{I}_3 \otimes \mathbf{J}_3] + \sigma_{BT}^2\left[\mathbf{I}_3 \otimes \left(\mathbf{I}_3 - \frac{1}{3}\mathbf{J}_3\right)\right]\right\}\mathbf{M}' &= \sigma_B^2 \text{ A1SIGM1} \\ &\quad + \sigma_{BT}^2 \text{ A1SIGM2} \\ &= \left(2\sigma_B^2 + \frac{1}{3}\sigma_{BT}^2\right)\text{A1} \end{aligned}$$

$$\mathbf{A}_2\mathbf{M}[\mathbf{I}_3 \otimes \mathbf{J}_3]\mathbf{M}' = \text{A2SIGM1} = 2\text{A2}$$

$$\mathbf{A}_2\mathbf{M}\left[\mathbf{I}_3 \otimes \left(\mathbf{I}_3 - \frac{1}{3}\mathbf{J}_3\right)\right]\mathbf{M}' = \text{A2SIGM2} = \frac{1}{3}\text{A2}$$

so

$$\begin{aligned}\mathbf{A}_2\mathbf{M}\left\{\sigma_B^2[\mathbf{I}_3 \otimes \mathbf{J}_3] + \sigma_{BT}^2\left[\mathbf{I}_3 \otimes \left(\mathbf{I}_3 - \frac{1}{3}\mathbf{J}_3\right)\right]\right\}\mathbf{M}' &= \sigma_B^2\text{A2SIGM1} \\ &\quad + \sigma_{BT}^2\text{A2SIGM2} \\ &= \left(2\sigma_B^2 + \frac{1}{3}\sigma_{BT}^2\right)\text{A2}\end{aligned}$$

$$\mathbf{A}_3\mathbf{M}[\mathbf{I}_3 \otimes \mathbf{J}_3]\mathbf{M}' = \text{A3SIGM1} = \mathbf{0}_{6\times 6}$$

$$\mathbf{A}_3\mathbf{M}\left[\mathbf{I}_3 \otimes \left(\mathbf{I}_3 - \frac{1}{3}\mathbf{J}_3\right)\right]\mathbf{M}' = \text{A3SIGM2} = \text{A3}$$

so

$$\begin{aligned}\mathbf{A}_3\mathbf{M}\left\{\sigma_B^2[\mathbf{I}_3 \otimes \mathbf{J}_3] + \sigma_{BT}^2\left[\mathbf{I}_3 \otimes \left(\mathbf{I}_3 - \frac{1}{3}\mathbf{J}_3\right)\right]\right\}\mathbf{M}' &= \sigma_B^2\text{A3SIGM1} \\ &\quad + \sigma_{BT}^2\text{A3SIGM2} \\ &= \sigma_{BT}^2\text{A3}\end{aligned}$$

$$\mathbf{A}_4\mathbf{M}[\mathbf{I}_3 \otimes \mathbf{J}_3]\mathbf{M}' = \text{A4SIGM1} = \mathbf{0}_{6\times 6}$$

$$\mathbf{A}_4\mathbf{M}\left[\mathbf{I}_3 \otimes \left(\mathbf{I}_3 - \frac{1}{3}\mathbf{J}_3\right)\right]\mathbf{M}' = \text{A4SIGM2} = \text{A4}$$

so

$$\begin{aligned}\mathbf{A}_4\mathbf{M}\left\{\sigma_B^2[\mathbf{I}_3 \otimes \mathbf{J}_3] + \sigma_{BT}^2\left[\mathbf{I}_3 \otimes \left(\mathbf{I}_3 - \frac{1}{3}\mathbf{J}_3\right)\right]\right\}\mathbf{M}' &= \sigma_B^2\text{A4SIGM1} \\ &\quad + \sigma_{BT}^2\text{A4SIGM2} \\ &= \sigma_{BT}^2\text{A4}\end{aligned}$$

$$\mathbf{X}'\mathbf{A}_1\mathbf{X} = \text{XA1X} = \frac{2}{3}\mathbf{J}_3$$

so

$$\beta'\mathbf{X}'\mathbf{A}_1\mathbf{X}\beta = \beta'\frac{2}{3}\mathbf{J}_3\beta = \frac{2}{3}(\beta_1 + \beta_2 + \beta_3)^2$$

$$\mathbf{X}'\mathbf{A}_2\mathbf{X} = \mathrm{XA2X} = \frac{1}{3}\left(\mathbf{I}_3 - \frac{1}{2}\mathbf{J}_3\right)$$

so

$$\beta'\mathbf{X}'\mathbf{A}_2\mathbf{X}\beta = \beta'\frac{1}{3}\left(\mathbf{I}_3 - \frac{1}{2}\mathbf{J}_3\right)\beta = \frac{1}{3}(\beta_1^2 + \beta_2^2 + \beta_3^2 - \beta_1\beta_2 - \beta_1\beta_3 - \beta_2\beta_3)^2$$

$$\mathbf{X}'\mathbf{A}_3\mathbf{X} = \mathrm{XA3X} = \left(\mathbf{I}_3 - \frac{1}{2}\mathbf{J}_3\right)$$

so

$$\beta'\mathbf{X}'\mathbf{A}_3\mathbf{X}\beta = \beta'\left(\mathbf{I}_3 - \frac{1}{2}\mathbf{J}_3\right)\beta = (\beta_1^2 + \beta_2^2 + \beta_3^2 - \beta_1\beta_2 - \beta_1\beta_3 - \beta_2\beta_3)^2$$

$$\mathbf{X}'\mathbf{A}_4\mathbf{X} = \mathrm{XA4X} = \mathbf{0}_{3\times 3}$$

so

$$\beta'\mathbf{X}'\mathbf{A}_4\mathbf{X}\beta = 0$$

$$\mathbf{X}'\mathbf{A}_3^*\mathbf{X} = \mathrm{XA3STARX} = \mathbf{0}_{3\times 3}$$

so

$$\beta'\mathbf{X}'\mathbf{A}_3^*\mathbf{X}\beta = 0$$

$$\begin{aligned}
&\mathrm{tr}[\mathbf{A}_3^*\sigma_B^2\mathbf{M}(\mathbf{I}_3 \otimes \mathbf{J}_3)\mathbf{M}' + \mathrm{tr}\left\{\mathbf{A}_3^*\sigma_{BT}^2\mathbf{M}\left[\mathbf{I}_3 \otimes \left(\mathbf{I}_3 - \frac{1}{3}\mathbf{J}_3\right)\right]\mathbf{M}'\right\} \\
&= \sigma_B^2\mathrm{TRA3ST1} + \sigma_{BT}^2\mathrm{TRA3ST2} \\
&= 3\sigma_B^2 + \sigma_{BT}^2
\end{aligned}$$

A2.2 COMPUTER OUTPUT FOR SECTION 7.3

Table A2.2 provides a summary of the notation used in section 7.3 of the text and the corresponding names used in the SAS PROC IML computer program. The program names XSTAR, SIG1, SIG2, X1, Q3, Z1, X2, Z2, and M are defined as in Section A.1 and therefore are not redefined in Table A2.2.

Table A2.2
Section 7.3 Notation and Corresponding PROC IML Program Names

Section 7.3 Notation	PROC IML Program Name
$\mathbf{R}, \mathbf{D}$	R, D
$\mathbf{RM}[\mathbf{I}_3 \otimes \mathbf{J}_3]\mathbf{M}'\mathbf{R}'$	SIGRM1
$\mathbf{RM}[\mathbf{I}_3 \otimes (\mathbf{I}_3 - \frac{1}{3}\mathbf{J}_3)]\mathbf{M}'\mathbf{R}'$	SIGRM2
$\mathbf{RMI}_9\mathbf{M}'\mathbf{R}'$	SIGRM3
$\mathbf{X} = \mathbf{RMX}^*$	X
$\mathbf{S}_1, \mathbf{T}_1, \mathbf{S}_2$	S1, T1, S2
$\mathbf{A}_1, \ldots, \mathbf{A}_5$	A1, ..., A5
$\mathbf{A}_1\mathbf{RM}[\mathbf{I}_3 \otimes \mathbf{J}_3]\mathbf{M}'\mathbf{R}'\mathbf{A}_1, \ldots, \mathbf{A}_5\mathbf{RM}[\mathbf{I}_3 \otimes \mathbf{J}_3]\mathbf{M}'\mathbf{R}'\mathbf{A}_5$	A1SIGRM1, ..., A5SIGRM1
$\mathbf{A}_1\mathbf{RM}[\mathbf{I}_3 \otimes (\mathbf{I}_3 - \frac{1}{3}\mathbf{J}_3)]\mathbf{M}'\mathbf{R}'\mathbf{A}_1, \ldots,$ $\mathbf{A}_5\mathbf{RM}[\mathbf{I}_3 \otimes (\mathbf{I}_3 - \frac{1}{3}\mathbf{J}_3)]\mathbf{M}'\mathbf{R}'\mathbf{A}_5$	A1SIGRM2, ..., A5SIGRM2
$\mathbf{A}_1\mathbf{I}_9\mathbf{A}_1, \ldots, \mathbf{A}_5\mathbf{I}_9\mathbf{A}_5$	A1SIGRM3, ..., A5SIGRM3
$\mathbf{X}'\mathbf{A}_1\mathbf{X}, \ldots, \mathbf{X}'\mathbf{A}_5\mathbf{X}$	XA1X, ..., XA5X
$\mathbf{S}_2^*, \mathbf{T}_1^*$	S2STAR, T1STAR
$\mathbf{A}_2^*, \mathbf{A}_3^*$	A2STAR, A3STAR
$\mathbf{A}_2^*\mathbf{RM}[\mathbf{I}_3 \otimes \mathbf{J}_3]\mathbf{M}'\mathbf{R}'\mathbf{A}_2, \mathbf{A}_3^*\mathbf{RM}[\mathbf{I}_3 \otimes \mathbf{J}_3]\mathbf{M}'\mathbf{R}'\mathbf{A}_3^*$	A2STARM1, A3STARM1
$\mathbf{A}_2^*\mathbf{RM}[\mathbf{I}_3 \otimes (\mathbf{I}_3 - \frac{1}{3}\mathbf{J}_3)]\mathbf{M}'\mathbf{R}'\mathbf{A}_2^*$	A2STARM2
$\mathbf{A}_3^*\mathbf{RM}[\mathbf{I}_3 \otimes (\mathbf{I}_3 - \frac{1}{3}\mathbf{J}_3)]\mathbf{M}'\mathbf{R}'\mathbf{A}_3^*$	A3STARM2
$\mathbf{A}_2^*\mathbf{I}_9\mathbf{A}_2^*, \mathbf{A}_3^*\mathbf{I}_9\mathbf{A}_3^*$	A2STARM3, A3STARM3
$\mathbf{X}'\mathbf{A}_2^*\mathbf{X}, \mathbf{X}'\mathbf{A}_3^*\mathbf{X}$	XA2STARX, XA3STARX
$\text{tr}\{\mathbf{A}_3^*\mathbf{RM}[\mathbf{I}_3 \otimes \mathbf{J}_3]\mathbf{M}'\mathbf{R}'\}$	TRA3ST1
$\text{tr}\{\mathbf{A}_3^*\mathbf{RM}[\mathbf{I}_3 \otimes (\mathbf{I}_3 - \frac{1}{3}\mathbf{J}_3)]\mathbf{M}'\mathbf{R}'\}$	TRA3ST2
$\text{tr}\{\mathbf{A}_3^*\mathbf{I}_9\}$	TRA3ST3

The SAS PROC IML program for Section 7.3 follows:

```
PROC IML;
XSTAR=J(3, 1, 1)@I(3);
SIG1=I(3)@J(3, 3, 1);
SIG2=I(3)@(I(3)−(1/3)#J(3, 3, 1));
X1=J(9, 1, 1);
Q3={ 1    1,
    −1    1,
     0   −2};
Z1=Q3@J(3, 1, 1);
X2=J(3, 1, 1)@Q3;
Z2=Q3@Q3;
M={0  1  0  0  0  0  0  0  0,
   0  0  1  0  0  0  0  0  0,
   0  0  0  1  0  0  0  0  0,
   0  0  0  0  0  1  0  0  0,
   0  0  0  0  0  0  1  0  0,
   0  0  0  0  0  0  0  1  0};
```

```
R=BLOCK(1,J(2, 1, 1), J(2, 1, 1), 1, 1, J(2, 1, 1));
D=T(R)*R;
X=R*M*XSTAR;
SIGRM1=R*M*SIG1*T(M)*T(R);
SIGRM2=R*M*SIG2*T(M)*T(R);
SIGRM3=I(9);
S1=R*M*X1;
T1=R*M*(XI||Z1);
S2=R*M*(X1||Z1||X2);
A1=S1*INV(T(S1)*S1)*T(S1);
A2=T1*INV(T(T1)*T1)*T(T1)−A1;
A3=S2*INV(T(S2)*S2)*T(S2)−A1 − A2;
A4=R*INV(D)*T(R) − A1 −A2 −A3;
A5=I(9) − R*INV(D)*T(R);
A1SIGRM1=A1*SIGRM1*A1;
A1SIGRM2=A1*SIGRM2*A1;
A1SIGRM3=A1*SIGRM3*A1;
A2SIGRM1=A2*SIGRM1*A2;
A2SIGRM2=A2*SIGRM2*A2;
A2SIGRM3=A2*SIGRM3*A2;
A3SIGRM1=A3*SIGRM1*A3;
A3SIGRM2=A3*SIGRM2*A3;
A3SIGRM3=A3*SIGRM3*A3;
A4SIGRM1=A4*SIGRM1*A4;
A4SIGRM2=A4*SIGRM2*A4;
A4SIGRM3=A4*SIGRM3*A4;
A5SIGRM1=A5*SIGRM1*A5;
A5SIGRM2=A5*SIGRM2*A5;
A5SIGRM3=A5*SIGRM3*A5;
XA1X = T(X)*A1*X;
XA2X = T(X)*A2*X;
XA3X = T(X)*A3*X;
XA4X = T(X)*A4*X;
XA5X = T(X)*A5*X;
S2STAR=R*M*(X1||X2);
T1STAR=R*M*(X1||X2||Z1);
A2STAR=S2STAR*INV(T(S2STAR)*S2STAR)*T(S2STAR)−A1;
A3STAR=T1STAR*INV(T(T1STAR)*T1STAR)*T(T1STAR)−A2STAR−A1;
A2STARM1=A2STAR*SIGRM1*A2STAR;
A2STARM2=A2STAR*SIGRM2*A2STAR;
A2STARM3=A2STAR*SIGRM3*A2STAR;
A3STARM1=A3STAR*SIGRM1*A3STAR;
```

```
A3STARM2=A3STAR*SIGRM2*A3STAR;
A3STARM3=A3STAR*SIGRM3*A3STAR;
XA2STARX = T(X)*A2STAR*X;
XA3STARX = T(X)*A3STAR*X;
TRA3ST1=TRACE(A3STAR*SIGRM1);
TRA3ST2=TRACE(A3STAR*SIGRM2);
TRA3ST3=TRACE(A3STAR*SIGRM3);
PRINT  A1 A2 A3 A4 A5 A1SIGRM1 A1SIGRM2 A1SIGRM3
       A2SIGRM1 A2SIGRM2 A2SIGRM3 A3SIGRM1 A3SIGRM2
       A3SIGRM3 A4SIGRM1 A4SIGRM2 A4SIGRM3 A5SIGRM1
       A5SIGRM2 A5SIGRM3 XA1X XA2X XA3X XA4X XA5X
       XA3STARX TRA3ST1 TRA3ST2 TRA3ST3;
QUIT;
RUN;
```

Since the program output is lengthy, it is omitted. In the following discussion, the computer output is translated into the results that appear in Section 7.3.

$$\begin{aligned}
\mathbf{A}_1\mathbf{RM}[\mathbf{I}_3 \otimes \mathbf{J}_3]\mathbf{M}'\mathbf{R}'\mathbf{A}_1 &= \text{A1SIGRM1} = 3\mathbf{A}_1 \\
\mathbf{A}_1\mathbf{RM}\left[\mathbf{I}_3 \otimes \left(\mathbf{I}_3 - \frac{1}{3}\mathbf{J}_3\right)\right]\mathbf{M}'\mathbf{R}'\mathbf{A}_1 &= \text{A1SIGRM2} = \frac{2}{3}\mathbf{A}_1 \\
\mathbf{A}_1\mathbf{I}_9\mathbf{A}_1 &= \text{A1SIGRM3} = \mathbf{A}_1
\end{aligned}$$

so

$$\mathbf{A}_1\boldsymbol{\Sigma}\mathbf{A}_1 = \left(3\sigma_B^2 + \frac{2}{3}\sigma_{BT}^2 + \sigma_{R(BT)}^2\right)\mathbf{A}_1$$

$$\begin{aligned}
\mathbf{A}_2\mathbf{RM}[\mathbf{I}_3 \otimes \mathbf{J}_3]\mathbf{M}'\mathbf{R}'\mathbf{A}_2 &= \text{A2SIGRM1} = 3\mathbf{A}_2 \\
\mathbf{A}_2\mathbf{RM}\left[\mathbf{I}_3 \otimes \left(\mathbf{I}_3 - \frac{1}{3}\mathbf{J}_3\right)\right]\mathbf{M}'\mathbf{R}'\mathbf{A}_2 &= \text{A2SIGRM2} = \frac{2}{3}\mathbf{A}_2 \\
\mathbf{A}_2\mathbf{I}_9\mathbf{A}_2 &= \text{A2SIGRM3} = \mathbf{A}_2
\end{aligned}$$

so

$$\mathbf{A}_2\boldsymbol{\Sigma}\mathbf{A}_2 = \left(3\sigma_B^2 + \frac{2}{3}\sigma_{BT}^2 + \sigma_{R(BT)}^2\right)\mathbf{A}_2$$

$$\begin{aligned}
\mathbf{A}_3\mathbf{RM}[\mathbf{I}_3 \otimes \mathbf{J}_3]\mathbf{M}'\mathbf{R}'\mathbf{A}_3 &= \text{A3SIGRM1} = \mathbf{0}_{9\times 9} \\
\mathbf{A}_3\mathbf{RM}\left[\mathbf{I}_3 \otimes \left(\mathbf{I}_3 - \frac{1}{3}\mathbf{J}_3\right)\right]\mathbf{M}'\mathbf{R}'\mathbf{A}_3 &= \text{A3SIGRM2} = \frac{4}{3}\mathbf{A}_3 \\
\mathbf{A}_3\mathbf{I}_9\mathbf{A}_3 &= \text{A3SIGRM3} = \mathbf{A}_3
\end{aligned}$$

so

$$\mathbf{A}_3 \Sigma \mathbf{A}_3 = \left(\frac{4}{3}\sigma^2_{BT} + \sigma^2_{R(BT)}\right)\mathbf{A}_3$$

$$\mathbf{A}_4\mathbf{RM}[\mathbf{I}_3 \otimes \mathbf{J}_3]\mathbf{M}'\mathbf{R}'\mathbf{A}_4 = \text{A4SIGRM1} = \mathbf{0}_{9\times 9}$$

$$\mathbf{A}_4\mathbf{RM}\left[\mathbf{I}_3 \otimes \left(\mathbf{I}_3 - \frac{1}{3}\mathbf{J}_3\right)\right]\mathbf{M}'\mathbf{R}'\mathbf{A}_4 = \text{A4SIGRM2} = \frac{4}{3}\mathbf{A}_4$$

$$\mathbf{A}_4\mathbf{I}_9\mathbf{A}_4 = \text{A4SIGRM3} = \mathbf{A}_4$$

so

$$\mathbf{A}_4 \Sigma \mathbf{A}_4 = \left(\frac{4}{3}\sigma^2_{BT} + \sigma^2_{R(BT)}\right)\mathbf{A}_4$$

$$\mathbf{A}_5\mathbf{RM}[\mathbf{I}_3 \otimes \mathbf{J}_3]\mathbf{M}'\mathbf{R}'\mathbf{A}_5 = \text{A5SIGRM1} = \mathbf{0}_{9\times 9}$$

$$\mathbf{A}_5\mathbf{RM}\left[\mathbf{I}_3 \otimes \left(\mathbf{I}_3 - \frac{1}{3}\mathbf{J}_3\right)\right]\mathbf{M}'\mathbf{R}'\mathbf{A}_5 = \text{A5SIGRM2} = \mathbf{0}_{9\times 9}$$

$$\mathbf{A}_5\mathbf{I}_9\mathbf{A}_5 = \text{A5SIGRM3} = \mathbf{A}_5$$

so

$$\mathbf{A}_5 \Sigma \mathbf{A}_5 = \sigma^2_{R(BT)}\mathbf{A}_5$$

$$\mathbf{X}'\mathbf{A}_1\mathbf{X} = \text{XA1X} = \mathbf{J}_3$$

so

$$\boldsymbol{\beta}'\mathbf{X}'\mathbf{A}_1\mathbf{X}\boldsymbol{\beta} = \boldsymbol{\beta}'\mathbf{J}_3\boldsymbol{\beta} = (\beta_1 + \beta_2 + \beta_3)^2$$

$$\mathbf{X}'\mathbf{A}_2\mathbf{X} = \text{XA2X} = \frac{2}{3}\left(\mathbf{I}_3 - \frac{1}{2}\mathbf{J}_3\right)$$

so

$$\boldsymbol{\beta}'\mathbf{X}'\mathbf{A}_2\mathbf{X}\boldsymbol{\beta} = \boldsymbol{\beta}'\frac{2}{3}\left(\mathbf{I}_3 - \frac{1}{2}\mathbf{J}_3\right)\boldsymbol{\beta} = \frac{2}{3}(\beta_1^2 + \beta_2^2 + \beta_3^2 - \beta_1\beta_2 - \beta_1\beta_3 - \beta_2\beta_3)^2$$

$$\mathbf{X}'\mathbf{A}_3\mathbf{X} = \text{XA3X} = \frac{4}{3}\left(\mathbf{I}_3 - \frac{1}{2}\mathbf{J}_3\right)$$

so

$$\boldsymbol{\beta}'\mathbf{X}'\mathbf{A}_3\mathbf{X}\boldsymbol{\beta} = \boldsymbol{\beta}'\frac{4}{3}\left(\mathbf{I}_3 - \frac{1}{2}\mathbf{J}_3\right)\boldsymbol{\beta} = \frac{4}{3}(\beta_1^2 + \beta_2^2 + \beta_3^2 - \beta_1\beta_2 - \beta_1\beta_3 - \beta_2\beta_3)^2$$

$$\mathbf{X}'\mathbf{A}_4\mathbf{X} = \text{XA4X} = \mathbf{0}_{3\times 3}$$

so

$$\boldsymbol{\beta}'\mathbf{X}'\mathbf{A}_4\mathbf{X}\boldsymbol{\beta} = 0$$

$$\mathbf{X}'\mathbf{A}_5\mathbf{X} = \text{XA5X} = \mathbf{0}_{3\times 3}$$

so

$$\boldsymbol{\beta}'\mathbf{X}'\mathbf{A}_5\mathbf{X}\boldsymbol{\beta} = 0$$

$$\mathbf{X}'\mathbf{A}_3^*\mathbf{X} = \text{XA3STARX} = \mathbf{0}_{3\times 3}$$

so

$$\boldsymbol{\beta}'\mathbf{X}'\mathbf{A}_3^*\mathbf{X}\boldsymbol{\beta} = 0$$

$$\begin{aligned}
\text{tr}[\mathbf{A}_3^*\sigma_B^2\mathbf{RM}(\mathbf{I}_3 \otimes \mathbf{J}_3)\mathbf{M}'\mathbf{R}'] + \text{tr}\left\{\mathbf{A}_3^*\sigma_{BT}^2\mathbf{RM}\left[\mathbf{I}_3 \otimes \left(\mathbf{I}_3 - \frac{1}{3}\mathbf{J}_3\right)\right]\mathbf{M}'\mathbf{R}'\right\}& \\
+ \text{tr}\{\mathbf{A}_3^*\sigma_{R(BT)}^2\mathbf{I}_9\}& \\
= \sigma_B^2 \text{ TRA3ST1} + \sigma_{BT}^2 \text{ TRA3ST2}& \\
+ \sigma_{R(BT)}^2 \text{ TRA3ST3}& \\
= 4\sigma_B^2 + \frac{4}{3}\sigma_{BT}^2 + 2\sigma_{R(BT)}^2.&
\end{aligned}$$

APPENDIX 3: COMPUTER OUTPUT FOR CHAPTER 8

Table A3.1 provides a summary of the notation used in Section 8.3 of the text and the corresponding names used in the SAS PROC IML computer programs and outputs.

Table A3.1
Section 8.3 Notation and Corresponding PROC IML Program Names

Section 8.3 Notation	PROC IML Program Names
b, t, k	B, T, K
$\mathbf{N}, \mathbf{M}$,	N, M

The SAS PROC IML programs and the outputs for Section 8.3 follow:

```
PROC IML;
THIS PROGRAM USES THE DIMENSIONS B, T, AND K AND THE T × B
INCIDENCE MATRIX N TO CREATE THE BK * BT PATTERN MATRIX M
FOR A BALANCED INCOMPLETE BLOCK DESIGN. THE PROGRAM
IS RUN FOR THE EXAMPLE IN FIGURE 7.2.1 WITH B=T=3, K=R=2,
LAMBDA=1. THE PROGRAM CAN BE GENERALIZED TO PRODUCE
THE BK × BT PATTERN MATRIX M FOR ANY BALANCED INCOM-
PLETE DESIGN BY SIMPLY INPUTING THE APPROPRIATE DIMEN-
SIONS B, T, K AND THE APPROPRIATE INCIDENCE MATRIX N.
B=3; T=3; K=2;
N={ 0 1 1,
    1 0 1,
    1 1 0 };
BK=B#K; BT=B#T;
M=J(BK, BT, 0);
ROW=1; COL=0;
  DO I=1 TO B;
  DO J=1 TO T;
     COL=COL+1;
     IF N(|J, I|)=1 THEN M(|ROW, COL|)=1;
     IF N(|J, I|)=1 THEN ROW=ROW+1;
  END;
  END;
PRINT M;
```

```
M
0  1  0  0  0  0  0  0  0
0  0  1  0  0  0  0  0  0
0  0  0  1  0  0  0  0  0
0  0  0  0  0  1  0  0  0
0  0  0  0  0  0  1  0  0
0  0  0  0  0  0  0  1  0
QUIT;
RUN;

PROC IML;
THIS PROGRAM USES THE DIMENSIONS B, T, AND K AND THE BK ×
BT PATTERN MATRIX M TO CREATE THE T * B INCIDENCE MATRIX
N FOR A BALANCED INCOMPLETE BLOCK DESIGN. THE PROGRAM
IS RUN FOR THE EXAMPLE IN FIGURE 7.2.1 WITH B=T=3, K=R=2,
LAMBDA=1 . THE PROGRAM CAN BE GENERALIZED TO PRODUCE
THE T × B INCIDENCE MATRIX N FOR ANY BALANCED INCOMPLETE
BLOCK DESIGN BY SIMPLY INPUTING THE APPROPRIATE DIMEN-
SIONS B, T, K AND THE APPROPRIATE PATTERN MATRIX M.
B=3; T=3; K=2;
M = {  0  1  0  0  0  0  0  0  0,
       0  0  1  0  0  0  0  0  0,
       0  0  0  1  0  0  0  0  0,
       0  0  0  0  0  1  0  0  0,
       0  0  0  0  0  0  1  0  0,
       0  0  0  0  0  0  0  1  0   };
BK=B# K;
N=J(T, B, 0);
ROW=1; COL=0;
  DO I=1 TO B;
  DO J=1 TO T;
     COL=COL+1;
     IF M(|ROW, COL|)=1 THEN N(|J, I|)=1;
     IF M(|ROW, COL|)=1 & ROW < BK  THEN ROW=ROW+1;
  END;
  END;
PRINT N;
N
0  1  1
1  0  1
1  1  0
QUIT;
RUN;
```

APPENDIX 4: COMPUTER OUTPUT FOR CHAPTER 9

Table A4.1 provides a summary of the notation used in Section 9.4 of the text and the corresponding names used in the SAS PROC GLM computer program and output.

Table A4.1
Section 9.4 Notation and Corresponding PROC GLM Program Names

Section 9.4 Notation	Proc GML Program Names
Levels of factor *A*	A
Replication number	R
Observed *Y*	Y

The SAS PROC GLM program and the output for Section 9.4 follow:

```
DATA A;
INPUT A R Y;
CARDS;
1  1  101
1  2  105
1  3   94
2  1   84
2  2   88
3  1   32
PROC GLM DATA=A;
CLASSES A;
MODEL Y=A/P SOLUTION;
PROC GLM DATA=A;
CLASSES A;
MODEL Y=A/P NOINT SOLUTION;
QUIT;
RUN;
```

General Linear Models Sums of Squares and Estimates
for the Less Than Full Rank Model

Dependent Variable: Y

Source	DF	Sum of Squares	Mean Square	F Value	Pr>F
Model	2	3480.00	1740.00	74.57	0.0028
Error	3	70.00	23.33		
Cor. Total	5	3550.00			

Parameter		Estimate	T for H0: Parameter=0	Pr > |T|	Std Error of Estimate
INTERCEPT		32.00 B	6.62	0.0070	4.83045892
A	1	68.00 B	12.19	0.0012	5.57773351
	2	54.00 B	9.13	0.0028	5.91607978
	3	0.00 B	.	.	.

NOTE: The X′X matrix has been found to be singular and a generalized inverse was used to solve the normal equations. Estimates followed by the letter 'B' are biased, and are not unique estimators of the parameters.

General Linear Models Estimates
for the Full Rank Mean Model

Parameter		Estimate	T for H0: Parameter=0	Pr > |T|	Std Error of Estimate
A	1	100.00	35.86	0.0001	2.78886676
	2	86.00	25.18	0.0001	3.41565026
	3	32.00	6.62	0.0070	4.83045892

In the following discussion, the preceding computer output is translated into the results that appear in Section 9.4.

The less than full rank parameter estimates in the output are titled INTERCEPT, A 1, 2, 3. These four estimates are SAS's solution for $\hat{\beta}_0 = (32, 68, 54, 0)'$.

The mean model estimates in the output are titled A 1, 2, 3. These three estimated equal $\hat{\mu} = (100, 86, 32)'$.

Note that the ANOVA table given in the SAS output is generated by the PROC GLM statement for the less than full rank model. However, the sums of squares presented are equivalent for the mean model and the less than full rank model. Therefore, the sum of squares labeled MODEL in the output is the sum of squares due to the treatment combinations for the data in Figure 9.4.1. Likewise, the sums of squares labeled error and cor. total in the output are the sums of squares due to the residual and the sum of squares total minus the sum of squares due to the overall mean for the data in Figure 9.4.1, respectively.

As a final comment, for a problem with more than one fixed factor, the classes statement should list the names of all fixed factors and the model statement should list the highest order fixed factor interaction between the '=' sign and the '/'. For example, if there are three fixed factors A, B, and C, then the CLASSES statement is

```
CLASSES A B C;
```

and the model statement is

```
MODEL Y=A*B*C/P SOLUTION; (for the less than full rank model)
MODEL Y=A*B*C/P NOINT SOLUTION; (for the mean model).
```

APPENDIX 5: COMPUTER OUTPUT FOR CHAPTER 10

A5.1 COMPUTER OUTPUT FOR SECTION 10.2

Table A5.1 provides a summary of the notation used in Section 10.2 of the text and the corresponding names used in the SAS PROC IML computer program and output.

Table A5.1

Section 10.2 Notation and Corresponding PROC IML Program Names

Section 10.2 Notation	PROC IML Program Names
$\mathbf{I}_3 \otimes \mathbf{J}_2, \mathbf{I}_3 \otimes (\mathbf{I}_2 - \frac{1}{2}\mathbf{J}_2), \mathbf{X}_1, \mathbf{Q}_2, \mathbf{Q}_3$	SIG1, SIG2, X1, Q2, Q3
$\mathbf{X}_2, \mathbf{Z}_1, \mathbf{Z}_2, \mathbf{M}, \mathbf{R}, \mathbf{D}, \mathbf{C}$	X2, Z1, Z2, M, R, D, C
$\mathbf{RM}(\mathbf{I}_3 \otimes \mathbf{J}_2)\mathbf{M}'\mathbf{R}', \mathbf{RM}[\mathbf{I}_3 \otimes (\mathbf{I}_2 - \frac{1}{2}\mathbf{J}_2)]\mathbf{M}'\mathbf{R}', \mathbf{I}_8$	SIGRM1, SIGRM2, SIGRM3
$\mathbf{T}_1, \mathbf{A}_1, \mathbf{A}_2, \mathbf{A}_3, \mathbf{A}_4, \mathbf{A}_5$	T1, A1, A2, A3, A4, A5
$\mathrm{tr}[\mathbf{A}_1\mathbf{RM}(\mathbf{I}_3 \otimes \mathbf{J}_2)\mathbf{M}'\mathbf{R}'], \ldots, \mathrm{tr}[\mathbf{A}_5\mathbf{RM}(\mathbf{I}_3 \otimes \mathbf{J}_2)\mathbf{M}'\mathbf{R}']$	TRA1SIG1, . . . , TRA5SIG1
$\mathrm{tr}\{\mathbf{A}_1\mathbf{RM}[\mathbf{I}_3 \otimes (\mathbf{I}_2 - \frac{1}{2}\mathbf{J}_2)]\mathbf{R}'\mathbf{M}'\}, \ldots, \mathrm{tr}[\mathbf{A}_5[\mathbf{I}_3 \otimes (\mathbf{I}_2 - \frac{1}{2}\mathbf{J}_2)]\}$	TRA1SIG2, . . . , TRA5SIG2
$\mathrm{tr}(\mathbf{A}_1\mathbf{I}_8), \ldots, \mathrm{tr}(\mathbf{A}_5\mathbf{I}_8)$	TRA1SIG3, . . . , TRA5SIG3
$\mathbf{C}'\mathbf{R}'\mathbf{A}_1\mathbf{RC}, \ldots, \mathbf{C}'\mathbf{R}'\mathbf{A}_5\mathbf{RC}$	RCA1RC, . . . , RCA5RC

The SAS PROC IML program and output for Section 10.2 follow:

```
PROC IML;
SIG1=I(3)@J(2, 2, 1);
SIG2=I(3)@(I(2)−(1/2)#J(2, 2, 1));
X1=J(6, 1, 1);
Q2={1,   −1};
Q3={ 1     1,
    −1     1,
     0    −2};
X2=J(3, 1, 1)@Q2;
Z1=Q3@J(2, 1, 1);
Z2=Q3@Q2;
M={1  0  0  0  0  0,
   0  1  0  0  0  0,
   0  0  1  0  0  0,
   0  0  0  0  1  0,
   0  0  0  0  0  1};
R=BLOCK (1, 1, J(2, 1, 1), 1, J(3, 1, 1));
D=T(R)*R;
```

```
C={1  0,
   0  1,
   1  0,
   1  0,
   0  1};
SIGRM1=R*M*SIG1*T(M)*T(R);
SIGRM2=R*M*SIG2*T(M)*T(R);
SIGRM3=I(8);
S1=R*M*X1;
S2=R*M*(X1||X2);
T1=R*M*(X1||X2||Z1);
A1=S1*INV(T(S1)*S1)*T(S1);
A2=S2*INV(T(S2)*S2)*T(S2) − A1;
A3=T1*INV(T(T1)*T1)*T(T1) − A1 − A2;
A4=R*INV(D)*T(R) − A1 −A2 −A3;
A5=I(8) − R*INV(D)*T(R);
TRA1SIG1=TRACE(A1*SIGRM1);
TRA1SIG2=TRACE(A1*SIGRM2);
TRA1SIG3=TRACE(A1*SIGRM3);
TRA2SIG1=TRACE(A2*SIGRM1);
TRA2SIG2=TRACE(A2*SIGRM2);
TRA2SIG3=TRACE(A2*SIGRM3);
TRA3SIG1=TRACE(A3*SIGRM1);
TRA3SIG2=TRACE(A3*SIGRM2);
TRA3SIG3=TRACE(A3*SIGRM3);
TRA4SIG1=TRACE(A4*SIGRM1);
TRA4SIG2=TRACE(A4*SIGRM2);
TRA4SIG3=TRACE(A4*SIGRM3);
TRA5SIG1=TRACE(A5*SIGRM1);
TRA5SIG2=TRACE(A5*SIGRM2);
TRA5SIG3=TRACE(A5*SIGRM3);
RCA1RC = T(C)*T(R)*A1*R*C;
RCA2RC = T(C)*T(R)*A2*R*C;
RCA3RC = T(C)*T(R)*A3*R*C;
RCA4RC = T(C)*T(R)*A4*R*C;
RCA5RC = T(C)*T(R)*A5*R*C;
PRINT  TRA3SIG1 TRA3SIG2 TRA3SIG3 TRA4SIG1 TRA4SIG2
       TRA4SIG3 TRA5SIG1 TRA5SIG2 TRA5SIG3 RCA1RC
       RCA2RC RCA3RC RCA4RC RCA5RC;
```

```
TRA3SIG1  TRA3SIG2  TRA3SIG3  TRA4SIG1  TRA4SIG2  TRA4SIG3
       4       0.8         2         0       1.2         1

TRA5SIG1  TRA5SIG2  TRA5SIG3
       0         0         3

RCA1RC      RCA2RC      RCA3RC     RCA4RC     RCA5RC
2     2     2    -2     0     0    0     0    0     0
2     2    -2     2     0     0    0     0    0     0

QUIT;
RUN;
```

In the following discussion, the computer output is translated into the results that appear in Section 10.2.

$$
\begin{aligned}
\mathrm{E}[\mathbf{Y}'\mathbf{A}_3\mathbf{Y}] &= \sigma_B^2 \mathrm{tr}[\mathbf{A}_3\mathbf{RM}(\mathbf{I}_3 \otimes \mathbf{J}_2)\mathbf{M}'\mathbf{R}'] \\
&\quad + \sigma_{BT}^2 \mathrm{tr}\left\{\mathbf{A}_3\mathbf{RM}\left[\mathbf{I}_3 \otimes \left(\mathbf{I}_2 - \frac{1}{2}\mathbf{J}_2\right)\right]\mathbf{R}'\mathbf{M}'\right\} + \sigma_{R(BT)}^2 \mathrm{tr}(\mathbf{A}_3\mathbf{I}_8) \\
&= \sigma_B^2 \mathrm{TRA3SIG1} + \sigma_{BT}^2 \mathrm{TRA3SIG2} + \sigma_{R(BT)}^2 \mathrm{TRA3SIG3} \\
&= 4\sigma_B^2 + 0.8\sigma_{BT}^2 + 2\sigma_{R(BT)}^2 \\
\mathrm{E}[\mathbf{Y}'\mathbf{A}_4\mathbf{Y}] &= \sigma_B^2 \mathrm{tr}[\mathbf{A}_4\mathbf{RM}(\mathbf{I}_3 \otimes \mathbf{J}_2)\mathbf{M}'\mathbf{R}'] \\
&\quad + \sigma_{BT}^2 \mathrm{tr}\left\{\mathbf{A}_4\mathbf{RM}\left[\mathbf{I}_3 \otimes \left(\mathbf{I}_2 - \frac{1}{2}\mathbf{J}_2\right)\right]\mathbf{R}'\mathbf{M}'\right\} + \sigma_{R(BT)}^2 \mathrm{tr}(\mathbf{A}_4\mathbf{I}_8) \\
&= \sigma_B^2 \mathrm{TRA4SIG1} + \sigma_{BT}^2 \mathrm{TRA4SIG2} + \sigma_{R(BT)}^2 \mathrm{TRA4SIG3} \\
&= 1.2\sigma_{BT}^2 + \sigma_{R(BT)}^2 \\
\mathrm{E}[\mathbf{Y}'\mathbf{A}_5\mathbf{Y}] &= \sigma_B^2 \mathrm{tr}[\mathbf{A}_5\mathbf{RM}(\mathbf{I}_3 \otimes \mathbf{J}_2)\mathbf{M}'\mathbf{R}'] \\
&\quad + \sigma_{BT}^2 \mathrm{tr}\left\{\mathbf{A}_5\mathbf{RM}\left[\mathbf{I}_3 \otimes \left(\mathbf{I}_2 - \frac{1}{2}\mathbf{J}_2\right)\right]\mathbf{R}'\mathbf{M}'\right\} + \sigma_{R(BT)}^2 \mathrm{tr}(\mathbf{A}_5\mathbf{I}_8) \\
&= \sigma_B^2 \mathrm{TRA5SIG1} + \sigma_{BT}^2 \mathrm{TRA5SIG2} + \sigma_{R(BT)}^2 \mathrm{TRA5SIG3} \\
&= 3\sigma_{R(BT)}^2
\end{aligned}
$$

Note that $\mathbf{C}'\mathbf{R}'\mathbf{A}_3\mathbf{RC} = \mathrm{RCA3RC} = \mathbf{0}_{2\times 2}$, $\mathbf{C}'\mathbf{R}'\mathbf{A}_4\mathbf{RC} = \mathrm{RCA4RC} = \mathbf{0}_{2\times 2}$, and $\mathbf{C}'\mathbf{R}'\mathbf{A}_5\mathbf{RC} = \mathrm{RCA5RC} = \mathbf{0}_{2\times 2}$, which implies $\boldsymbol{\mu}'\mathbf{C}'\mathbf{R}'\mathbf{A}_3\mathbf{RC}\boldsymbol{\mu} = \boldsymbol{\mu}'\mathbf{C}'\mathbf{R}'\mathbf{A}_4\mathbf{RC}\boldsymbol{\mu} = \boldsymbol{\mu}'\mathbf{C}'\mathbf{R}'\mathbf{A}_5\mathbf{RC}\boldsymbol{\mu} = 0$. Therefore, the $\mathrm{E}[\mathbf{Y}'\mathbf{A}_3\mathbf{Y}]$, $\mathrm{E}[\mathbf{Y}'\mathbf{A}_4\mathbf{Y}]$, and $\mathrm{E}[\mathbf{Y}'\mathbf{A}_5\mathbf{Y}]$ do not involve $\boldsymbol{\mu}$.

A5.2 COMPUTER OUTPUT FOR SECTION 10.4

Table A5.2 provides a summary of the notation used in Section 10.4 of the text and the corresponding names used in the SAS PROC VARCOMP computer program and output.

Table A5.2
Section 10.4 Notation and Corresponding PROC VARCOMP Program Names

Section 10.4 Notation	PROC VARCOMP Program/Output Names
$B, T, R, Y, R(BT)$	B, T, R, Y, Error
$\sigma^2_{B*}, \sigma^2_{BT*}, \sigma^2_{R(BT)*}, \Phi(T)$	var(B), var($T * B$), var(Error), $Q(T)$

The SAS program and output for Section 10.4 follow:

```
DATA A;
INPUT B T R Y;
CARDS;
1 1 1 237
1 2 1 178
2 1 1 249
2 1 2 268
3 1 1 186
3 2 1 183
3 2 2 182
3 2 3 165
PROC VARCOMP METHOD=TYPE1;
CLASS T B;
MODEL Y=T|B / FIXED=1;
RUN;
PROC VARCOMP METHOD=REML;
CLASS T B;
MODEL Y=T|B / FIXED=1;
QUIT;
RUN;
```

```
          Variance Components Estimation Procedure

                  Class Level Information
               Class     Levels     Values
               T         2          1 2
               B         3          1 2 3
          Number of observations in data set = 8
                  Dependent Variable : Y
```

TYPE 1 SS Variance Component Estimation Procedure

Source	DF	Type I SS	Type I MS
T	1	6728.0000	6728.0000
B	2	2770.8000	1385.4000
T*B	1	740.0333	740.0333
Error	3	385.1667	128.3889
Corrected Total	7	10624.0000	

Source	Expected Mean Square
T	Var(Error) + 2 Var(T*B) + Var(B) + Q(T)
B	Var(Error) + 1.4 Var(T*B) + 2 Var (B)
T*B	Var(Error) + 1.2 Var(T*B)
Error	Var(Error)

Variance Component	Estimate
Var(B)	271.7129
Var(T*B)	509.7037
Var(Error)	128.3889

RESTRICTED MAXIMUM LIKELIHOOD
Variance Components Estimation Procedure

Iteration	Objective	Var(B)	Var(T*B)	Var(Error)
0	35.946297	30.899	616.604	158.454
1	35.875323	51.780	663.951	144.162
2	35.852268	66.005	693.601	136.830
3	35.845827	81.878	701.484	133.406
4	35.843705	86.686	711.185	131.422
5	35.843069	87.877	718.188	130.334
6	35.842890	88.250	722.237	129.757
7	35.842842	88.421	724.385	129.458
8	35.842830	88.509	725.494	129.305
9	35.842827	88.553	726.060	129.227
10	35.842826	88.576	726.348	129.187
11	35.842825	88.587	726.493	129.167
12	35.842825	88.593	726.567	129.157
13	35.842825	88.596	726.605	129.152
14	35.842825	88.598	726.624	129.149

Convergence criteria met

In the following discussion, the computer output is translated into the results that appear in Section 10.4.

$\mathbf{Y}'\mathbf{A}_2\mathbf{Y}$ = Type I SS due to T = 6728.0

$\mathbf{Y}'\mathbf{A}_3\mathbf{Y}$ = Type I SS due to B = 2770.7

$\mathbf{Y}'\mathbf{A}_4\mathbf{Y}$ = Type I SS due to T*B = 740.03

$\mathbf{Y}'\mathbf{A}_5\mathbf{Y}$ = Type I SS due to Error = 385.17

$$
\begin{aligned}
E[\mathbf{Y}'\mathbf{A}_3\mathbf{Y}/2] &= 2\text{Var(B)} + 1.4\ \text{Var(T*B)} + \ \text{Var(Error)} \\
&= 2\sigma^2_{B^*} + 1.4\sigma^2_{BT^*} + \sigma^2_{R(BT)^*} \\
E[\mathbf{Y}'\mathbf{A}_4\mathbf{Y}/1] &= 1.2\ \text{Var(T*B)} + \ \text{Var(Error)} \\
&= 1.2\sigma^2_{BT^*} + \sigma^2_{R(BT)^*} \\
E[\mathbf{Y}'\mathbf{A}_5\mathbf{Y}/3] &= \ \text{Var(Error)} = \sigma^2_{R(BT)^*}
\end{aligned}
$$

A5.3 COMPUTER OUTPUT FOR SECTION 10.6

Table A5.3.1 provides a summary of the notation used in Section 10.6 of the text and the corresponding names used in the SAS PROC IML computer program and output.

Table A5.3

Section 10.6 Notation and Corresponding PROC IML Program Names

Section 10.6 Notation	PROC IML Program Names
$\mathbf{I}_3 \otimes \mathbf{J}_2, \mathbf{I}_3 \otimes (\mathbf{I}_2 - \frac{1}{2}\mathbf{J}_2), \mathbf{X}_1, \mathbf{Q}_2, \mathbf{Q}_3$	SIG1, SIG2, X1, Q2, Q3
$\mathbf{X}_2, \mathbf{Z}_1, \mathbf{Z}_2, \mathbf{M}, \mathbf{R}, \mathbf{D}, \mathbf{C}$	X2, Z1, Z2, M, R, D, C
$\mathbf{RM}(\mathbf{I}_3 \otimes \mathbf{J}_2)\mathbf{M}'\mathbf{R}', \mathbf{RM}[\mathbf{I}_3 \otimes (\mathbf{I}_2 - \frac{1}{2}\mathbf{J}_2)]\mathbf{M}'\mathbf{R}', \mathbf{I}_8$	SIGRM1, SIGRM2, SIGRM3
$\hat{\Sigma}, \hat{\boldsymbol{\mu}}_w, \widehat{\text{cov}}(\hat{\boldsymbol{\mu}}_w)$	COVHAT, MUHAT, COVMUHAT
$\mathbf{t}, \mathbf{t}'\hat{\boldsymbol{\mu}}, \widehat{\text{var}}(\mathbf{t}', \hat{\boldsymbol{\mu}}_w)$	T, TMU, VARTMU
$\mathbf{t}'\hat{\boldsymbol{\mu}}_w \pm Z_{0.025}\sqrt{\widehat{\text{var}}(\mathbf{t}'\hat{\boldsymbol{\mu}}_w)}$	LOWERLIM, UPPERLIM
$\mathbf{S}_1, \mathbf{S}_2, \mathbf{T}_1, \mathbf{A}_1, \mathbf{A}_2, \mathbf{A}_3, \mathbf{A}_4, \mathbf{A}_5$	S1, S2, T1, A1, A2, A3, A4, A5
$W, \text{f}, \mathbf{Y}'\mathbf{A}_2\mathbf{Y}/W$	W, F, TEST

The SAS PROC IML program and output for Section 10.6 follow:

```
PROC IML;
Y={237, 178, 249, 268, 186, 183, 182, 165};
SIG1=I(3)@J(2, 2, 1);
SIG2=I(3)@(I(2)–(1/2)#J(2, 2, 1));
X1=J(6, 1, 1);
```

```
Q2={1,   −1};
Q3={ 1    1,
    −1    1,
     0   −2};
X2=J(3, 1, 1)@Q2;
Z1=Q3@J(2, 1, 1);
Z2=Q3@Q2;
M={1  0  0  0  0  0,
   0  1  0  0  0  0,
   0  0  1  0  0  0,
   0  0  0  0  1  0,
   0  0  0  0  0  1};
R=BLOCK(1, 1, J(2, 1, 1), 1, J(3, 1, 1));
D=T(R)*R;
C={1  0,
   0  1,
   1  0,
   1  0,
   0  1};
SIGRM1=R*M*SIG1*T(M)*T(R);
SIGRM2=R*M*SIG2*T(M)*T(R);
SIGRM3=I(8);
COVHAT=526.56#SIGRM1 + 509.70#SIGRM2 + 128.39#SIGRM3;
MUHAT=INV(T(C)*T(R)*INV(COVHAT)*R*C)*T(C)*T(R)
   *INV(COVHAT)*Y;
COVMUHAT=INV(T(C)*T(R)*INV(COVHAT)*R*C);
T={1, −1)};
TMU=T(T)*MUHAT;
VARTMU=T(T)*COVMUHAT*T;
LOWERLIM=TMU−1.96#(VARTMU**.5);
UPPERLIM=TMU+1.96#(VARTMU**.5);
S1=R*M*X1;
S2=R*M*(X1||X2);
T1=R*M*(XI||X2||Z1);
A1=S1*INV(T(S1)*S1)*T(S1);
A2=S2*INV(T(S2)*S2)*T(S2) − A1;
A3=T1*INV(T(T1)*T1)*T(T1) − A1 − A2;
A4=R*INV(D)*T(R) − A1 −A2 −A3;
A5=I(8) − R*INV(D)*T(R);
W=.25#T(Y)*A3*Y + (1.3/1.2)#T(Y)*A4*Y − (0.7/3.6)#T(Y)*A5*Y;
F=W**2/ ((((.25#T(Y)*A3*Y)**2)/2) + (((1.3/1.2)#T(Y)*A4*Y)**2) +
   (((0.7#T(Y)*A5*Y/3.6)**2)/3));
```

```
TEST=T(Y)*A2*Y/W;
PRINT    MUHAT COVMUHAT TMU VARTMU LOWERLIM
     UPPERLIM W F TEST;

       MUHAT      COVMUHAT                  TMU          VARTMU
      227.94      295.7818     88.33466     45.318467    537.25697
   182.62153      88.33466    418.14449


LOWERLIM      UPPERLIM     W            F            TEST
−0.111990     90.748923    1419.5086    2.2780974    4.739663

QUIT;
RUN;
```

References and Related Literature

Anderson, R. L., and Bancroft, T. A. (1952), *Statistical Theory in Research*, McGraw-Hill Book Co., New York.

Anderson, T. W. (1958), *An Introduction to Multivariate Statistical Analysis*, John Wiley & Sons, New York.

Anderson, V. L., and McLean, R. A. (1974), *Design of Experiments, A Realistic Approach*, Marcel Dekker, New York.

Arnold, S. F. (1981), *The Theory of Linear Models and Multivariate Analysis*, John Wiley & Sons, New York.

Bhat, B. R. (1962), "On the distribution of certain quadratic forms in normal variates," *Journal of the Royal Statistical Society, Ser. B* **24**, 148–151.

Box, G. E. P., Hunter, W. G., and Hunter, J. S. (1978), *Statistics for Experimenters*, John Wiley & Sons, New York.

Chew, V. (1970), "Covariance matrix estimation in linear models," *Journal of the American Statistical Society* **65**, 173–181.

Cochran, W. D., and Cox, G. M. (1957), *Experimental Designs*, John Wiley & Sons, New York.

Davenport, J. M. (1971), "A comparison of some approximate F-tests," *Technometrics* **15**, 779–790.

Draper, N. R., and Smith, H. (1966), *Applied Regression Analysis*, John Wiley & Sons, New York.

Federer, W. T. (1955), *Experimental Design*, Macmillan, New York.

Graybill, F. A. (1954), "On quadratic estimates of variance components," *Annals of Mathematical Statistics*, **25**, No. 2, 367–372.

Graybill, F. A. (1961), *An Introduction to Linear Statistical Models, Volume 1*, McGraw-Hill Book Co., New York.
Graybill, F. A. (1969), *An Introduction to Matrices with Applications in Statistics*, Wadsworth Publishing Co., Belmont, CA.
Graybill, F. A. (1976), *Theory and Application of the Linear Model*, Wadsworth & Brooks/Cole, Pacific Grove, CA.
Guttman, I. (1982), *The Analysis of Linear Models*, John Wiley & Sons, New York.
Harville, D. A. (1977), "Maximum likelihood approaches to variance component estimation and to related problems," *Journal of the American Statistical Association* **72**, No. 358, 320–338.
Hocking, R. R. (1985), *The Analysis of Linear Models*, Brooks/Cole, Publishing Co., Belmont, CA.
Hogg, R. V., and Craig, A. T. (1958), "On the decomposition of certain χ^2 variables," *Annals of Mathematical Statistics* **29**, 608–610.
John, P. W. M. (1971), *Statistical Design and Analysis of Experiments*, Macmillan, New York.
Kempthorne, O. (1952, 1967), *The Design and Analysis of Experiments*, John Wiley & Sons, New York.
Kshirsagar, A. M. (1983), *A Course in Linear Models*, Marcel Dekker, New York.
Milliken, G. A., and Johnson, D. E. (1992), *Analysis of Messy Data, Volme 1: Designed Experiments*, Von Nostrand Reinhold, New York.
Morrison, D. (1967), *Multivariate Statistical Methods*, McGraw-Hill Book Co., New York.
Moser, B. K. (1987), "Generalized F variates in the general linear model," *Communications in Statistics: Theory and methods* **16**, 1867–1884.
Moser, B. K., and McCann, M. H. (1996), "Maximum likelihood and restricted maximum likelihood estimators as functions of ordinary least squares and analysis of variance estimators," *Communications in Statistics: Theory and Methods*, **25**, No. 3.
Moser, B. K., and Lin, Y. (1992), "Equivalence of the corrected F test and the weighted least squares procedure," *The American Statistician* **46**, No. 2, 122–124.
Moser, B. K., and Marco, V. R. (1988), "Bayesian outlier testing using the predictive distribution for a linear model of constant intraclass form," *Communications in Statistics: Theory and Methods* **17**, 849–860.
Moser, B. K., Stevens, G. R., and Watts, C. L. (1989), "The two sample T test versus Satterthwaite's approximate F test," *Communications in Statistics: Theory and Methods*, **18**, 3963–3976.
Myers, R. H., and Milton, J. S. (1991), *A First Course in the Theory of Linear Statistical Models*, PWS-Kent Publishing Co., Boston.
Nel, D. G., van der Merwe, C. A., and Moser, B. K. (1990), "The exact distributions of the univariate and multivariate Behrens–Fisher statistic with a comparison of several solutions in the univariate case," *Communications in Statistics: Theory and Methods* **19**, 279–298.
Neter, J. Wasserman, W., and Kutner, M. H. (1983), *Applied Linear Regression Models*, Richard D. Irwin, Homewood, IL.
Patterson, H. D., and Thompson, R. (1971), "Recovery of interblock information when block sizes are unequal," *Biometrika* **58**, 545–554.
Patterson, H. D., and Thompson, R. (1974), "Maximum likelihood estimation of components of variance," *Proceedings of the 8th International Biometric Conference*, pp. 197–207.
Pavur, R. J., and Lewis, T. O. (1983), "Unbiased F tests for factorial experiments for correlated data," *Communications in Statistics: Theory and Methods* **13**, 3155–3172.
Puntanen, S., and Styan, G. P. (1989), "The equality of the ordinary least squares estimator and the best linear unbiased estimator," *The American Statistician* **43**, No. 3, 153–161.
Rao, C. R. (1965), *Linear Statistical Inference and Its Applications*, John Wiley & Sons, New York.
Russell, T. S., and Bradley, R. A. (1958), "One-way variances in a two-way classification," *Biometrika*, **45**, 111–129.
Satterthwaite, F. E. (1946), "An approximate distribution of estimates of variance components," *Biometrics Bulletin* **2**, 110–114.

Scariano, S. M., and Davenport, J. M. (1984), "Corrected F-tests in the general linear model," *Communications in Statistics: Theory and Methods* **13**, 3155–3172.

Scariano, S. M., Neill, J. W., and Davenport, J. M. (1984), "Testing regression function adequacy with correlation and without replication," *Communications in Statistics: Theory and Methods*, **13**, 1227–1237.

Scheffé, H. (1959), *The Analysis of Variance*, John Wiley & Sons, New York.

Searle, S. R. (1971), *Linear Models*, John Wiley & Sons, New York.

Searle, S. R. (1982), *Matrix Algebra Useful for Statistics*, John Wiley & Sons, New York.

Searle, S. R. (1987), *Linear Models for Unbalanced Data*, John Wiley & Sons, New York.

Seber, G. A. F. (1977), *Linear Regression Analysis*, John Wiley & Sons, New York.

Smith, J. H., and Lewis, T. O. (1980), "Determining the effects of intraclass correlation on factorial experiments," *Communications in Statistics: Theory and Methods*: **9**, 1353–1364.

Smith, J. H., and Lewis, T. O. (1982), "Effects of intraclass correlation on covariance analysis," *Communications in Statistcs: Theory and Methods*, **11**, 71–80.

Snedecor, W. G., and Cochran, W. G. (1978), *Statistical Methods*, The Iowa State University Press, Ames, IA.

Steel, R. G., and Torrie, J. H. (1980), *Principles and Procedures of Statistics*, McGraw-Hill Book Co., New York.

Thompson, W. A. (1962), "The problem of negative estimates of variance components," *Annals of Mathematical Statistics* **33**, 273–289.

Weeks, D. L., and Trail, S. M. (1973), "Extended complete block designs generated by BIBD," *Biometrics* **29**, No. 3, 565–578.

Weeks, D. L., and Urquhart, N. S. (1978), Linear models in messy data: Some problems and alternatives," *Biometrics*, **34**, No. 4, 696–705.

Weeks, D. L., and Williams, D. R. (1964), "A note on the determination of correctness in an N-way cross classification," *Technometrics*, **6**, 319–324.

Zyskind, G. (1967), "On canonical forms, non-negative covariance matrices and best and simple least squares linear estimators in linear models," *Annals of Mathematical Statistics* **38**, 1092–1109.

Zyskind, G., and Martin, F. B. (1969), "On best linear estimation and a general Gauss–Markov theorem in linear models with arbitrary non-negative covariance structure," *SIAM, Journal of Applied Mathematics* **17**, 1190–1202.

Subject Index

s

t

u

v

PROBABILITY AND MATHEMATICAL STATISTICS

Thomas Ferguson, *Mathematical Statistics: A Decision Theoretic Approach*
Howard Tucker, *A Graduate Course in Probability*
K. R. Parthasarathy, *Probability Measures on Metric Spaces*
P. Révész, *The Laws of Large Numbers*
H. P. McKean, Jr., *Stochastic Integrals*
B. V. Gnedenko, Yu. K. Belyayev, and A. D. Solovyev, *Mathematical Methods of Reliability Theory*
Demetrios A. Kappos, *Probability Algebras and Stochastic Spaces*
Ivan N. Pesin, *Classical and Modern Integration Theories*
S. Vajda, *Probabilistic Programming*
Robert B. Ash, *Real Analysis and Probability*
V. V. Fedorov, *Theory of Optimal Experiments*
K. V. Mardia, *Statistics of Directional Data*
H. Dym and H. P. McKean, *Fourier Series and Integrals*
Tatsuo Kawata, *Fourier Analysis in Probability Theory*
Fritz Oberhettinger, *Fourier Transforms of Distributions and Their Inverses: A Collection of Tables*
Paul Erdös and Joel Spencer, *Probabilistic Methods of Statistical Quality Control*
K. Sarkadi and I. Vincze, *Mathematical Methods of Statistical Quality Control*
Michael R. Anderberg, *Cluster Analysis for Applications*
W. Hengartner and R. Theodorescu, *Concentration Functions*
Kai Lai Chung, *A Course in Probability Theory, Second Edition*
L. H. Koopmans, *The Spectral Analysis of Time Series*
L. E. Maistrov, *Probability Theory: A Historical Sketch*
William F. Stout, *Almost Sure Convergence*
E. J. McShane, *Stochastic Calculus and Stochastic Models*
Robert B. Ash and Melvin F. Gardner, *Topics in Stochastic Processes*
Avner Friedman, *Stochastic Differential Equations and Applications, Volume 1, Volume 2*
Roger Cuppens, *Decomposition of Multivariate Probabilities*
Eugene Lukacs, *Stochastic Convergence, Second Edition*
H. Dym and H. P. McKean, *Gaussian Processes, Function Theory, and the Inverse Spectral Problem*
N. C. Giri, *Multivariate Statistical Inference*
Lloyd Fisher and John McDonald, *Fixed Effects Analysis of Variance*
Sidney C. Port and Charles J. Stone, *Brownian Motion and Classical Potential Theory*

Konrad Jacobs, *Measure and Integral*
K. V. Mardia, J. T. Kent, and J. M. Biddy, *Multivariate Analysis*
Sri Gopal Mohanty, *Lattice Path Counting and Applications*
Y. L. Tong, *Probability Inequalities in Multivariate Distributions*
Michel Metivier and J. Pellaumail, *Stochastic Integration*
M. B. Priestly, *Spectral Analysis and Time Series*
Ishwar V. Basawa and B. L. S. Prakasa Rao, *Statistical Inference for Stochastic Processes*
M. Csörgö and P. Révész, *Strong Approximations in Probability and Statistics*
Sheldon Ross, *Introduction to Probability Models, Second Edition*
P. Hall and C. C. Heyde, *Martingale Limit Theory and Its Applications*
Imre Csiszár and János Körner, *Information Theory: Coding Theorems for Discrete Memoryless Systems*
A. Hald, *Statistical Theory of Sampling Inspection by Attributes*
H. Bauer, *Probability Theory and Elements of Measure Theory*
M. M. Rao, *Foundations of Stochastic Analysis*
Jean-Rene Barra, *Mathematical Basis of Statistics*
Harald Bergström, *Weak Convergence of Measures*
Sheldon Ross, *Introduction to Stochastic Dynamic Programming*
B. L. S. Prakasa Rao, *Nonparametric Functional Estimation*
M. M. Rao, *Probability Theory with Applications*
A. T. Bharucha-Reid and M. Sambandham, *Random Polynomials*
Sudhakar Dharmadhikari and Kumar Joag-dev, *Unimodality, Convexity, and Applications*
Stanley P. Gudder, *Quantum Probability*
B. Ramachandran and Ka-Sing Lau, *Functional Equations in Probability Theory*
B. L. S. Prakasa Rao, *Identifiability in Stochastic Models: Characterization of Probability Distributions*
Moshe Shaked and J. George Shanthikumar, *Stochastic Orders and Their Applications*
Barry K. Moser, *Linear Models: A Mean Model Approach*

ISBN 0-12-508465-X